IFMBE Proceedings

Volume 78

The IFMBE Proceedings Book Series is an official publication of *the International Federation for Medical and Biological Engineering* (IFMBE). The series gathers the proceedings of various international conferences, which are either organized or endorsed by the Federation. Books published in this series report on cutting-edge findings and provide an informative survey on the most challenging topics and advances in the fields of medicine, biology, clinical engineering, and biophysics.

The series aims at disseminating high quality scientific information, encouraging both basic and applied research, and promoting world-wide collaboration between researchers and practitioners in the field of Medical and Biological Engineering.

Topics include, but are not limited to:

- Diagnostic Imaging, Image Processing, Biomedical Signal Processing
- Modeling and Simulation, Biomechanics
- Biomaterials, Cellular and Tissue Engineering
- Information and Communication in Medicine, Telemedicine and e-Health
- Instrumentation and Clinical Engineering
- Surgery, Minimal Invasive Interventions, Endoscopy and Image Guided Therapy
- Audiology, Ophthalmology, Emergency and Dental Medicine Applications
- Radiology, Radiation Oncology and Biological Effects of Radiation

IFMBE proceedings are indexed by SCOPUS and EI Compendex. They are also submitted for ISI proceedings indexing.

Proposals can be submitted by contacting the Springer responsible editor shown on the series webpage (see "Contacts"), or by getting in touch with the series editor Ratko Magjarevic.

More information about this series at http://www.springer.com/series/7403

Muhamed Brka · Enisa Omanović-Mikličanin · Lutvija Karić · Vedad Falan · Almir Toroman
Editors

30th Scientific-Experts Conference of Agriculture and Food Industry

Answers for Forthcoming Challenges in Modern Agriculture

Springer

Editors
Muhamed Brka
Faculty of Agriculture and Food Sciences
University of Sarajevo
Sarajevo, Bosnia and Herzegovina

Enisa Omanović-Mikličanin
Faculty of Agriculture and Food Sciences
University of Sarajevo
Sarajevo, Bosnia and Herzegovina

Lutvija Karić
Faculty of Agriculture and Food Sciences
University of Sarajevo
Sarajevo, Bosnia and Herzegovina

Vedad Falan
Faculty of Agriculture and Food Sciences
University of Sarajevo
Sarajevo, Bosnia and Herzegovina

Almir Toroman
Faculty of Agriculture and Food Sciences
University of Sarajevo
Sarajevo, Bosnia and Herzegovina

ISSN 1680-0737 ISSN 1433-9277 (electronic)
IFMBE Proceedings
ISBN 978-3-030-40048-4 ISBN 978-3-030-40049-1 (eBook)
https://doi.org/10.1007/978-3-030-40049-1

This Springer imprint is published by the registered company Springer Nature Switzerland AG
The registered company address is: Gewerbestrasse 11, 6330 Cham, Switzerland

Preface

This book gathers the proceedings of the 30th jubilee International Scientific-Expert Conference of Agriculture and Food Sciences, which was held on 25–27 September 2019, in Sarajevo, Bosnia and Herzegovina.

The conference covered five areas:

1. Plant production,
2. Animal production,
3. Agricultural economics,
4. Food production,
5. Innovative technologies in agriculture and human health.

Based on that, this book is organized into five parts.

Organization

Organizing Committee

Rifat Škrijelj (Rector)	University of Sarajevo, Bosnia and Herzegovina
Necdet Budak (Rector)	Ege University, Turkey
Muhamed Brka	Dean of the faculty of Agriculture and Food Sciences, University of Sarajevo, Bosnia and Herzegovina
Nedim Kosum	Dean of the Faculty of Agriculture, Ege University, Turkey
Dušan Živković	Dean of the Faculty of Agriculture, Faculty of Agriculture, University of Belgrade, Serbia
Almir Badnjević	President of the Bosnia and Herzegovina Medical and Biological Engineering Society
Enisa Omanović-Mikličanin	President of the Organizing Committee, Faculty of Agriculture and Food Sciences, University of Sarajevo, Bosnia and Herzegovina
Senaid Memić	Mayor of municipality Ilidža, Faculty of Agriculture and Food Sciences, University Sarajevo, Bosnia and Herzegovina
Ševal Muminović	Faculty of Agriculture and Food Sciences, University of Sarajevo, Bosnia and Herzegovina
Sabahudin Bajramović	Faculty of Agriculture and Food Sciences, University of Sarajevo, Bosnia and Herzegovina
Zlatan Sarić	Faculty of Agriculture and Food Sciences, University of Sarajevo, Bosnia and Herzegovina
Enver Karahmet	Faculty of Agriculture and Food Sciences, University of Sarajevo, Bosnia and Herzegovina

Mirsad Kurtović	Faculty of Agriculture and Food Sciences, University of Sarajevo, Bosnia and Herzegovina
Selim Škaljić	Faculty of Agriculture and Food Sciences, University of Sarajevo, Bosnia and Herzegovina
Nedžad Karić	Faculty of Agriculture and Food Sciences, University of Sarajevo, Bosnia and Herzegovina
Drena Gadžo	Faculty of Agriculture and Food Sciences, University of Sarajevo, Bosnia and Herzegovina
Halil Omanović	Faculty of Agriculture and Food Sciences, University of Sarajevo, Bosnia and Herzegovina
Almir Toroman	Faculty of Agriculture and Food Sciences, University of Sarajevo, Bosnia and Herzegovina
Banu Yucel	Faculty of Agriculture, Ege University, Turkey
Eftal Duzyaman	Faculty of Agriculture, Ege University, Turkey
Emre Ilker	Faculty of Agriculture, Ege University, Turkey
Hakan Geren	Faculty of Agriculture, Ege University, Turkey
Hakan Bayraktar	Faculty of Agriculture, Ege University, Turkey
Ozlem Tuncay	Faculty of Agriculture, Ege University, Turkey
Fulsen Ozen	Faculty of Agriculture, Ege University, Turkey
Cigdem Seremet Tugalay	Faculty of Agriculture, Ege University, Turkey
Erkan Urkan	Faculty of Agriculture, Ege University, Turkey

Local Organizing Committee

Petar Glamočlija
Alen Mujčinović
Adel Salihović
Emina Sijahović
Munevera Begić
Fejzo Bašić
Anera Kazlagić
Berina Borovac

Scientific Committee

Johann Wollmann	BOKU University, Vienna, Austria
Michael Murkovic	Graz Technical University, Austria
Hamid El Bilali	International Centre for Advanced Mediterranean Agronomic Studies (CIHEAM Bari), Italy

Paride D'Ottavio	Università Politecnica delle Marche Department of Agricultural, Food and Environmental Sciences, Ancona, Italy
Ivan Kreft	Biotechnical Faculty, University of Ljubljana, Slovenia
Selim Škaljić	Faculty of Agriculture and Food Sciences, University of Sarajevo, Bosnia and Herzegovina
Mirha Đikić	Faculty of Agriculture and Food Sciences, University of Sarajevo, Bosnia and Herzegovina
Drena Gadžo	Faculty of Agriculture and Food Sciences, University of Sarajevo, Bosnia and Herzegovina
Pakeza Drkenda	Faculty of Agriculture and Food Sciences, University of Sarajevo, Bosnia and Herzegovina
Milenko Blesić	Faculty of Agriculture and Food Sciences, University of Sarajevo, Bosnia and Herzegovina
Senada Čengić-Džomba	Faculty of Agriculture and Food Sciences, University of Sarajevo, Bosnia and Herzegovina
Ervin Zečević	Faculty of Agriculture and Food Sciences, University of Sarajevo, Bosnia and Herzegovina
Emir Džomba	Faculty of Agriculture and Food Science, University of Sarajevo, Bosnia and Herzegovina
Admir Dokso	Faculty of Agriculture and Food Sciences, University of Sarajevo, Bosnia and Herzegovina
Nedžad Karić	Faculty of Agriculture and Food Sciences, University of Sarajevo, Bosnia and Herzegovina
Enisa Omanović-Mikličanin	Faculty of Agriculture and Food Sciences, University of Sarajevo, Bosnia and Herzegovina
Sanja Oručević Žuljević	Faculty of Agriculture and Food Sciences, University of Sarajevo, Bosnia and Herzegovina
Asima Akagić	Faculty of Agriculture and Food Sciences, University of Sarajevo, Bosnia and Herzegovina

Blažo Lalević	Faculty of Agriculture, University of Belgrade, Serbia
Milena Savić	Faculty of Agriculture, University of Belgrade, Serbia
Aleksandar Kostić	Faculty of Agriculture, University of Belgrade, Serbia
Aleksandar Simić	Faculty of Agriculture, University of Belgrade, Serbia
Vladan Bogdanović	Faculty of Agriculture, University of Belgrade, Serbia
Radica Đedović	Faculty of Agriculture, University of Belgrade, Serbia
Branka Krstić	Faculty of Agriculture, University of Belgrade, Serbia
Zoran Popović	Faculty of Agriculture, University of Belgrade, Serbia
Miroljub Barać	Faculty of Agriculture, University of Belgrade, Serbia
Blažo Lalević	Faculty of Agriculture, University of Belgrade, Serbia
Dragan Stanojević	Faculty of Agriculture, University of Belgrade, Serbia
Jelena Jovičić.Petrović	Faculty of Agriculture, University of Belgrade, Serbia
Snežana Trivunović	Faculty of Agriculture, University of Novi Sad, Serbia
Branka Pilic	Faculty of Agriculture, University of Novi Sad, Serbia
Denis Kučević	Faculty of Agriculture, University of Novi Sad, Serbia
Aleksandar Joksimović	Faculty of Sport and Physical Education, University of Niš, Serbia
Nebojša Savić	Faculty of Economics, Finance and Administration, Metropolitan University, Serbia
Milka Popović	University of Novi Sad Faculty of Medicine, Department of Hygiene
Snežana Popov	Faculty of Science, University of Novi Sad
Vesna Vujasinović	Faculty of Science, University of Novi Sad
Vesna Gantner	Faculty of Bio-Agricultural Sciences, University of Osijek, Croatia
Mirela Kopjar	Faculty of Bio-Agricultural Sciences, University of Osijek, Croatia
Vlado Guberac	Faculty of Bio-Agricultural Sciences, University of Osijek, Croatia

Karolina Vrandečić	Faculty of Bio-Agricultural Sciences, University of Osijek, Croatia
Jadranka Frece	Faculty of Food Technology and Biotechnology, University of Zagreb, Croatia
Zvonko Antunović	Faculty of Agriculture, University of Osijek, Croatia
Ante Ivanković	Faculty of Agriculture, University of Zagreb, Croatia
Sandra Pedišić	Faculty of Food Technology and Biotechnology, University of Zagreb, Croatia
Damir Ježek	Faculty of Food Technology and Biotechnology, University of Zagreb, Croatia
Elda Vitanović	Institute for Adriatic Crops and Karst Reclamation, Split, Croatia
Zoran Popovski	Faculty of Agricultural Sciences and Food, Cyril and Methodius University of Skopje, North Macedonia
Zoran Dimov	Faculty of Agricultural Sciences and Food, Cyril and Methodius University of Skopje, North Macedonia
Ordan Cukaliev	Faculty of Agricultural Sciences and Food, Cyril and Methodius University of Skopje, North Macedonia
Snežana Barjaktarović-Labović	Specialist in Hygiene, Subspecialists Dietotherapy at Primary Health Care Center, Bar Montenegro

Contents

Plant Production

Ex-Situ Remediation of Heavy Metals Contaminated Soils Using Natural Aluminosilicate Minerals

Emina Sijahović[1], Hamdija Čivić[1], and Senad Murtić[2](✉)

[1] Department of Plant Nutrition, Faculty of Agriculture and Food Science, University of Sarajevo, Zmaja od Bosne 8, 71 000 Sarajevo, Bosnia and Herzegovina

[2] Department of Plant Physiology, Faculty of Agriculture and Food Science, University of Sarajevo, Zmaja od Bosne 8, 71 000 Sarajevo, Bosnia and Herzegovina

s.murtic@ppf.unsa.ba

Abstract. The effects of zeolite and pyrophyllite on decreasing heavy metals availability in contaminated soil were evaluated in pot experiments. The zeolite and pyrophyllite materials were added to contaminated soil at four rates i.e. 0, 200, 400 and 600 kg ha^{-1}. The available forms of heavy metals from soil samples were extracted using EDTA solution (0.01 mol dm^{-3} $C_{10}H_{16}N_2O_8$ and 1 mol dm^{-3} $(NH_4)_2CO_3$, adjusted to pH 8.6) and their concentrations in the obtained extract were determined using atomic absorption spectrophotometry.

Zeolite and pyrophyllite incorporation in soil was found to reduce the availability of heavy metals in soil under experimental conditions. The addition of zeolite was more effective in decreasing Cu and Pb availability, while the pyrophyllite showed a better effect on reducing Mn and Cd availability. These results indicated that addition of these clay minerals could be a suitable technique for reducing heavy metals availability in soils. However, the further research is necessary to confirm this hypothesis across different experimentation and soil ecosystems.

Keywords: Availability · Pyrophyllite · Zeolite

1 Introduction

Preserving agricultural soil, as the primary resource for food production, is the basis of the progress and stability of each society [1]. Unfortunately, most soils in the central and northeastern part of Bosnia and Herzegovina are contaminated, to a greater or lesser extent, with harmful substances especially with toxic heavy metals due to use of pesticides and fertilizers that contain a certain amount of these substances and gas emissions and residues from different manufacturing processes. This problem is particularly pronounced in Zenica region where industrial and agricultural activity is constantly rising.

Considering the soils contaminated by heavy metals may pose risks and hazards to groundwater, food crops and consequently human health, the remediation of

M. Brka et al. (Eds.): AgriConf 2019, IFMBE Proceedings 78, pp. 3–10, 2020.
https://doi.org/10.1007/978-3-030-40049-1_1

contaminated soils has become increasingly important [2]. During the past decades, many studies have been conducted to find appropriate techniques for the removal of heavy metals from soils [3–5]. As a result, different techniques have been established for the removal of heavy metals from contaminated soils: physical, biological and chemical techniques, but most are costly and complicated in implementation [6]. Accordingly, in the last years, researchers and the community in general invested great efforts in the development of simple and less expensive techniques to eliminate or reduce the negative effects of toxic heavy metals in soils. Applying some natural aluminosilicate minerals such as zeolite and pyrophyllite are certainly some of these techniques [7].

Due to specific structure based on AlO_4 and SiO_4 tetrahedra with numerous channels and cavities, these minerals possess high sorption and expansion properties which make them very useful in decreasing bioavailability of heavy metals in soils. Many scientists agree that the ability of natural aluminosilicate minerals to decrease the mobility of heavy metals in soils is the result of adsorption of heavy metals in their inter layers or on its surface through ion exchange reaction [8–10]. However, the efficiency of clay minerals to decrease the mobility of heavy metals in soils depends on its sorption capacity, soil pH, soil physical properties as well as the distribution coefficient of heavy metals in soils [11].

The aim of this study was to evaluate the efficacy of applying two natural aluminosilicate minerals i.e. zeolite and pyrophyllite as a technique for reducing heavy metals availability in soils through pot experiments.

2 Materials and Methods

2.1 Materials

Zeolite: The natural zeolite clinoptilolite materials, which was used in this study, were obtained by grinding and sieving zeolite ore from deposits in Slanci - Veliko Selo, near Belgrade, Serbia. The zeolites were grinded to <100 μm prior to experimentation and were characterized by high cation exchange capacity (CEC), more than 180 meq 100 g^{-1}. Median total SiO_2, Al_2O_3, Fe_2O_3, CaO, MgO, Na_2O, K_2O, Cu, Cd, Co, Cr, Ni, Pb, Zn were 64.55%, 14.49%, 2.3%, 4.9%, 0.88%, 0.88%, 1.04%, 6.2 mg kg^{-1}, 3 mg kg^{-1}, 28 mg kg^{-1}, 5.8 mg kg^{-1}, 21 mg kg^{-1}, 35 mg kg^{-1} and 42 mg kg^{-1}, respectively.

Pyrophyllite: Pyrophyllite materials used in this study were obtained by grinding and sieving pyrophyllite ore from deposits in Parsovići - Konjic, Bosnia and Herzegovina. The pyrophyllites were grinded to <100 μm prior to experimentation and were characterized by following properties: very soft mineral, white to grey in color, hydrophobic interaction, and cation exchange capacity (CEC) between 50 and 70 meq 100 g^{-1}. Median total SiO_2, Al_2O_3, K, Ca, Mg, Cu, Ni, Zn, Co, Mn, Pb and Cr contents were 67.55%, 19.10%, 0.3%, 6.65%, 0.14%, 1.40 mg kg^{-1}, 2.74 mg kg^{-1}, 25.68 mg kg^{-1}, 0.4 mg kg^{-1}, 93.14 mg kg^{-1}, 7.97 mg kg^{-1} and 0.76 mg kg^{-1}, respectively. The above-mentioned results of pyrophyllite chemical analysis are obtained in the laboratory of Faculty of Agriculture and Food Sciences University of Sarajevo.

2.2 Study Area

The five soil plots in the immediate vicinity of steel factory in Zenica formed the study area (44° 22′5″ N, 17° 89′85″ E). The soil plots had approximately 100 m^2 in area and were located at a very close distance from each other (up to 500 m). According to Soil Taxonomy, all examined soils were classified as Eutric Cambisol. Base saturation of more than 50% and pH value > 5.5 is a typical characteristic of this type of soil [12].

The soil average samples were taken from the selected soil plots in March 2018, then they were transported to the laboratory of the Faculty of Agriculture and Food Science University of Sarajevo where the determination of heavy metals in soil samples was carried out. Concentrations of heavy metals (Cu, Zn, Mn, Pb, Cd) were highest in the soil plot at the local community 'Gradišće', and therefore the soil from this plot was collected for performing the ex situ soil remediation technique.

2.3 Soil Sampling and Analysis

Soil samples for the pot experiment were collected from the surface layers (depth 0–30 cm) of selected soil plot using plastic shovel and then transferred to the laboratory of Faculty of Agriculture and Food Sciences University of Sarajevo. Soil samples were cleared of impurities, air-dried at room temperature, ground and sieved through a 2 mm sieve, homogenized and then uniformly added into the treatment pots.

Before the start of the experiment, the soil chemical analysis was performed and the following parameters were analyzed: soil acidity (pH), organic matter content, content of available forms of phosphorus (P_2O_5) and potassium (K_2O) and content of total and available forms of heavy metals (Cu, Zn, Mn, Pb, Cd).

Soil pH was determined according to ISO 10390 method [13], organic matter according to ISO 14235 method [14] and the content of available phosphorus and potassium according to AL – method [15]. Total and available forms of heavy metals in soil sample were extracted using aqua regia (mixture of nitric and hydrochloric acid in 1:3 ratio) and EDTA solution (0.01 mol dm^{-3} ethylenediaminetetraacetic acid (EDTA) and 1 mol dm^{-3} $(NH_4)_2CO_3$, adjusted to pH 8.6), respectively [16, 17]. The concentration of heavy metals in the obtained extract were determined using Shimadzu AA-7700 atomic absorption spectrophotometer [18].

2.4 Pot Experiments

Plastic pots (height 13 cm and diameter 20) with a capacity of 2 kg of soil were used to conduct the experiment. The experiment involved 21 pots comprised 7 treatments with three replicates in a completely randomized design.

Experiment treatments were as follows:

T_1 - soil without clay minerals i.e. control treatment
T_2 - soil with zeolite at rate of 200 kg ha^{-1}
T_3 - soil with zeolite at rate of 400 kg ha^{-1}
T_4 - soil with zeolite at rate of 600 kg ha^{-1}
T_5 - soil with pyrophyllite at rate of 200 kg ha^{-1}
T_6 - soil with pyrophyllite at rate of 400 kg ha^{-1}
T_7 - soil with pyrophyllite at rate of 600 kg ha^{-1}

Recommended clay minerals rate was recalculated based on pot area. The treated soils in pots were incubated at 60% of their water holding capacity for 21 days under laboratory conditions.

2.5 Statistical Analysis

All experimental measurements were made in triplicate and the results were given as means ± standard errors. Obtained data were analyzed by a one-way analysis of variance (ANOVA) using Microsoft Excel 2013 software program, and the significant differences between the variants were determined using Least Significant Differences test at 0.05 level of probability ($P \leq 0.05$).

3 Results

3.1 Basic Chemical Properties of the Studied Soil

The results of the soil chemical analysis are presented in Table 1.

Table 1. Chemical analysis of soil sample

Parameter	Unit	Measured value
pH H_2O	pH unit	7.4
pH KCl	pH unit	6.7
Organic matter	%	2.81
P_2O_5	mg 100 g^{-1}	20.84
K_2O	mg 100 g^{-1}	60.9
Cu	mg kg^{-1}	49.3
Zn	mg kg^{-1}	108.7
Mn	mg kg^{-1}	1061.9
Pb	mg kg^{-1}	121.4
Cd	mg kg^{-1}	0.45

Analyzed soil was slightly acidic to neutral with moderate levels of organic matter and high content of available forms of phosphorus (P_2O_5) and potassium (K_2O).

The ranking order of occurrence of heavy metals in the studied soil was Mn > Pb > Zn > Cu > Cd. The mean concentration of Pb and Zn exceeded, while Cu and Cd did not exceed the limit value for agricultural soils prescribed by legislation in Bosnia and Herzegovina [19]. Accordingly, the limit value for Pb, Zn, Cu and Cd in soil is 100 mg kg^{-1}, 100 mg kg^{-1}, 80 mg kg^{-1} and 1.5 mg kg^{-1} respectively. The limit value of Mn is not prescribed by legislation because Mn is not considered as hazardous heavy metal. However, the concentration of Mn in studied soil exceeded the toxic level of Mn in soils (850 mg kg^{-1}) reported by Pais and Jones [20]. Obtained data suggests that the studied soil is contaminated by Pb, Zn and Mn and is therefore selected for performing ex-situ remediation technique through pot experiments.

3.2 Concentrations of Available Heavy Metals in the Soil After Pot Experiments

Determined values of available forms of Cu, Zn, Mn, Pb and Cd in the studied soil, depending on the experimental treatment are presented in Table 2.

Table 2. Concentrations of available heavy metals (mg kg^{-1})

Treatment**	Cu	Zn	Mn	Pb	Cd
T_1	6.5 ± 1.0^a	7.5 ± 0.3	11.9 ± 1.8^a	24.68 ± 4.3	0.12 ± 0.02^a
T2	4.5 ± 0.8^{cd}	7.7 ± 0.9	11.2 ± 1.6^{abcd}	19.52 ± 3.5	0.10 ± 0.02^{ab}
T_3	4.4 ± 1.4^d	7.5 ± 0.8	11.2 ± 3.0^{abcde}	18.82 ± 5.7	0.09 ± 0.03^b
T_4	4.3 ± 0.1^d	7.5 ± 0.1	11.3 ± 1.5^{abc}	18.72 ± 0.4	0.10 ± 0.01^b
T_5	4.7 ± 0.5^{bcd}	7.4 ± 0.2	11.4 ± 2.0^{ab}	19.63 ± 2.5	0.11 ± 0.01^{ab}
T_6	5.4 ± 2.0^{bc}	7.5 ± 1.3	10.0 ± 0.6^{bcdef}	21.50 ± 6.0	0.10 ± 0.02^{ab}
T_7	5.6 ± 0.3^b	7.2 ± 0.1	8.3 ± 0.4^g	19.40 ± 0.4	0.10 ± 0.01^b
$Lsd_{0.05}$	0.96	–	1.61	3.67	0.017

* Means denoted by the same letter indicate no significant difference ($P \leq 0.05$)
Experimental treatment: T_1 - control treatment; T_2 - zeolite at rate of 200 kg ha^{-1}; T_3 - zeolite at rate of 400 kg ha^{-1}; T_4 - zeolite at rate of 600 kg ha^{-1}; T_5 - pyrophyllite at rate of 200 kg ha^{-1}; T_6 - pyrophyllite at rate of 400 kg ha^{-1}; T_7 - pyrophyllite at rate of 600 kg ha^{-1}

As shown in Table 2, the concentrations of available Cu, Mn, Pb and Cd were lower in pots where zeolite and pyrophyllite were added. The statistical analysis does not confirm a significant effect of the added clay minerals on the reduction of Zn availability between experimental treatment.

4 Discussion

The pot experiments were undertaken to evaluate the effect of the zeolite and pyrophyllite aluminosilicate minerals on decreasing availability of studied heavy metals in soils. The results showed that applying these clay minerals at different amounts into the soil, in principle, decreases the heavy metals availability. The addition of zeolite at rate of 600 kg ha^{-1} reduced available forms of Mn by 5.3%, Pb by 24.1%, Cd by 17.6% and Cu by 34.9%, while the addition of pyrophyllite at the same rate significantly reduced available forms of Mn by 28%, Pb by 19.4%, Cd by19.3% and Cu by 15% compared to the untreated soil. Furthermore, the addition of pyrophyllite and zeolite at rate of 200 kg ha^{-1} and 400 kg ha^{-1} also reduced the concentration of available forms for Mn, Cd, Pb and Cu, indicating that these treatments could also be effective in remediation of soil polluted by heavy metals.

The above-mentioned results implied that the addition of zeolite was more effective in decreasing Cu and Pb availability, while the pyrophyllite showed a better effect on reducing Mn and Cd availability. Brännvall [21] reported a similar finding related to the effect of zeolite on reducing Cu and Pb availability in the soil. In this paper is also noted that zeolite could adsorb high Zn amount from solution, but this observation was not confirmed in our experiment. The high Zn concentrations in studied soil could be one of the reasons for the lower efficiency of the applied zeolite in our study. Accordingly, it is assumed that the amount of applied zeolite as well as pyrophyllite used in our study was not sufficient to significantly reduce mobility of Zn in the soil.

Jiménez-Castañeda and Medina [22] reported that zeolite efficiency in removal of heavy metals is result of its high ion exchange and adsorption molecular sieving capacity. This finding consistent with the results of other studies [23–25].

The cation exchange capacity for pyrophyllite is lower and thus its ability to tightly bond heavy metals through ion exchange process. Regardless, the pyrophyllite also has high efficiency in removing heavy metals from the soil and results of this study as well as many other studies confirm this hypothesis [26–28]. Although pyrophyllite has high efficiency in removing heavy metals from the soil, its adsorption mechanism for heavy metals is still not fully understood. It is well known that pyrophyllite ($Al_2[Si_4O_{10}](OH)_2$) is the typical 2:1 dioctahedral phyllosilicate that consists of an aluminium hydroxide layer ($Al(OH)_2$) sandwiched between two silicate layers, where all tetrahedral sites are filled with Si^{4+}, two-thirds of the octahedral sites are filled with Al^{3+}, and the remaining one-third of the sites is vacant. Except vacant sites, pyrophyllite possess a large number of voids in the structure, especially between adjacent layers, representing potential binding sites for heavy metals.

Interesting finding in this study was the fact that the efficiency of zeolite did not change significantly with increasing their amount in the soil. Namely, in the present study there was no statistically significant difference between efficiency of zeolite at a rate of 200 kg ha^{-1}, 400 kg ha^{-1} and 600 kg ha^{-1} on decreasing the availability of tested heavy metals in soil. This inconsistent result could be potentially attributed to the uneven distribution of added zeolite within soil, resulting in its lower exposure area and thus efficiency. However, further investigation is necessary to confirm or denied this observation as well as other observations presented in this study.

5 Conclusions

Zeolite and pyrophyllite incorporation in soil was found to reduce the availability of tested heavy metals in soil under experimental conditions, indicating that addition of these aluminosilicate minerals is a suitable technique for reducing heavy metals availability in soils.

Acknowledgments. The authors would like to express thanks to ZEO Invest Ltd and Harbi Ltd for donation materials used in this research.

Funding. The research was supported by the Environmental Fund of the Federation of Bosnia and Herzegovina.

Conflicts of Interest. The authors declare no conflict of interest. The funders had no role in the design of the study; in the writing of the manuscript, and in the decision to publish the results.

References

1. Ball, B.C.: A framework of connections between soil and people can help improve sustainability of the food system and soil functions. AMBIO J. Hum. Environ. **47**(3), 269–283 (2018)
2. Doabi, S.A., Karami, M., Afyni, M., Yeganeh, M.: Pollution and health risk assessment of heavy metals in agricultural soil, atmospheric dust and major food crops in Kermanshah province, Iran. Ecotoxicol. Environ. Saf. **163**, 153–164 (2018)
3. Khalid, S., Shahid, M., Niazi, N.K., Murtaza, B., Bibi, I., Dumat, C.: A comparison of technologies for remediation of heavy metal contaminated soils. J. Geochem. Explor. **182**, 247–268 (2016)
4. Agnello, A.C., Bagard, M., van Hullebusch, E.D., Esposito, G., Huguenot, D.: Comparative bioremediation of heavy metals and petroleum hydrocarbons co-contaminated soil by natural attenuation, phytoremediation, bioaugmentation and bioaugmentation-assisted phytoremediation. Sci. Total Environ. **564**, 693–703 (2015)
5. Ali, H., Khan, E., Sajad, M.A.: Phytoremediation of heavy metals-concepts and applications. Chemosphere **91**, 869–881 (2013)
6. Liu, L., Li, W., Song, W., Guo, M.: Remediation techniques for heavy metal-contaminated soils: principles and applicability. Sci. Total Environ. **633**, 206–219 (2018)
7. Mosa, K.A., Saadoun, I., Kumar, K., Helmy, M., Dhankher, O.P.: Potential biotechnological strategies for the cleanup of heavy metals and metalloids. Front. Plant Sci. **7**, 303 (2016)
8. Uddin, M.K.: A review on the adsorption of heavy metals by clay minerals, with special focus on the past decade. Chem. Eng. J. **308**, 438–462 (2017)
9. Ghorbel-Abid, I., Trabelsi-Ayadi, M.: Competitive adsorption of heavy metals on local landfill clay. Arab. J. Chem. **8**(1), 25–31 (2015)
10. Adeyemo, A.A., Adeoye, I.O., Bello, O.S.: Adsorption of dyes using different types of clay: a review. Appl. Water Sci. **7**(2), 543–568 (2017)
11. Bradl, H.B.: Adsorption of heavy metal ions on soils and soils constituents. J. Colloid Interface Sci. **277**, 1–18 (2004)
12. FAO: World reference base for soil resources. Food and agriculture organization of the United Nations. http://www.fao.org/3/i3794en/I3794en.pdf. Accessed 23 May 2019
13. ISO: International Standard ISO 10390: soil quality - determination of pH. International Organization for Standardization, Geneva, Switzerland (2005). https://www.sis.se/api/document/preview/905787/. Accessed 23 May 2019
14. ISO: International Standard ISO 14235: Soil quality - determination of organic carbon in soil by sulfochromic oxidation, International Organization for Standardization, Geneva, Switzerland (1998). https://www.iso.org/standard/23140.html. Accessed 23 May 2019
15. Egnér, H., Riehm, H., Domingo, W.R.: Untersuchungen über die chemische Boden analyse als Grundlage für die Beurteilung de Nährstoffzustandes der Böden. II. Chemische Extraktions methoden zur Phosphor und Kaliumbestimmung. Kungliga Lantbrukshögskolans Annaler **26**, 199–215 (1960)
16. ISO: International Standard ISO 11466: Soil quality - extraction of trace elements soluble in aqua regia, International Organization for Standardization, Geneva, Switzerland (1995). https://www.iso.org/standard/19418.html. Accessed 23 May 2019
17. Trierweiler, J.E., Lindsay, W.L.: EDTA-ammonium carbonate soil test for zinc. Soil Sci. Soc. Am. Proc. **39**, 49–54 (1969)

18. ISO: International Standard ISO 11047: Soil quality - determination of cadmium, chromium, cobalt, copper, lead, manganese, nickel and zinc - flame and electrothermal atomic absorption spectrometric methods, International Organization for Standardization, Geneva, Switzerland (1998). https://www.iso.org/standard/24010.html. Accessed 23 May 2019
19. Official Gazette of FBiH: Rulebook on determination of allowable quantities of harmful and hazardous substances in soils of Federation of Bosnia and Herzegovina and methods for their testing, No. 72/09. Edit. Official Gazette of FBiH, Sarajevo, Bosnia and Herzegovina (2009). http://www.fzofbih.org.ba/userfiles/file/izmj_zakon_otpad.pdf
20. Pais, I., Jones, J.B.: The Handbook of Trace Elements, 1st edn. Lucie Press, Boca Raton (1997)
21. Brannvall, E.: An experimental study on the use of natural zeolite for Cu, Pb and Zn immobilization in soil. Geologija **56**, 1–4 (2006)
22. Jiménez-Castañeda, M.E., Medina, D.I.: Use of surfactant-modified zeolites and clays for the removal of heavy metals from water. Water **9**(4), 235 (2017)
23. Taamneh, Y., Sharadqah, S.: The removal of heavy metals from aqueous solution using natural Jordanian zeolite. Appl. Water Sci. **7**(4), 2021–2028 (2017)
24. Gorimbo, J., Taenzana, B., Muleja, A.A., Kuvarega, A.T., Jewell, L.L.: Adsorption of cadmium, nickel and lead ions: equilibrium, kinetic and selectivity studies on modified clinoptilolites from the USA and RSA. Environ. Sci. Pollut. **25**(31), 30962–30978 (2018)
25. Colella, C.: Ion exchange equilibria in zeolite minerals. Miner. Deposita **31**(6), 554–562 (1996)
26. Prasad, M., Saxena, S.: Attenuation of divalent toxic metal ions using natural sericitic pyrophyllite. J. Environ. Manag. **88**(4), 1273–1279 (2008)
27. Scheidegger, A.M., Sparks, D.L., Fendorf, M.: Mechanisms of nickel sorption on pyrophyllite: macroscopic and microscopic approaches. Soil Sci. Soc. Am. J. **60**(6), 1763–1772 (1996)
28. Caporale, A.G., Vioalante, A.: Chemical processes affecting the mobility of heavy metals and metalloids in soil environments. Curr. Pollut. Rep. **2**(1), 15–27 (2016)

Soil Water Balance Response to Climate Change in Posavina Region

Sabrija Čadro[1(✉)], Monika Marković[2], Selman Edi Kaloper[1], Marija Ravlić[2], and Jasminka Žurovec[1]

[1] Faculty of Agriculture and Food Sciences, University of Sarajevo, Zmaja Od Bosne 8, Sarajevo, Bosnia and Herzegovina
s.cadro@ppf.unsa.ba

[2] Faculty of Agrobiotechnical Sciences, Josip Juraj Strossmayer University of Osijek, Vladimira Preloga 1, 31000 Osijek, Croatia

Abstract. Posavina region represents the area with the most favorable natural conditions for agricultural production in Croatia and Bosnia and Herzegovina. Increase in the air temperature and changes in the amount of rainfall results in changes in evapotranspiration and the values of main water balance elements indicating the need for a precise determination of the current and future state of basic water balance elements, such as soil moisture deficit and the amount of runoff. Linear regression was applied to determine soil water balance response to climate change in Posavina region. The air temperature, precipitation, reference evapotranspiration and water balance components: actual evapotranspiration, total runoff, soil moisture deficit and amount of snow trends were analyzed. Monthly weather data from 4 weather stations, two in Croatia: Slavonski Brod and Gradište; and two in Bosnia and Herzegovina: Doboj and Gradačac, for the time period of 58 years (1961–2018) were used. The results obtained show increasing trends in annual air temperature (0.032 °C–0.057 °C year^{-1}), sum of precipitation (1.424 mm–2.317 mm year^{-1}), reference evapotranspiration (0.462 mm–4.640 mm year^{-1}), actual evapotranspiration (0.019 mm–2.190 mm year^{-1}), soil moisture deficit (0.443 mm–2.672 mm year^{-1}) and total runoff (0.286–2.46 9 mm year^{-1}) series and decreasing trend in the annual amount of snow (0.676–1.664 mm year^{-1}). Results obtained are showing an urgent need to start with climate change adaptation measures and actions to combat the negative impact of climate change in the Posavina region.

Keywords: Water balance · Climate change · Posavina region · Sava River

1 Introduction

The Sava River Basin is located in the Balkan Peninsula and it extends to six countries: Slovenia, Croatia, Bosnia and Herzegovina, Serbia, Montenegro and Albania. The Sava River Basin is a major drainage basin of the South Eastern Europe, covering the total area of approximately 97,713.20 km^2 and represents one of the most significant sub-basins of the Danube River Basin, with the share of 12% [1]. Most of this area is located in BiH (39.25%) and Croatia (25.97%) while the remaining four countries

M. Brka et al. (Eds.): AgriConf 2019, IFMBE Proceedings 78, pp. 11–24, 2020.
https://doi.org/10.1007/978-3-030-40049-1_2

account for 34.79% [2]. Posavina region represents the inner regions of the Sava River Basin, located in Croatia and Bosnia and Herzegovina (BiH).

As for Croatian territory, Posavina area is placed in continental part of the Republic of Croatia and it covers Vukovar-Srijem and Brod-Posavina Counties. Vukovar-Srijem County belongs to Slavonija and Srijem geographic region of eastern Republic of Croatia. It has geostrategic importance because of the boundary with BiH and Republic of Serbia. Regarding the geography, as the part of Panonnian basin the Vukovar-Srijem County is mostly plain with two main types of arable land, automorphic soil (45.38%) and hydromorphic (54.62%). Above 60% of land in Vukovar-Srijem County is arable land (126,968 ha) with average farm size below 10 ha while organic farming in Vukovar-Srijem County counts 1.944 ha [3]. Given that, improving of agricultural strategy in the area of Vukovar-Srijem County is a foundation for achieving the social economic development of this area [4]. The climate of the Vukovar-Srijem County is moderate continental with average annual temperature of 11.4 °C, average maximum air temperature is 16.5 °C, and average annual minimum temperature is 6.2 °C. The average annual rainfall is approximately 660 mm with summer maximum while the lowest amount of rainfall is recorded during the winter time. The average air humidity is 75% [5].

The area of Brod-Posavina County covers 46 soil unit [6] with the most suitable soils for cultivation located near the Sava river. Most of the arable land (63,702 ha) is sown with crops (57.6%). Climate of Brod-Posavina County is moderately warm and rainy continental climate with average annual rainfall 778 mm (spring and autumn maximum). The average annual air temperature is 10.4 °C. The lowland part of the County by the river Sava is mainly the flood-hit area because of the high amount of rainfall in this area and the type of soil (sandy loam and clay) that increases the surface drainage because of the low infiltration rate [7]. The results of RegCM (A2 scenario) simulation showed the increase of air temperature in two periods as follows: for 0.6 °C during the winter time and 0.8 °C during the summer time for 2011–2040 period, furthermore 2 °C during the winter time and 2.4 °C during the summer time for 2041–2070 period [8]

According to Branković, Srnec [9] in the forthcoming period 2041–2070 the expected increase in amplitude and air temperature during the winter time in continental part of the Croatia is approximately 2 °C, while during the summer months up to 2.4 °C. This is important for the future agricultural production in this area since both of the Counties are the regions with the most intensive crop and production in the Republic of Croatia [10].

The region of northern Bosnia, more precisely the Peripanese Bosnia (Bosnian Posavina) represents the area with the most favorable natural conditions for agricultural production. In the classification of land surface structures, BiH has 5% of lowland areas predominantly in the region of northern Bosnia [11]. Growing grain, vegetable-crops as well as livestock production are largely represented in this area. Of the total cultivable land in BiH, currently less than 20% of the land is suitable for intensive agriculture and is mostly located in the lowlands areas of northern Bosnia. In this part of the country, in recent years, due to global climate change, there has been an increase in air temperature, precipitation and wind speed [12–19]. Precipitation does not fully meet the needs of cultivated plants for water, which negatively affects the yield of most agricultural crops. In April and May 2014, record rainfall, over 420 mm [20], resulting in large floods, and in northern Bosnia caused, among other things, enormous damage to agricultural-food production.

The most vulnerable municipalities to climate change in BiH are located in the north part of the country [21], with few of them in Posavina region (Orašje, Bosanski Brod, Modriča). Based on available data and climate projections, exposure to climate change threats will continue to grow, suggesting the need for adaptation to both current and future climate change. This is particularly true for the agricultural sector, which is, due to its socio-economic importance in BiH, highly vulnerable to the negative impact of climate change [22].

The Global Climate Risk Index 2018 (CRI) analyses to what extent countries have been affected by the impacts of weather-related loss events (storms, floods, heat waves etc.). Less developed countries are generally more affected than industrialized countries [23]. The most recent data available from 1998 to 2017, places B&H at 67th place, and Croatia at 35th on the climate risk index list (CRI rank) [24].

Increase in the air temperature and changes in the amount of rainfall results in changes in evapotranspiration and the values of main water balance elements [25] indicating the need for a precise determination of the current and future state of basic water balance elements, such as soil moisture deficit and the amount of runoff.

Thus, the main objective of this study is to determine soil water balance response to climate change in Posavina region using linear regression method, in order to analyze severity of climate change influence to soil water balance and enable better understanding of the impact of climate change on agriculture in this region.

2 Materials and Methods

2.1 Study Area and Data Availability

Study area is represented by Posavina region, the narrow area along the right (BiH) and left (Republic of Croatia) side of river Sava. Administratively this region is part of Vukovar-Srijem and Brod-Posavina County (2,030 km^2) in Croatia and Posavina canton (325 km^2) in BiH.

Because of the low density of weather stations (WS) and the insufficient data they collect, apart from the WS that are located in the very Posavina region, consideration has been given to those in its immediate vicinity. Thus, 4 weather stations (WS), two in Croatia: Slavonski Brod and Gradište; and two in BiH: Doboj and Gradačac, were selected for this study. Daily climatic data, including mean maximum and minimum air temperature (°C), sum of precipitation (mm), mean relative humidity (%), wind speed (m s^{-1}) and sunshine hours (h) for the period 1961–2018 (58 years) were collected and averaged over each month.

Basic location characteristics and number of months used are shown in Table 1.

Table 1. Location, climate characteristics and observation periods of 4 used weather stations (WS) in Croatia and BiH.

WS	Country	A (m)	°E	°N	Period	No. of months	T_{max} (°C)	T_{min} (°C)	RH_{mean} (%)	u (m s^{-1})	n (h)
Slavonski Brod	Croatia	88	17.995	45.159	1963–2018	672	16.8	5.6	78	1.7	5.2
Gradište	Croatia	97	18.704	45.159	1981–2018	456	17.0	6.8	74	2.2	6.0
Doboj	BiH	147	18.083	44.733	1961–2016	672	17.0	6.1	79	1.5	4.6
Gradačac	BiH	225	18.417	44.883	1981–2018	444	16.8	7.7	74	2.3	5.6

Note: A – altitude; °E – longitude; °N – latitude; T_{max} – mean maximum air temperature; T_{min} – mean minimum air temperature; RH_{mean} – mean relative humidity; u – mean wind speed; n – actual duration of sunshine.

2.2 Reference Evapotranspiration (ET_o)

Reference evapotranspiration (ET_0) was calculated using standard FAO-PM equation, given by Allen et al. [26]:

$$ET_0 = \frac{0.408\Delta \cdot (R_n - G) + \gamma \cdot \frac{900}{T_{mean} + 273} \cdot u_2 \cdot (e_s - e_a)}{\Delta + \gamma \cdot (1 + 0.34 \cdot u_2)} \tag{1}$$

where ET_o is the reference evapotranspiration (mm day^{-1}), R_n the net radiation at the crop surface (MJ m^{-2} day^{-1}), G the soil heat flux density (MJ m^{-2} day^{-1}), T_{mean} the mean daily air temperature at 2 m height (°C), u_2 the wind speed at 2 m height (m s^{-1}), e_s the saturation vapor pressure, e_a the actual vapor pressure, $e_s - e_a$ the saturation vapor pressure deficit, Δ the slope of the vapor pressure curve (kPa $°C^{-1}$) and γ is the psychrometric constant (kPa $°C^{-1}$).

All necessary parameters required for calculation of *ETo* where computed following the procedure developed in FAO-56 [26].

Since reflected solar radiation (R_s) is required for R_n calculation and this parameter is not measured on WS in Croatia and BiH, it was estimated from the measured sunshine hours data (The Campbell–Stokes sunshine recorder) with the Ångström [27] equation:

$$R_s = (a_s + b_s \cdot \mathrm{n/N}) \cdot R_a \tag{2}$$

where R_a is the extraterrestrial radiation (MJ m^{-2} day^{-1}) calculated for each day of the year and for different latitudes, from the solar constant (G_{sc} = 0.0820 MJ m^{-2} min^{-1}), the solar declination (δ) and the time of the year (J) and then by selecting the R_a for 15th day of each month converted to monthly values, n is the actual duration of sunshine (h), N is the maximum possible duration of sunshine or daylight hours (h), as is the regression constant, expressing the fraction of extraterrestrial radiation reaching the earth on overcast days (n = 0) and $a_s + b_s$ is the fraction of extraterrestrial radiation reaching the earth on clear days ($n = N$). In the absence of actual solar radiation (R_s) measurements, the values $a_s = 0.25$ and $b_s = 0.5$ were used as suggested by Allen and Pereira [26].

For the WS where measured solar radiation data (R_s) or sunshine hours data were missing for a certain month, solar radiation was estimated using Hargreaves' formula [28] (Eq. 3), as suggested in Allen, Pereira [26]:

$$R_s = k_{R_s}\sqrt{(T_{max} - T_{min})} \times R_a \tag{3}$$

where R_a is extra-terrestrial radiation (MJ m^{-2} d^{-1}), T_{max} maximum air temperature (°C), T_{min} minimum air temperature (°C), k_{Rs} adjustment coefficient (°C$^{-0.5}$).

In the case of used WS, since they are located inland ("interior" location) where land mass dominates and air masses are not strongly influenced by a large water body, value of $k_{Rs} = 0.13$ was used [17], as suggested by Cadro, Cherni-Cadro [13] and Čadro, Uzunović [17].

Actual vapor pressure (e_a) was derived from relative humidity data [26] as:

$$e_a = \frac{e^0(T_{min})\frac{RH_{max}}{100} + e^0(T_{max})\frac{RH_{min}}{100}}{2} \tag{4}$$

where e_a is actual vapor pressure (kPa), $e^0(T_{min})$ saturation vapor pressure at daily minimum temperature (kPa), $e^0(T_{max})$ saturation vapor pressure at daily maximum temperature (kPa), RH_{max} maximum relative humidity (%), RH_{min} minimum relative humidity (%). In the absence of relative humidity data, e_a was estimated by assuming that the dew pint temperature (T_{dew}) is close to daily minimum temperature (T_{min}) [26].

When wind speed was not available, the average regional wind speed value was used.

Monthly values of FAO-PM ET_0 were calculated using REF-ET: Reference Evapotranspiration Calculator [29].

2.3 Soil Water Balance

Monthly water balance was calculated by using Thornthwaite-Mather method [30, 31] that was modified and later described in Dingman [32]. Except data on monthly precipitation (P) and evapotranspiration (ET_o) applied water balance requires data on soil available water content ($SOIL_{max}$). The value $SOIL_{max}$ = 100 mm was used [33] since this is the most commonly used value for the types of soil that are found on the study locations.

After calculation of annual means (μ) and the standard deviation (σ) for all analyzed water balance components a statistical measure of the dispersion of data points, the coefficient of variation (CV) was calculated using equation:

$$CV = \frac{\sigma}{\mu} \tag{5}$$

To detect the trends within time series of water balance components (annual precipitation, reference evapotranspiration, actual evapotranspiration, soil moisture deficit, total runoff and snow) parametric method of linear regression was used, as shown in following equation:

$$y = a + b \times x \quad (6)$$

where x is the explanatory variable, y the dependent variable, b the slope of the line and a the intercept. The slope indicates the mean temporal change of the studied variable. Positive values of the slope show increasing trends, while negative values of the slope indicate decreasing trends [34, 35].

3 Results and Discussion

3.1 Descriptive Statistics

Descriptive statistics (mean and coefficient of variation) for the air temperature (T), precipitation (P), ET_0 and via water balance calculated actual evapotranspiration (AET), soil moisture deficit (SMD), total runoff (TRO) and snow for the 4 WS for the period 1961–2018 are summarized in Table 2.

Table 2. Annual statistics for the climate and water balance (WB) elements in Posavina during the period 1961–2018.

	Croatia				Bosnia and Herzegovina			
WB Element	Slavonski Brod		Gradište		Doboj		Gradačac	
	μ	CV	μ	CV	μ	CV	μ	CV
T_{mean}	11.1	0.2	11.8	0.2	11.1	0.2	12.1	0.2
P	752	34.3	680	43.2	927	52.1	835	46.3
ET_0	765	15.1	843	20.3	730	13.9	857	21.1
AET	599	16.9	575	25.2	624	15.0	644	20.4
SMD	166	25.9	268	35.6	106	21.8	213	32.0
TRO	154	17.7	106	14.5	304	34.4	192	21.9
$SNOW$	113	14.8	99	13.4	155	19.8	117	17.6

Note: T_{mean} – mean air temperature; P – Precipitation; ET_0 – reference evapotranspiration; AET – Actual evapotranspiration; SMD – Soil moisture deficit; TRO – Total runoff; CV – Coefficient of variation

Higher values of almost all climate and water balance elements are recorded in WS at the right side of the Sava River, i.e. in BiH part of Posavina. The mean annual air temperature is very similar, ranging from 11.1 °C (Slavonski Brod, Doboj) to 12.1 °C (Gradačac). Annual sum of precipitation is highest in Doboj (927 mm), or in BiH in general.

These differences in precipitation resulted in differences between water balance elements. Mean ET_o is higher in Slavonski Brod, Gradište and Gradačac, while in Doboj is lower than mean annual precipitation. As the result, at all observed WS, except Doboj, the high values of the *SMD* were obtained (166–268 mm), while in Doboj, because of high annual precipitation, higher TRO values were obtained (304 mm).

These data point to priorities when addressing climate-related issues. Under the conditions of *SMD*, the priorities are the fight against droughts, i.e. the construction of irrigation systems, while in high TRO conditions it is necessary to solve the problems of high water through the construction of embankment, water regulation and drainage system.

The highest coefficient of variation (CV) is for annual precipitation (*P*) ranging from 34.3% to 52.1%. Also, high variations are found for *SMD* (21.8%–35.6%). These results are in line with previous studies of water scarcity [36–38] and high vulnerability [21] of this region to climate extremes, which is, among other things, caused by its great importance, especially when it comes to agricultural production.

3.2 Regression Analysis

Results of the regression analysis for annual air temperature (T_{mean}), precipitation (*P*), ET_0 and via water balance calculated actual evapotranspiration (*AET*), soil moisture deficit (*SMD*), total runoff (*TRO*) and snow for the 4 WS for the period 1961–2018 are presented in Table 3.

Table 3. Results for the statistical tests for the annual climate and water balance elements in Posavina during the period 1961–2018.

WB Element	Croatia				Bosnia and Herzegovina			
	Slavonski Brod		Gra dište		Do boj		Gradačac	
	b	**R**	**b**	**R**	**b**	**R**	**b**	**R**
T	0.037	0.503	0.057	0.534	0.032	0.465	0.048	0.452
P	-0.004	0.005	2.317	0.026	2.162	0.034	1.424	0.014
ET_0	1.451	0.183	4.640	0.483	0.462	0.022	4.019	0.542
AET	0.080	0.001	2.190	0.070	0.019	0.001	1.340	0.064
SMD	1.371	0.056	2.450	0.043	0.443	0.008	2.679	0.104
TRO	0.152	0.001	0.286	0.003	2.469	0.102	0.119	0.004
SNOW	-1.082	0.106	-0.676	0.002	-1.158	0.068	-1.664	0.133

Note: T_{mean} – mean air te mperature; *P* – Precipitation; *ETo* – reference eva potranspiration; *AET* – Actual evapotranspiration; *SMD* – Soil moisture deficit; *TRO* – Total runoff; b – slope, R - Correlation coefficient

The results obtained show increasing trends in *T, P*, ET_0, *AET, SMD* and *TRO* series and decreasing trend in the amount of the *SNOW*. The increasing trends of the air temperature (*T*) are similar between 4 analyzed WS in Croatia and BiH, with values ranging from 0.032 °C $year^{-1}$ to 0.057 °C $year^{-1}$. Increasing trends were also detected for the annual amount of precipitation, ranging from 1.424 mm $year^{-1}$ (Gradačac) to

2.317 mm $year^{-1}$ (Gradište). The exception is WS Slavonski Brod where the amount of precipitation does not change. Based on this data, the annual sum of precipitation for a period of 50 years (1969–2018) in Posavina has increased by 100 mm. The observed air temperature and precipitation change patterns in Croatia and B&H are consistent with the predominant trends in other areas of East Europe [14–16, 39–45] as well as trends observed globally [46–49].

The annual increase of ET_0 vary depending on WS. They are high in Gradište and Gradačac (4.019–4.640 mm $year^{-1}$), and low in Slavonski Brod and Doboj (0.462–1.451 mm $year^{-1}$). Similarly, the magnitude of trend for the annual *AET* is higher in Gradište and Gradačac and lower in Slavonski Brod and Doboj. *AET* has a smaller increasing trend, which will in the future cause an increasing gap between atmospheric potential for receiving water through evapotranspiration and actual evapotranspiration.

Such trends of ET_0 and *AET* point to the fact that in the future, for the intensive agricultural production in the Posavina region, it will be necessary to provide increasing amount of water, since crops will not be able to get this water from precipitation, for serious agriculture production it will be an obligation to build adequate irrigation systems.

SMD is result of difference between ET_0 and available soil moisture [50, 51]. *SMD* has increasing trend at all 4 analyzed WS, ranging from 0.443 mm $year^{-1}$ (Doboj) to 2.672 mm $year^{-1}$ (Gradačac). Such positive trend in *SMD* in the future will cause more severe long-lasting droughts and yield reduction. Similar results were found in other studies all over Croatia [52–54].

Annual amount of *TRO* has low increasing trend at all WS except in Doboj, which is about ten times higher, 2.469 mm $year^{-1}$ compared to 0.286 mm $year^{-1}$.

The only decreasing trends was found for the amount of *SNOW*. Trend values are similar between 4 analyzed WS in Croatia and BiH, with values ranging from −0.676 mm $year^{-1}$ to 1.664 mm $year^{-1}$.

The obtained results indicate an increasing amount of surface waters that can cause soil erosion, landslides, floods and cause great damage both to the areas, infrastructure and agriculture. The Doboj region suffered heavy damage during the 2014 floods [55]. In addition to Doboj, within Posavina region, the areas of Orašje, Bosanski Brod, Bosanski Šamac and Odžak are at the top of the list of Municipalities affected by floods [56].

Average annual T_{max}, T_{mean}, T_{min}, *P*, ET_0, *AET, SMD, TRO* and *SNOW* values together with their linear trend, correlation coefficient (R) and coefficient of determination (R^2) for the Posavina region and time period 1981–2018 are shown in Figs. 1, 2, 3 and 4. Period 1981−2018 was selected since all 4 WS had measurements during that period, so it was possible to determine the averages.

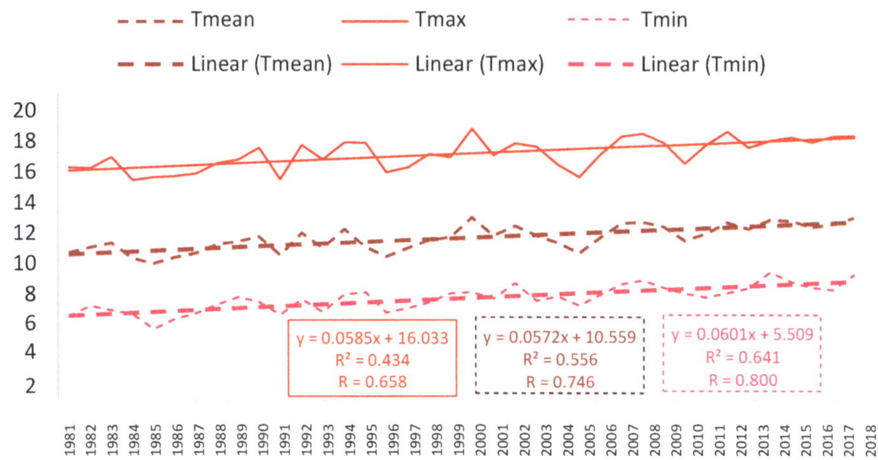

Fig. 1. Linear trend, correlation coefficient (R) and coefficient of determination (R^2) of annual maximum, mean and minimum air temperature for the Posavina region for the period 1981–2018.

The average results for the Posavina region show similar values of the trend as in the case of individual WS. There is an increasing trend in T_{max}, T_{mean} and T_{min} for this area (Fig. 1), ranging from 0.057 to 0.060 °C $year^{-1}$.

Also, this Posavina region is getting more precipitation, where the amount of rain increases as the amount of snow decreases (Fig. 2). It is also interesting to note that the variations in annual precipitation are higher in the period after 2000, when we had 2 extremely dry years 2000 and 2011 (P < 471 mm) and 3 extremely wet: 2001, 2010, and 2014 (P > 1100 mm).

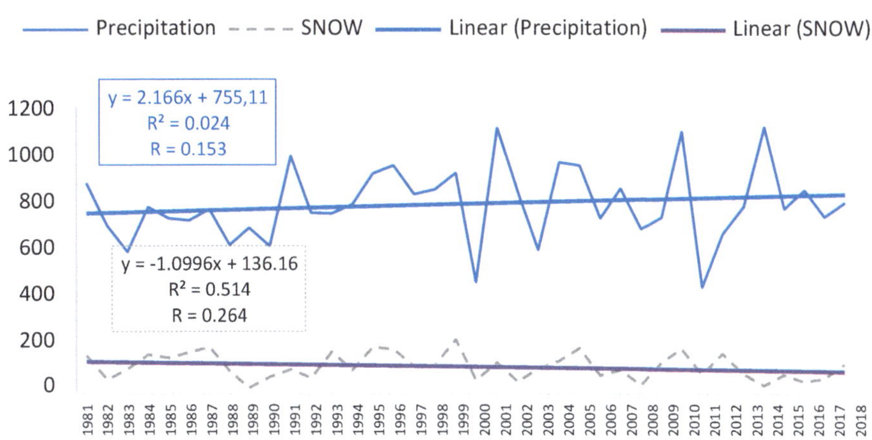

Fig. 2. Linear trend, correlation coefficient (R) and coefficient of determination (R^2) of annual precipitation (*P*) and snow amount (*SNOW*) for the Posavina region for the period 1981–2018.

Such high variations in recent years are also causing high variations in *TRO* and *SMD*, that eventually have devastating effects on Croatia and BiH economies. In Fig. 3. it is possible to notice the shift of extremely wet and dry years, i.e. larger extremes and less and less "normal" years.

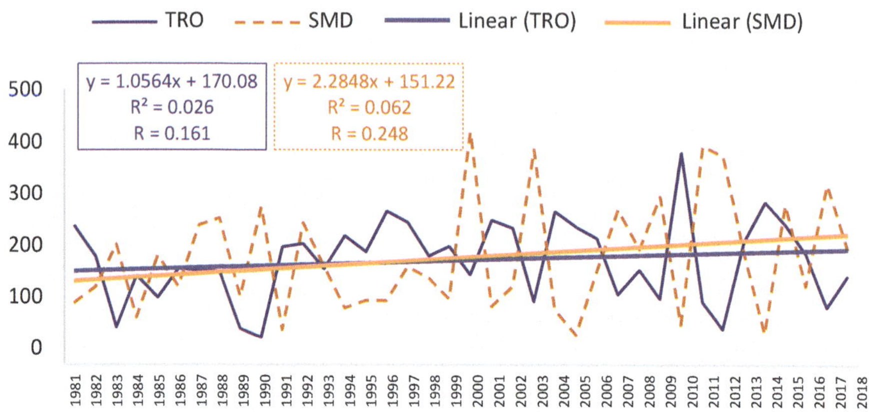

Fig. 3. Linear trend, correlation coefficient (R) and coefficient of determination (R^2) of annual total runoff (*TRO*) and soil moisture deficit (*SMD*) for the Posavina region for the period 1981–2018.

In addition, the gap between ET_0 and *AET* is also increasing (Fig. 4).

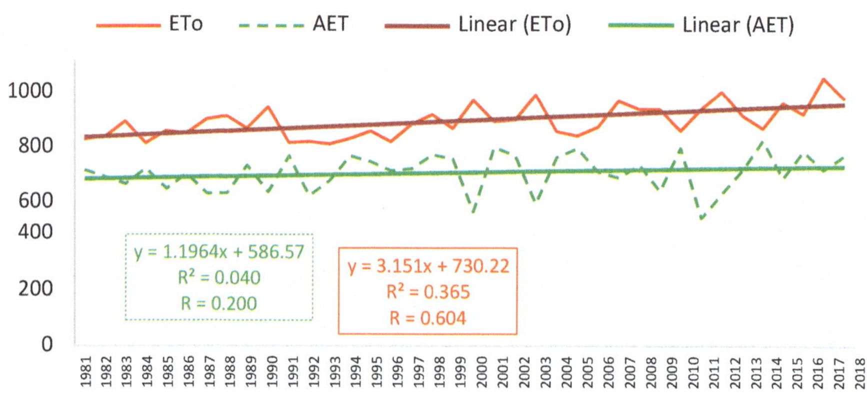

Fig. 4. Linear trend, correlation coefficient (R) and coefficient of determination (R2) of annual reference (*ET0*) and actual evapotranspiration (*AET*) for the Posavina region for the period 1981–2018.

4 Conclusion

Linear regression was applied to determine soil water balance response to climate change in Posavina region. The air temperature, precipitation, FAO-56 PM reference evapotranspiration and water balance components: actual evapotranspiration, total runoff, soil moisture deficit and amount of snow trends were analyzed.

Monthly weather data from four weather stations, two in Croatia: Slavonski Brod and Gradište; and two in BiH: Doboj and Gradačac, for the time period of 58 years (1961–2018) were used.

The results obtained showed that higher values of almost all climate and water balance elements are in BiH part of Posavina. This is especially true for annual sum of precipitation. These differences in precipitation resulted in differences between main water balance elements. Mean reference evapotranspiration is higher in Slavonski Brod, Gradište and Gradačac, while in Doboj is lower than mean annual precipitation. As the result, at all observed locations, except Doboj, the high values of the soil moisture deficit were obtained (166–268 mm), indicating need for irrigation, while in Doboj, because of high annual precipitation, higher total runoff values were obtained (304 mm) pointing to a greater need for drainage.

The results obtained show an increasing amount of surface waters that can cause soil erosion, landslides and floods while in the same time there is an increasing trend in temperature and soil moisture deficit causing the prolonged hot and dry conditions, thus droughts that can seriously affect agricultural production in Posavina.

High variations in amount of precipitation, together with increasing trend of temperature and evapotranspiration, are causing high variations and more frequent occurrence of extreme amount of total runoff and severe droughts. In addition, as the result of such climate variations, the possibility of occurrence of years with extreme rainfall in spring and extreme droughts during the summer, becomes certain.

Based on the obtained results for the Posavina region, we can say that the problems caused by climate change are only partly the result of the increase in trend of climate elements, which is primarily related to temperature increase, the bigger problem is increased variability and frequent occurrence of extreme weather events.

There is an urgent need to start with climate change adaptation measures, with coordinated actions to combat the negative impact of climate change otherwise, agriculture of the Posavina area and therefore the people engaged in it, will in the near future face significant negative consequences.

References

1. ISRBC: Sava River Basin Analysis Summary. The International Sava River Basin Commission, Zagreb, Croatia (2010)
2. ISRBC: Annual Report on Work and Activities of the International Sava River Basin Commission for the period April 1, 2017 - March 31, 2018. International Sava River Basin Commission, Zagreb, Croatia (2018)
3. Deže, J., et al.: Possibilities for entrepreneurship development in organic production of Vukovar-Srijem County. Poljoprivreda **22**(2), 64–69 (2016)

4. Jurić, M.: European Funds in Function of Agricultural Development in Vukovar-srijem County. University in Zadar, Zadar (2016)
5. IMO and Hrast, Razvojna strategija Vukovarsko-srijemske županije 2011. - 2013. Institute for international relationships and Agencija za razvoj Vukovarsko-srijemske županije (2011)
6. Bogunović, M., Husnjak, S.: Primjena GIS tehnologije na primjeru višenamjenskog vrednovanja prostora Brodsko-posavske županije. Agronomski glasnik 1–2 (2000)
7. Petrić-Stjepanović, V.: Zaštita Slavonskog Broda od poplava bujičnih tokova. Građevinar **60** (8), 695–703 (2008)
8. SEEFCCA: Regional climate vulnerability assessment - Synthesis report Croatia, FYR Macedonia, Montenegro, Serbia (2012)
9. Branković, Č., Srnec, L., Patarčić, M.: An assessment of global and regional climate change based on EH5OM climate models ensemble. Clim. Change **98**, 21–49 (2010)
10. CBS: Prirodna bogatstva hrvatskih regija. Croatian Bureau of Statistics Zagreb (2018)
11. Radusin, S., et al.: Third National Communication and Second Biennial Update Report on Greenhouse Gas Emissions of Bosnia And Herzegovina under the United Nations Framework Convention on Climate Change, Sarajevo (2016)
12. Žurovec, J., Čadro, S.: Klimatske promjene, potreba i značaj navodnjavanja poljoprivrednih kultura na području sjeveroistočne BiH. In: 21st Scientific-Expert Conference in Agriculture and Food Industry, Faculty of Agriculture and Food Science, Sarajevo, Neum, Bosnia and Herzegovina (2010)
13. Cadro, S., et al.: A reference evapotranspiration map for Bosnia and Herzegovina. Int. Soil Water Conserv. Res. **7**(1), 89–101 (2019)
14. Miseckaite, O., et al.: Climate and aridity change. In: Utsav Bhattarai, V.P.P. (eds.) 8th Asian Regional Conference (8ARC): Irrigation in Support of Evergreen Revolution, Nepal National Committee of ICID (NENCID), Kathmandu, Nepal (2018)
15. Čadro, S., et al.: Impact of climate change on the annual water balance in a humid climate. In: Agriculture & Forestry, Podgorica, pp. 129–143 (2018)
16. Miseckaite, O., Čadro, S.: Climate change impact on the main conponents of Thornthwaite-Mather water balance. In: 8th International Scientific and Technical Conference "Modern Problems of Water Managmenet, Environmental Protection, Architecture and Construction", Tbilsi, Georgia (2018)
17. Čadro, S., et al.: Validation and calibration of various reference evapotranspiration alternative methods under the climate conditions of Bosnia and Herzegovina. Int. Soil Water Conserv. Res. **5**(4), 309–324 (2017)
18. Čadro, S., et al.: Effect of climate change on agro hydrological balance for some regions in Bosnia and Herzegovina. In: VII International Scientific Agricultural Symposium "Agrosym 2016", Jahorina, pp. 1863–1869 (2016)
19. Žurovec, J., Čadro, S.: Temporal drought and soil moisture variability in the arable land of Spreča valley. In: 26th International Scientific expert Conference in Agriculture and Food Industry. Faculty of Agriculture and Food Scence, Ilidža, Sarajevo (2015)
20. Radusin, S., et al.: Second National Communication of Bosnia and Herzegovina under the United Nations Framework Convention on Climate Change (2013)
21. Žurovec, O., Čadro, S., Sitaula, B.K.: Quantitative assessment of vulnerability to climate change in rural municipalities of Bosnia and Herzegovina. Sustainability **9**(1208), 18 (2017)
22. Žurovec, O., Vedeld, P.O., Sitaula, B.K.: Agricultural sector of Bosnia and Herzegovina and climate change—challenges and opportunities. Agriculture **5**(2), 245–266 (2015)
23. Eckstein, D., Künzel, V., Schäfer, L.: Global Climate Risk Index 2018. Germanwatch, Bonn (2018). 36 p
24. Eckstein, D., Hutfils, M.-L., Winges, M.: Global Climate Risk Index 2019. Germanwatch, Berlin (2018)

25. Giugliano, R.P., et al.: Edoxaban versus Warfarin in Patients with Atrial Fibrillation. N. Engl. J. Med. **369**(22), 2093–2104 (2013)
26. Allen, R.G., et al.: Crop evapotranspiration: guidelines for computing crop water requirements. FAO Irrigation and Drainage Paper. United Nations FAO, Rome, p. 326 (1998)
27. Ångström, A.: Solar and terrestrial radiation. Quaterly J. R. Meteorol. Soc. **50**, 121–125 (1924)
28. Hargreaves, G.H., Samani, Z.A.: Reference crop evapotranspiration from temperature. Trans. ASAE **1**(2), 96–99 (1985)
29. Allen, R.G., Zhenguli, L.G.: REF-ET Reference Evapotranspiration Calculator. University of Idaho Research and Extension Center Kimberly, Idaho (2016)
30. Thornthwaite, C.W., Mather, J.R.: The Water Balance. Publications in Climatology. Drexel Institute of Technology, Laboratory of Climatology, Centerton (1955)
31. Thornthwaite, C.W., Mather, J.R.: Instructions and Tables for Computing Potential Evapotranspiration and the Water Balance. Publications in Climatology, vol. 10, no. 3, p. 311. Laboratory of Climatology, Centerton (1957)
32. Dingman, S.L.: Physical Hydrology, 2nd edn. Prentice Hall, Upper Saddle River (2002). 646 p
33. McBean, E.A., Rovers, F.A., Farquhar, G.J.: Soil Waste Landfill Engineering and Design. Prentice Hall PRT, New Jersey (1995)
34. Gocic, M., Trajkovic, S.: Analysis of changes in meteorological variables using Mann-Kendall and Sen's slope estimator statistical tests in Serbia. Global and Planet. Change **100**, 172–182 (2013)
35. Gocic, M., Trajkovic, S.: Analysis of trends in reference evapotranspiration data in a humid climate. Hydrol. Sci. J.-Journal Des Sciences Hydrologiques **59**(1), 165–180 (2014)
36. Čustović, H., Ljuša, M., Situala, B.K.: Adaptacija na klimatske promjene u sektoru poljoprivrede (vrijeme je da djelujemo odmah), Sarajevo (2015)
37. Vlahinić, M., Čustović, H., Alagić, E.: Prijeti li kriza vode potrebne za navodnjavanje u poljoprivredi Bosne i Hercegovine. Radovi poljoprivredno- prehrabenog fakulteta **58**(2), 25–36 (2006)
38. Hodžić, S., Marković, M., Čustović, H.: Drought Conditions and Management. Strategies in Bosnia and Herzegovina - Concise Country Report. UNW-DPC Proceedings (2013)
39. Bukantis, A., Rimkus, E.: Climate variability and change in Lithuania. Acta Zoologica Lituanica **15**(2), 100–104 (2005)
40. Branković, Č., et al.: Sixth National Communication of the Republic of Croatia under the United Nations Framework Convention on the Climate Change (UNFCCC) Selected Sections in Chapters: 7. Climate Change Impacts and Adaptation Measures 8. Resarch, Systematic Observation and Monitoring. Meteorological and hydrological service of Croatia, Zagreb, Croati (2013)
41. Burić, D., et al.: Recent extreme air temperature changes in Montenegro. Bull. Serb. Geogr. Soc. **XCV**(4), 53–66 (2013)
42. Unkasevic, M., Tosic, I.: Trends in temperature indices over Serbia: relationships to large-scale circulation patterns. Int. J. Climatol. **33**(15), 3152–3161 (2013)
43. Jaagus, J., et al.: Precipitation pattern in the Baltic countries under the influence of large-scale atmospheric circulation and local landscape factors. Int. J. Climatol. **30**(5), 705–720 (2009)
44. Rutgersson, A., et al.: Observed changes and variability of atmospheric parameters in the Baltic Sea region during the last 200 years. Clim. Res. **61**(2), 177–190 (2014)

45. Tripolskaja, L., Pirogovskaja, G.: Impact of climate variability in Lithuania and Belarus on atmospheric precipitation infiltration: lysimetric study. Zemdirbyste Agric. **100**(4), 369–376 (2013)
46. Popov, T., et al.: Recent trends in extreme temperature indices in Bosnia and Herzegovina. Carpathian J. Earth and Environ. Sci. **13**(1), 211–224 (2018)
47. Trenberth, K.E., et al.: Global warming and changes in drought. Nat. Clim. Change **4**, 17 (2013)
48. Kharin, V.V., et al.: Changes in temperature and precipitation extremes in the CMIP5 ensemble. Clim. Change **119**(2), 345–357 (2013)
49. Jacob, D., et al.: Climate impacts in Europe under +1.5°C global warming. Earth's Future. **6** (2), 264–285 (2018)
50. Kos, Z., et al.: Priručnik za hidrotehničke melioracije. II Kolo Navodnjavanje: potrebe vode za navodnjavanje, vol. 2. Građevinski fakultet Seučilišta u Rijeci, Hrvatsko društvo za odvodnjavanje i navodnjavanje, Rijeka (1993)
51. Žurovec, J.: Melioracije i uređenje poljoprivrednog zemljišta, p. 245. Univerzitet u Sarajevu, Poljoprivredno-prehrambeni fakultet, Sarajevo (2012)
52. Cindrić, K., Pasarić, Z., Gajić-Čapka, M.: Spatial and temporal analysis of dry spells in Croatia. Theor. Appl. Climatol. **102**(1–2), 171–184 (2010)
53. Perčec Tadić, M., et al.: Drought vulnerability in Croatia. Agriculturae Conspectus Scientificus **79**(1), 31–38 (2014)
54. Whan, K., et al.: Impact of soil moisture on extreme maximum temperatures in Europe. Weather Clim. extremes **9**, 57–67 (2015)
55. EC: Bosnia and Herzegovina Floods, 2014 Recovery Needs Assessment. European Commission, Sarajevo (2014)
56. HEIS: Floods and Landslides Risk Assessment for the Housing Sector in Bosnia and Herzegovina. Hydro-Engineering Institute Sarajevo – HEIS, Sarajevo (2015)

The Influence of Changes in Way of Use, in Climate and in Hydrological Conditions on Histosol in Reclamation Area in Peatland Ždralovac

Possibilities of Re-establish of Agricultural Production in Abandoned Reclamation Area in Peatland Ždralovac in Livanjsko Polje After 25 yr.

Mirza Tvica[1(✉)], Hamdija Čivić[1], and Vesna Tunguz[2]

[1] Faculty of Agriculture and Food Sciences, University Sarajevo, Sarajevo, Bosnia and Herzegovina
m.tvica@ppf.unsa.ba
[2] Faculty of Agriculture, University Istočno Sarajevo, Lukavica, Bosnia and Herzegovina

Abstract. Fen type peatland Ždralovac in Livanjsko Polje (Livno's karsts field) in south-west Bosnia and Herzegovina traditionally were used for livestock grazing with mowing grass, agricultural production by the sides of the field and peat excavation only for own necessaries.

The 1970s at southern part peatland with shallow peat was built a reclamation area Table (1000 ha) used for agricultural production. The entire hydro meliorated system is designed based on shallow and controlled drainage with the system of water-gates that maintained the water table at a depth which enabled the continuous capillary wetting of the rhizosphere in the dry summer period. After the last war (1991–1995) in Bosnia and Herzegovina, these areas were abandoned over the past 20 years so that the entire system of water-gates is now devastated. Peatland use change and uncontrolled drainage together with climate change are lowered the water table level and prolonged drying of the surface layer in the summer period. This has further caused accelerates peat mineralization by changing its traits and reducing the peat depth to only 20 cm. Everything was followed numerous long-lasting fires resulting in the changes in vegetation cover on agricultural areas that overgrown into grassy surfaces, shrubs and trees.

The aim of this paper is to evaluate the possibility of renewal of agricultural production on these surfaces taking into account the results of the analysis of climate change, changes in hydrological conditions, soil water balance and the physical and chemical properties of Histosol.

The possibility of restoration of agricultural production on these surfaces primarily includes the continuous supply crops with the water in the summer period.

Keywords: Peatland · Drainage · Climate change · Histosol

M. Brka et al. (Eds.): AgriConf 2019, IFMBE Proceedings 78, pp. 25–35, 2020.
https://doi.org/10.1007/978-3-030-40049-1_3

1 Introduction

Livanjsko Polje, located in the southwestern part of Bosnia and Herzegovina, is one of the world largest karsts fields (65 km × 6 km), at the average height of 700 m above sea level. Karsts fields have an elongated shape where field sides are built from carbonate rocks. On the field margins, mostly at higher hypsometric levels, there are water springs. At some lower levels, there are swallow holes that receive surface watercourses which then flow underground toward a lower hypsographic level to the river Cetina and the Adriatic Sea.

Peatland Ždralovac (3.615 ha), located in the northwestern part of the field, is situated in two landscape depressions with impermeable bottom named Veliki Ždralovac and Mali Ždralovac. It was formed as a geogenous fen type peatland, fed by groundwater that has been in contact with limestone and mineral soils and with dominant vegetation reeds and sedges. Most important input water in peatland is by wellspring/watercourse Bastašica which is active average 160 days per year. Ždralovački channel located in the central part of peatland drains vast areas towards the swallow hole of Kazanci. In the rainy season, when the inflow of water into the swallow hole Kazanci is greater than the velocity of swelling through, around the swallow hole is formed water retention basin (Fig. 1) and along with fen Ždralovac, this entire northwestern part of the karsts field is flooded. In hydrological terms, peatland represents a separate part from the rest of the field from whose surface water is collected and used for hydroelectric purposes. Livanjsko Polje is of particular importance for birds, in 2008 was proclaimed the RAMSAR area, and since 2011 it has been included in the IBA- Important Birds Areas list.

People have always drained peatland for their needs. Traditionally using peatland was: livestock grazing with grass and young reeds mowing in the dry summertime, agricultural production by the sides of the field which has not been flooded and peat excavation only for their own needs. But, the 1970s in central part of peatland with deepest Histosol (200 cm) started peat excavation and in same time on the southern part of the peatland – Mali Ždralovac (1000 ha) with shallow Histosol, is converted for use in agricultural production, thus becoming the reclamation area named "Table" (Fig. 1). The entire system in "Table" is designed based on shallow and controlled drainage by water-gates, which ensured the continuous wetting of the root spreading layer during the dry summer months. After the last war in Bosnia (1991–1995), these areas were abandoned over the past 20 years so that the entire system of water-gates is now devastated.

Peatland use change and uncontrolled drainage together with climate change lowered the water table level and led to mineralization of peat, as well as to changes in vegetation cover [7–9, 13]. Also, once the peat surface has been desiccated during the summer months it becomes more prone to fire and erosion by wind.

The aim of this paper is to point to the main limiting factors of renewal of agricultural production on these surfaces. In fact, the impact of changes in the use of peatland, in the hydrological condition after drainage and climate change of this area on the properties of Histosol should be evaluated and the possibilities of re-establishing agricultural production should be assessed.

2 Material and Method

For climate analysis of the Livno area especially in the context of climate change we used data from Meteorological station Livno for the period 1961–2014 [11]. Vegetation period is calculated by [6], ds $= ((T_{trsh} - T_{below}) \cdot Dm)/(T_{above} - T_{below})$

- T_{trsh} - temperature threshold 10 °C,
- T_{below} and T_{above} - the average monthly temperature for two consecutive months, between which the temperature threshold is,
- ds the number of days we need to add on the 15th day of the first month to get the date when it was overrun T trsh at the beginning of the season in the spring, or the number of days we should subtract in the fall of the 15th day of the second month to get the Ttrsh at the end of the season, and
- Dm - the number of days between both months (taken 30 days)

For the analysis of hydrological data in peatland before this research, we used Hydrological and ecological study of Livanjsko polje [14]. Measurement of groundwater level is done in piezometer. On two occasions during 2013, 28.02. and 20.03. We have performed the geodetic recording of the upper level of water in the area of the peatland and swallow hole Kazanci. As the floods in the hydro-meliorated area "Table" directly under the influence of water retention around the Kazanci swallow hole (Fig. 1) data about the beginning and end of the duration water retention Kazanci were collected from the local population during the 2011–2014 research period.

For the analysis of the morphological, physical and chemical properties of Histosol, a profile was opened up to the appearance of an impermeable layer of carbonate clay from which the peat samples were taken in the disordered condition and in the Kopecky ring (100 cm^3) by layers in 3 repeats.

As a standard method for moisture, ash and organic matter of peat and other organic soils, ASTM Standards D 2974-87, 1990 [3] was used. Method A for determining moisture of peat dried in the oven at 110 ± 5 °C (Ms), according to the formula: ws = ((M − Ms) • 100)/Ms (%), where M is mass of the wet peat and Ms is mass of the peat dried in the oven at 105°C. The ash content was determined according to Method C by incineration in the oven at a temperature of 440 ± 40 °C of the dried sample. The residual substance after incineration is ash. The ash percentage represents the weight of ash related to the weight of the sample dried in the oven at 105 °C. The percentage of the organic part is obtained by subtracting the ash percentage of 100. ASTM Standard D 4531-86, reapproved 2002, [2] was used as the Standard Test Method for Bulk Density (BD). Bulk density (BD) - the mass of dried peat (Ms) dried in the oven at 105 °C per unit of total peat volume (V) defined by the volume of the sampling cylinder (100 cm^3); is calculated according to the formula:

$$BD = Ms/V\,(g/cm^3).$$

The specific gravity Sg, was made according to the mass of dried peat at 105 °C (Ms) in relation to the volume of the same dried peat (Vs). The volume of dried peat was measured using the standard air pycnometer method.

Porosity, P, is calculated by the formula P = ((SDBD) × 100)/SD.

Available forms of phosphorus (P_2O_5) and potassium (K_2O) were made by the standard AL-method Enger-Riehm-Domingo [5].

The pH value of the peat was determined electrometrically in water suspension. The preparation was taken from ASTM 2976-71, 1990 [4].

The content (%) of total carbon (C) and total nitrogen (N) in the peat sample was determined by dry combustion in the elemental analyzer (CHNS Elemental Analyzer: CHNS Elementar Vario EL III, Hanau, Germany, 2001).

Soil water balance for the area Table calculated by [21]: Knowing the annual precipitation (P - mm) and reserve water in the soil (R - mm) which is associated with the water capacity of the soil, it is possible by introducing potential evapotranspiration (PET - mm) [12] into the equilibrium of the soil water balance to achieve real evapotranspiration (RET - mm) which is limited by the amount of water available in the soil, and obtaining data of water deficits (D = PET–RET) in the vegetative season as well as data on surplus of water as a potential outflow. 50% of the total available water was used as a water reserve (R) in the soil. The largest water flow is in 50 cm deep of soil and for these case the available water is usually taken 100 mm, but if it is a shallow soil (like in "Table", this value (R) may be lower than 50 mm [22].

Fig. 1. Water phenomena in the area of the swalow hol Kazanci with a series of piezometers (red point 1–11)

3 Results and Discussion

Climate: Many scientists point out that global warming also has impact on reducing the water level on peatlands [18, 19].

Data show that the average annual temperature in four years of the study (10,6 °C) is higher than average annual temperature in the series from 1961–2014 (9,4 °C) by 1.2 °C, while the average annual rainfall for the period of investigation (1260 mm) exceeds by 105 mm average annual rainfall for the same series (1155 mm) (Table 1).

Table 1. Climate data for the years research 2011–2014 and in the series 1961–2014 from meteorological station Livno [11]

Month	Precipitation P (mm)		Temperature (°C)	
	Average 2011–2014	Average 1961–2014	Average 2011–2014	Average 1961–2014
I	110	97	1,6	−0,1
II	101	86	1,2	1,0
III	86	91	5,7	4,4
IV	115	99	10,1	8,6
V	105	78	13,4	13,5
VI	92	87	18,6	16,9
VII	86	52	20,7	19,3
VIII	46	66	20,6	18,7
XI	151	97	15,8	14,4
X	109	115	10,7	9,9
XI	114	154	7,1	5,1
XII	145	132	2,0	0,9
Year	1260	1155	10,6	9,4

In Bosnia and Herzegovina's first national report in accordance with the framework of climate change [10] the area of Livanjsko Polje was concluded to be in the region of increase of average annual air temperature in the range of 0.6 to 0.8 °C in the last 100 years. These changes are manifested in the more frequent occurrences of temperature extremes, above the absolute maximum, especially in the period 2000–2012. This is also apparent from the following picture (Fig. 2).

According to the same report [10], the precipitation in this wider area shows an increase of 2% and can be said to stagnate but noticeably increase the intensity of the particular rain while simultaneously extending the dry (no rain) periods. The first phenomenon results in an increased risk of flooding, and the second results in drying of the surface layer in the summer period.

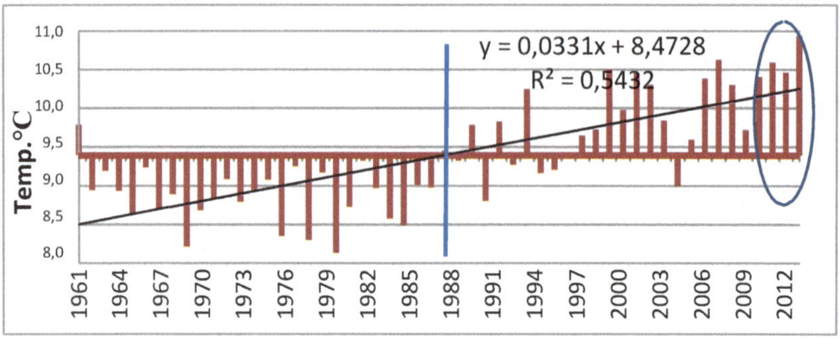

Fig. 2. Average annual temperatures for the period 1961-2014 and trend of growth, MS Livno

3.1 Duration of Water Retention Kazanci in the Research Period

As can be seen from Fig. 1, the floods in the hydro-meliorated area "Table" is directly under the influence of water retention around the Kazanci swallow hole. Table 2 shows that the average annual water retention period of Kazanci in the three-year period of research (2011–2014) is around 139 days, which corresponds to the average duration of retention in the rainy years of 140 days for the period 1959–1991 [14].

Table 2. Duration water retention Kazanci*

Start (dates)	End (dates)	Duration water retencion Kazanci (days)
	25.03.2011	
25.12.2011	12.04.2012	109*
20.12.2012	22.05.2013	154*
22.11.2013	29.03.2014	129*
05.05.2014	25.05.2014	25*

*±4 days

As in normal conditions, most plants begin to develop in spring when the peat temperature reaches 6–8 °C, so the total vegetation period begins to count when the mean air temperature reaches 10 °C and the soil temperature is about 5 °C. Calculated vegetation period [6] in an average year for temperature threshold of 10 °C is in the duration of 174 days and started 24.04. and finished at 14.10. As can see from Table 2, during the intense rainy season as it was in 2012/2013. and due to the heavy rainfall in May 2014, the water retention disappeared on May 22–25 so that the "Table" area was flooded about one month during the vegetation period.

3.2 Histosol

Especially low peatlands are unusually grateful for reclamation because good regulation of hydrological conditions and rational agro-techniques can convert them into much more fertile soil than mineral soil [17]. For use in agricultural purpose, peatland Mali Ždralovac covered with drainage channels and the surfaces between the channels represent the production area. However, the intensive cultivation of peat in agricultural production increases the aeration, and thus the intensive decomposition of organic matter as well as the change of the physical properties of peat [16]. Agricultural exploitation of the peat usually converts the top layer of peat into a special mineralized layer. This layer called "muck" [1] represents an organic soil containing more than 35% of inorganic material (as opposed to peat with 75% organic matter) and in which the botanical origin of organic matter can no longer be easily detrmined (Fig. 3).

Fig. 3. Profile 1, Peatland used in agriculture, mineralized "muck" layer

During peat cultivation, a specific moorsh forming process (MFP) is a major step in the degradation of organic materials that lead to physical deformation of colloid, peat degradation, and related processes such as CO_2 production, mineralization of organic N and P, nitrification and denitrification, as well as rinsing of inorganic phosphates, nitrates and soluble C [20]. The physical and chemical properties of Profiles 1 are presented in Tables 3 and 4.

Table 3. Profile 1. physical properties

Layer	Depth cm	Bulk density (BD) g/cm^3	Specific gravity g/cm^3	Porosity (P) %	Water retention (0,01 Mpa) %
nHpI	0–10	0,33	1,73	81,0	229,8
nHpII	10–20	0,53	2,18	75,8	133,1
clay	20–35	0,87	2,73	68,1	74,0

Table 4. Profile 1. chemical properties

Layer	Depth cm	pH H_2O	Organic content %	Mineral content %	N %	C %	C kg/m^2	mg/100 g	
								P_2O_5	K_2O
nHpI	0–10	6,9	43,7	56,3	1,6	20,5	6,7	2,8	22,7
nHpII	10–20	7,2	31,1	68,9	1,1	13,9	7,3	1,8	8

The depth of peat body is only 20 cm. The 10–20 cm layer is more mineralized than the surface layer 0–10 cm and as such it is characterized by higher values ofbulk density and specific gravity, less porosity and a lower water retention capacity.

Both layers are highly mineralized with a percentage of an organic part of less than 75%, so they are classified into organic soil and not peat [12, 15].

The pH value is moderately acidic to neutral.

The surface layer with a higher proportion of organic material (43.72%) is characterized by higher concentrations of C and N. A total of 14 kg C/m2 is stored in profile 1.

The layer 0–10 cm is also characterized by the higher content of available potassium (high content) compared to the deeper layer. The available phosphorus values are in the range of low content.

3.3 Soil Water Balance

During the summer period in the nearby piezometer 10, during 2012 (the average year with less precipitation in the vegetation period), for a continuous duration of five months, the dry bottom was recorded. During the rainy 2013, was also a dry bottom for a total duration of about four months recorded.

Since in the largest part of the vegetation period (summer period) there is no underground water influence, using the soil water balance it is possible to monitor the conditions of moisture of the shallow peat layer on the "Table" and to assess the possibility for the restart of the plant production in the existing conditions.

Table 5. Water balance of soil for "Table" in average year (1961–2014), R = 50 mm

Mth.	1	2	3	4	5	6	7	8	9	10	11	12	Yea.
P*	97	86	91	99	78	87	52	66	97	115	154	132	1155
PET*	2	5	19	44	81	105	123	111	71	43	18	4	626
R*	50	50	50	50	47	29	**0**	**0**	26	50	50	50	452
RET*	2	5	19	44	81	105	81	66	71	43	18	4	539
D*	0	0	0	0	0	0	**42**	**45**	0	0	0	0	**87**
S	95	81	72	55	0	0	0	0	0	48	136	128	615

*P - precipitation (mm), PET - potential evapotranspiration (mm), R - water reserve in soil, RET - real avapotranspiration (mm); D - Deficit water in soil (mm); S - surplus water (mm)

Table 6. Water balance of soil for Table in 2012 (driest year), R = 50 mm

Mth	1	2	3	4	5	6	7	8	9	10	11	12	Yea.
P	37	102	1	177	102	35	34	0	194	212	102	202	1197
PET	0	0	26	39	72	127	147	130	80	42	26	1	690
R	50	50	25	50	50	**0**	**0**	**0**	50	50	50	50	425
RET	0	0	26	39	72	85	34	0	80	42	26	1	405
D	0	0	0	0	0	**42**	**113**	**130**	0	0	0	0	**285**
S	37	102	0	113	30	**0**	**0**	**0**	64	170	76	201	**793**

In the Tables 5 and 6 are given the soil water balance for the average year for series 1961–2014 and 2012 which was the hottest and the driest year in vegetation period.

The deficit (D) of water i.e. the amount of water for which real evapotranspiration (RET) was less than potential evapotranspiration (PET) in the vegetation period of the average year (1961–2014) of 87 mm was expressed in VII (42 mm) and VIII months (45 mm). This is the period when available water reserves in the soil have been discharged. Water deficiency (D) in 2012 recorded in three months VI (42 mm), VII (113) and VIII (130), which made a total deficit of 285 mm. In the VIII month, no precipitation was recorded and the water reserves in the soil were 0 mm, so that in that month in the water bilance the value of real evapotranspiration RET was 0.

4 Conclusions

The hydrological regime of the hydro-meliorated part of the peat Ždralovac was changed under the influence of climate change, drainage and peat ploughing, which caused lower groundwater level, shortening the duration of the wet condition and drying the surface layer of peat.

The changed hydrological regime influenced the intensification of peat mineralization and the change in the properties of Peat soils - Histosol, which is mostly reflected in the reduction of the peat depth, in the increased the mineral part of the peat so they are classified into organic soil and not peat.

The further discussed results suggest that it is possible to use the "Table" in agriculture primarily with the continuous supply of water to the rhizosphere layer during the dry summer months. This implies the reconstruction of a system of controlled drainage i.e. reconstruction of a system of water-gates which can provide capillary wetting of the rhizosphere in dry periods or the introduction of some irrigation systems. Deep tillage and mixing a layer of organic soil and layer of clay would also deepen the rhizosphere layer and increase the retention capacity of the water, i.e. the water reserves that would be consumed during the summer dry periods.

Conflict of Interest. The authors declare that there are no conflicts of interest regarding the publication of this paper.

References

1. Allison, F.E.: Soil Organic Matter and its Role in Crop Production. Elsevier, Amsterdam (1973). 637 p.
2. ASTM Designation: D 4531–86: Standard Test Methods for Bulk Density of Peat and Peat Products, These test methods are under the jurisdiction of ASTM Committee D18 on Soil and Rock and are the direct responsibility of Subcommittee D18.22 on Soil as a Medium for Plant Growth. Current edition approved Feb. 24, 1986. Published May 1986. Reapproved 2002 (1986)
3. ASTM Standards D 2974–87: Test methods for moisture, ash and organic matter of peat and other organic soil, Active Standard ASTM D2980 | Developed by Subcommittee: D18.22 Book of Standards Volume: 04.08 (1990)
4. ASTM Standards 2976–71: Test methods for pH of peat materials, (Withdrawn 2013), Withdrawn Standard: ASTM D2976-71, 2004 | Developed by Subcommittee: D18.22 (1990)
5. Čivić, H., Šaćiragić, B., Elezi, Dž.: Agrohemija sa ishranom biljaka, Graforad Travnik (2004). ISBN 9958–9301-2-9
6. Črepinšek, Z.: AGROKLIMATOLOGIJA. Vaje, Ljubljana (2005)
7. Dite, D., Hajek, M., Hajkova, P.: Formal definitions of Slovakian mire plant associations and their application in regional research. Biologia **62**, 400–408 (2007)
8. Driessen, P., Deckers, J., Spaargaren, O., Nachtergaele, F.: Lecture notes on the major soils of the world. World Soil Resources Reports, no. 94, FAO, Rome (2001). 344 p.
9. Dupieux, N.: La gestion conservatoire des tourbieres de France: premiers elements scientifiques et techniques. Espaces Naturels de France, programme Life Tourbieres de France (1998). 244 p.
10. Federalni meteorološki zavod iz Sarajeva i Hidro meteorološki zavod RS-a iz Banja Luke: Prvi nacionalni izvještaj Bosne i Hercegovine, u skladu sa Okvirnom konvencijom UN o klimatskim promjenama (financirano od strane UNDP) (2012). www.msb.gov.ba/onama/struktura/zastita_spasavanje/publikacije/default.aspx?id=7288&langTag=bs-BA
11. Federalni meteorološki zavod iz Sarajeva, Meteorološki godišnjak za period 1961–2014. godina
12. Huat, B.B.K.: Organic and Peat Soils Engineering. Universiti Putra Malaysia Press, Serdang (2004). 146 p.
13. Holdena, J., Chapmana, P.J., Labadzb, J.C.: Artificial drainage of peatlands: hydrological and hydrochemical process and wetland restoration. Prog. Phys. Geogr. **28**(1), 95–123 (2004)
14. Institut za hidrotehniku Građevinskog fakulteta u Sarajevu: Hidrološka i ekološka studija Livanjskog polja, u okviru UNDP projekta: Integriranje smjernica za zaštitu kraških tresetišta u ključne ekonomske sektore - Lot 3, Sarajevo 2013 (2013)
15. Jarrett, P.M.: Geoguide 6. Site investigations for organic soils and peat. J.K.R. Document 20709-0341-95. Institut Kerja Raya Malaysia (1995)
16. Kemmers, R.H., Jansen, P.C.: Hydrochemistry of rich fen and water management. Agric. Water Manag. **14**, 399–412 (1998)
17. Kurtović, J.: Zemljište kao prirodni resurs i faktor razvoja: Mineralizacija treseta, njen uticaj na slijeganje i način iskorišćavanja, Akademija nauka i umjetnosti BiH, Posebna izdanja: U spomen na akademika Milivoja Ćirića, Knjiga XCVIII, Sarajevo 1991 (1991)
18. Laine, J., Silvola, J., Tolonen, K., Alm, J., Nykänen, H., Vasander, V., Sallantaus, T., Savolainen, I., Sinisalo, J., Martikainen, P.J.: Effect of water-Level drawdown on global climatic warming: Northern Peatlands. Source Ambio J. Hum. Environ. **25**(3), 179–184 (1996)

19. Minayeva, T., Yu, Sirin, A.A.: Peatland biodiversity and climate change. Bio. Bull. Rev. **2** (2), 164–175 (2012). ISSN 2079 0864
20. Parent, E.L., Ilnicki, P.: Organic Soils and Peat Materials for Sustainable Agriculture Includes Bibliographical References and Index. CRC Press LLC, Boca Raton (2010). ISBN 0-8493-1458-5
21. Thornthwaite, C.W.: An approach toward a rational classification of climate. Geogr. Rev. **38**, 55–94 (1948)
22. Vlahinić, M.: Vodno-fizičke osobine tresetišta Livanjskog polja, Radovi poljoprivrednog fakulteta Univerziteta u Sarajevu, godina XVIII, broj 20, Sarajevo 1969 (1969)

Survival of Soil Microbial Population After Glyphosate Application

B. Lalević[1(✉)], S. Hamidović[2], T. Gavrić[2], A. Sunulahpašić[3], B. Borovac[2], M. Halilović[2], I. Jusić[2], A. Kazlagić[2], and M. Delić[2]

[1] Faculty of Agriculture, University of Belgrade, Belgrade-Zemun, Serbia
blazol@agrif.bg.ac.rs

[2] Faculty of Agriculture and Food Sciences, University of Sarajevo, Sarajevo, Bosnia and Herzegovina

[3] Ministry of Agriculture, Water Management and Forestry of Central Bosna Canton, Travnik, Bosnia and Herzegovina

Abstract. Modern agricultural production is often associated with application of agrochemicals. One of them is glyphosate, which has become one of the most widely used herbicides in weed control. However, the widespread use of glyphosate may influence the living organism's activity. The aim of this paper was to estimate the soil microbial activity after glyphosate application. Glyphosate (Glifomark SG) was applied at the end of 2017 on two locations: Principovo polje, and Novi Travnik (Travnik municipality, Bosnia and Herzegovina), so that corn and raspberry residues can be removed. Soil sampling (0–10 and 10–30 cm) was conducted at the time of glyphosate application, also 15 and 45 days after application. On control soil glyphosate was not applied. Chemical soil analysis (pH, content of humus, available P and K) was performed using standard methodology, whilst microbial abundance (total number of bacteria, ammonification bacteria, fungi, actinomycetes and glyphosate-tolerant bacteria) was analyzed using agar plate method. The results of chemical soil analysis showed slightly acid to slightly alkaline pH value, moderate to high humus content, and low to high content of available P and K. In most of samples, bacterial number was reduced 15 days after glyphosate application, whilst at the end of experiment increase of bacterial abundance was recorded. Similar results were obtained for fungi. Rapid decrease of actinomycetes number was observed after application of glyphosate. Two bacterial strains (PP-23 and NT-11) were able to grow on mineral medium supplemented with glyphosate in concentrations of 1 and 2% (v/v), thus representing a promising candidates for bioremediation of soil contaminated with glyphosate.

Keywords: Microbial diversity · Glyphosate · Bacteria · Bioremediation of soil

1 Introduction

Modern plant production is based on pesticide application in order to suppress pathogens, pests and weeds on agricultural soil. However, pesticides may have potential toxic effects on environmental quality and non-target organisms (Partoazar

M. Brka et al. (Eds.): AgriConf 2019, IFMBE Proceedings 78, pp. 36–43, 2020.
https://doi.org/10.1007/978-3-030-40049-1_4

et al. 2011). Glyphosate is systemic herbicide which is commonly used to suppress weeds Zabaloy et al. (2008). Glyphosate was introduced in 1974 under the name Roundup (Lane 2011) and become one of the most used herbicides worldwide (Duke et al. 2012). Nevertheless, the widespread use of this herbicide may have detrimental effects on soil properties and living organisms (Yamada et al. 2009). Several reports have addressed the reduction of nutrient uptake (Eker et al. 2006) and soil microbial activity (Zhao et al. 2013) after glyphosate use. Šantrić et al. (2014a, b) suggest that incorporation of herbicides could have significant influence on microbial abundance in agricultural soils. Use of pesticides may have negative impact on activity of autochthonous microbial populations, causing soil disturbance, with possibility to enter in to the human food chain (Hussain et al. 2009). On the other hand, several microbial groups have capability to use herbicides as a unique carbon and energy sources (Johnsen et al. 2001). Thus, application of these chemical compounds reduces microbial prevalence, but enhances functional diversity of soil microorganisms (Wang et al. 2006).

Long-term application of glyphosate may lead to its accumulation in soil and ground waters, causing contamination (Battaglin et al. 2014). Even low concentrations of glyphosate (about 50 ppb) have adverse impact on living organisms in aquatic ecosystems (Cuhra et al. 2013). According to Albers et al. (2009), found that glyphosate has carcinogenic properties that affect humans (IARC 2016) and that half- life of glyphosate varied from 2 to 197 days, thus that it is necessary to find environmental friendly methods for removal of glyphosate from ecosystem. Compared to conventional techniques, use of microorganisms for removal of glyphosate is indicated as promising technology for improvement of environmental quality (Dzionek et al. 2016). The aim of this paper was to determine the impact of glyphosate on microbial activity of soil, and to isolate the indigenous glyphosate-tolerant bacterial strains capable of growth in the presence of glyphosate.

2 Materials and Methods

The experiment was conducted in 2017 on two locations in Travnik municipality (Bosnia and Herzegovina). Glyphosate (Glifomark SG) was applied at the end of 2017 on Principovo polje, and Novi Travnik in order to remove corn (3.5 l/ha), and raspberry (5 l/ha) residue. Soil sampling (0–10 and 10–30 cm) was conducted before application of glyphosate on 15th and 45th day after application. On control soil, glyphosate was not applied.

Chemical analysis of soil was performed using ISO 10390:2005 standard for determination of pH value in H_2O and KCl, whilst humus content was determined using Tyurin's method (Tyurin 1951), available phosphorus and potassium content using AL method (Egner et al. 1960) and carbonate content using Scheibler calcimeter.

Microbial prevalence was determined using agar plate method. Total number of bacteria was determined using 0.1 × TSA (Torlak, Serbia), ammonification bacteria using nutrient agar, fungi using Sabouraud maltose agar (Thermo Scientific, UK), and actinomycetes using starch-ammonia agar. Microbial number was expressed as colony forming units (CFU) per gram of absolutely dried soil.

Glyphosate-tolerant bacteria were determined using modification of Talaie et al. (2010) method, with addition of 0.5; 1.0; and 2.0% (v/v) of glyphosate in mineral salt medium. Growth of most glyphosate-tolerant bacteria was determined by inoculation of mineral salt medium (Talaie et al. 2010) supplemented with 25, 50 and 100 mg/l of glyphosate, and measured spectrophotometrically (OD_{600}) before application of glyphosate, and 6, 12, 24, 48 and 72 h of incubation.

3 Results and Discussion

Table 1 shows the basic chemical characteristics of soil. Neutral to slightly alkaline pH value was determined at Principovo polje, whilst at Novi Travnik slightly acid to neutral pH was detected. Humus content was higher at Principovo polje. High content of available P and K at Principovo polje was determined. At Novi Travnik, soil was had weak to moderate P and K content (Table 1).

Table 1. Chemical analysis of soil samples

Soil samples	Depth	pH value		Humus content	P_2O_5	K_2O
	(cm)	H_2O	KCl	(%)	(mg/100 g)	
Principovo polje - control	0–30	7.80	7.10	6.25	100.00	87.60
Principovo polje	0–10	7.70	7.00	3.69	76.75	55.40
Principovo polje	10–30	7.70	7.00	3.47	54.44	53.40
Novi Travnik - control	0–30	7.80	7.00	2.87	22.75	23.10
Novi Travnik	0–10	6.80	5.80	3.15	4.44	8.80
Novi Travnik	10–30	7.30	6.50	2.70	19.38	22.70

Impact of pesticides on soil microbial activity depends on various environmental factors, depending on their availability and toxicity (Abdel-Mallek et al. 1994). According to Katagi (2008), concentration of pesticides in soil is controlled by sorption processes; thus, pesticide toxicity is influenced by soil texture, vegetation cover and chemical properties of soil (Hussain et al. 2009). Despite the fact that herbicides are not primarily manufactured to suppress beneficial microorganisms, non-target effects may be achieved, especially in the case of sensitive soil microbiological parameters (Partoazar et al. 2011).

As can be seen from Table 2, total number of bacteria and prevalence of ammonification bacteria depends on the depth of soil and time of sampling. At Principovo polje, highest abundance of bacteria was detected in deepest layer (10–30 cm) and lowest abundance in control, whilst at Novi Travnik it was vice versa. On control soil, bacterial abundance at Principovo polje increased during experiment. Similar trend was detected at Novi Travnik, whilst prevalence of ammonification bacteria was highest before the application of glyphosate. In most of the samples, highest bacterial number, on both locations, was detected 45 days after glyphosate treatment, which indicates that glyphosate application increases soil microbial activity in the short term. Similar results were obtained in several earlier researches (Haney et al. 2000; Haney et al. 2002). Zabaloy et al. (2008) highlighted that glyphosate may serve as a source of nutrients to bacteria. In contrast, Ratcliff et al. (2006) and Weaver et al. (2007) found that glyphosate application has no effect on microbial activity of soil.

Table 2. Total number of bacteria and ammonification bacteria ($\times 10^4$ CFU/g) in soil

Soil	Total number of bacteria			Ammonification bacteria		
	Time zero (before glyphosate)	15 days after	45 days after	Time zero (before glyphosate)	15 days after	45 days after
Principovo polje (control)	30.5	74.6	112.3	37.5	46.6	51.4
Principovo polje (0–10 cm)	30.9	34.5	123.3	39.6	21.7	77.5
Principovo polje (10–30 cm)	35.8	29.9	50.6	46.3	13.6	45.6
Novi Travnik (control)	73.9	72.2	99.6	97.9	57.5	82.7
Novi Travnik (0–10 cm)	40.1	85.2	85.7	65.7	52.1	89.9
Novi Travnik (10–30 cm)	24.6	116.7	83.4	45.6	158.2	65.8

At both locations, number of fungi and actinomycetes was highest in control soils before glyphosate application (Table 3). In most of samples, fungal number decreased 15 days after the glyphosate application compared to same soil before application. Increase in fungal activity was also determined by other authors (Lane 2011). Fungi are most important microorganisms in glyphosate degradation (Krzysko-Lupicka et al. 1997). Laatikainen and Heinonen-Tanski (2002) found that glyphosate may stimulate the growth of several fungal populations *in vitro*. On the other hand, rapid decrease of actinomycetes was recorded in all treatments. Inhibition in actinomycetes count after glyphosate application was noticed by Šantrić et al. (2014b). In contrast, Araujo et al. (2003) found the enhancement of actinomycetes population after glyphosate soil treatment.

During screening of potential glyphosate-tolerant bacteria, five colony morphologies were found in mineral salt medium supplemented with 0.5% (v/v) of glyphosate. From these five morphologies, two bacterial isolates (PP-23 and NT-11) were able to

Table 3. Prevalence of fungi and actinomycetes ($\times 10^3$ CFU/g) in soil

Soil	Fungi			Actinomycetes		
	Time zero (before glyphosate)	15 days after	45 days after	Time zero (before glyphosate)	15 days after	45 days after
Principovo polje (control)	5.7	2.5	3.8	48.6	7.9	4.5
Principovo polje (0–10 cm)	2.4	2.1	3.5	39.5	3.8	3.8
Principovo polje (10–30 cm)	1.3	2.4	3.2	40.3	4.6	3.9
Novi Travnik (control)	6.8	1.9	3.3	63.6	4.5	3.3
Novi Travnik (0–10 cm)	5.5	2.8	9.4	55.4	6.6	5.6
Novi Travnik (10–30 cm)	5.9	4.6	6.9	52.8	13.7	2.5

grow on medium supplemented with glyphosate in concentrations of 1 and 2% (v/v). Both isolates were Gram-positive, rod-shape, spore forming bacteria, probably belonging to *Bacillus* genus. Previous researches showed that several *Bacillus* sp. may use glyphosate as carbon and energy source (Yu et al. 2015; Fan et al. 2012). Our results confirm that bacterial isolates may grow on mineral salt medium supplemented with glyphosate (Table 4). Growth rate depends on bacterial isolate and time of sampling.

Table 4. Growth of bacterial isolates (OD_{600}) in the presence of glyphosate

Isolate	Concentration (mg/l)	0 h	6 h	12 h	24 h	48 h	72 h
PP-23	25	0.023	0.132	0.127	0.312	0.522	0.432
	50	0.030	0.111	0.109	0.245	0.478	0.422
	100	0.032	0.055	0.054	0.062	0.070	0.074
NT-11	25	0.017	0.022	0.056	0.063	0.182	0.218
	50	0.018	0.020	0.051	0.054	0.139	0.204
	100	0.022	0.021	0.020	0.021	0.024	0.025

From data presented in Table 4, it is evident that growth of isolate PP-23 is more pronounced compared to NT-11. At concentrations of 25 and 50 mg/l, in PP-23 isolate, lag-phase is obvious after 6 h of incubation. Rapid increase of optical density (OD_{600}) was noticed between 24 and 48 h of incubation. Moderate decrease of OD_{600} was detected at the end of incubation. At highest glyphosate concentration (100 mg/l) moderate increase of bacterial growth was noticed during incubation period. On the other hand, in NT-11, slow bacterial growth was recorded up to 24 h of incubation; at the end of experiment, rapid increase of bacterial growth at 25 and 50 mg/l was achieved. At concentration of 100 mg/l, negligible bacterial growth was detected. *Escherichia coli* ATCC 25922 was able to grow on medium containing 0.08 mg/ml of glyphosate (Nielsen et al. 2018). Lima et al. (2014) found lag, log and stationary phases of *Pseudomonas aeruginosa* ATCC 15442 cultivated on medium containing from 0.850 to 1690 ppm of glyphosate.

4 Conclusion

Our results showed that highest bacterial and fungal prevalence was 45 days after glyphosate application. Rapid decrease of actinomycetes number was detected after treatment with glyphosate. Two bacterial isolates PP-23 and NT-11 represent a promising candidates for bioremediation of soil contaminated with glyphosate.

Acknowledgement. This research was partially supported by Ministry of Education, Science and Technological Development of Republic of Serbia (Grant No. TR31080).

References

Abdel-Mallek, A.Y., Moharram, A.M., Abdel-Kader, M.I., Omar, S.A.: Effect of soil treatment with the organophosphorus insecticide profenofos on the fungal flora and some microbial activities. Microbiol. Res. **149**, 167–171 (1994)

Albers, C.N., Banta, G.T., Hansen, P.E., Jacobsen, O.S.: The influence of organic matter on sorption and fate of glyphosate in soil—Comparing different soils and humic substances. Environ. Pollut. **157**, 2865–2870 (2009)

Araujo, A., Monteiro, R., Abarkeli, R.: Effect of glyphosate on the microbial activity of two Brazilian soils. Chemosphere **52**, 799–804 (2003)

Battaglin, W.A., Meyer, M.T., Kuivila, K.M., Dietze, J.E.: Glyphosate and its degradation product AMPA occur frequently and widely in U.S. soils, surface water, groundwater, and precipitation. JAWRA J. Am. Water Resour. Assoc. **50**(2), 275–290 (2014)

Cuhra, M., Traavik, T., Bohn, T.: Clone- and age-dependent toxicity of a glyphosate commercial formulation and its active ingredient in Daphnia magna. Ecotoxicology **22**(2), 251–262 (2013)

Dzionek, A., Wojcieszynska, D., Guzik, U.: Natural carriers in bioremediation: a review. Electron. J. Biotechnol. **23**, 28–36 (2016)

Duke, S.O., Lydon, L., Koskinen, W.C., Moorman, T.B., Chaney, R.L., Hammerschmidt, R.: Glyphosate effects on plant mineral nutrition, crop rhizosphere microbiota, and plant disease in glyphosate-resistant crops. J. Agric. Food Chem. **60**, 10375–10397 (2012)

Egner, H., Riehm, H., Domingo, W.R.: Untersuchungen über die chemische Boden-analyse als Grundlage für die Beurteilung des Nährstoffzustandes der Böden. II Chemische Extraktionsmethoden für Phosphor- und Kaliumbestimmung. Kungluga Lantbrukshoegsk Annaler, **26**, pp. 199–215 (1960)

Eker, S., Ozturk, L., Yazici, A., Erenoglu, B., Romheld, V., Cakmak, I.: Foliar- applied glyphosate substantially reduced uptake and transport of iron and manganese in sunflower (Helianthus annuus L.) plants. J. Agric. Food Chemi. **54**, 10019–10025 (2006)

Fan, J., Yang, G., Zhao, H., Shi, G., Geng, Y., Hou, T., Tao, K.: Isolation, identification and characterization of a glyphosate-degrading bacterium, Bacillus cereus CB4, from soil. J. Gen. Appl. Microbiol. **58**(4), 263–271 (2012)

Haney, R., Senseman, S., Hons, F.: Effect of roundup ultra on microbial activity and biomass from selected soils. J. Environ. Qual. **31**, 730–735 (2002)

Haney, R., Senseman, S., Hons, F., Zuberer, D.: Effect of glyphosate on soil microbial activity and biomass. Weed Sci. **48**, 89–93 (2000)

Hussain, S., Siddique, T., Saleem, M., Arshad, M., Khalid, A.: Impact of pesticides on soil microbial diversity, enzymes, and biochemical reactions. Adv. Agron. **102**, 159–200 (2009)

IARC: IARC Monographs on the evaluation of carcinogenic risks to humans – glyphosate, 2nd edn. International Agency for Research on Cancer, Lyon, France (2016)

Johnsen, K., Jacobsen, C.S., Torsvik, V.: Pesticides effects on bacterial diversity in agricultural soils-a review. Biol. Fertil. Soils **33**, 443–453 (2001)

Katagi, T.: Surfactant effects on environmental behavior of pesticides. Rev. Environ. Contam. Toxicol. **194**, 1–177 (2008)

Krzysko-Lupicka, T., Strof, W., Kubs, K., Skorupa, M., Wieczorek, P., Lejczak, B.: The ability of soil-borne fungi to degrade organophosphonate carbon-to- phosphorus bonds. Appl. Microbiol. Biotechnol. **48**, 549–552 (1997)

Laatikainen, T., Heinonen-Tanski, H.: Mycorrhizal growth in pure cultures in the presence of pesticides. Microbiol. Res. **157**, 127–137 (2002)

Lane, M.: The effect of glyphosate on soil microbial communities. Degree Master of Science in the Graduate School of The Ohio State University (2011)

Lima, I.S., Baumeier, N.C., Rosa, R.T., Stuelp Campelo, P.M., Ribeiro Rosa, E.A.: Influence of glyphosate in planktonic and biofilm growth of Pseudomonas aeruginosa. Braz. J. Microbiol. **45**(3), 971–975 (2014)

Nielsen, L.N., Roager, H.M., Casas, M.E., Frandsen, H.L., Gosewinkel, U., Bester, K., Liht, T. R., Hendriksen, B.N., Bahl, M.I.: Glyphosate has limited short-term effects on commensal bacterial community composition in the gut environment due to sufficient aromatic amino acid levels. Environ. Pollut. **233**, 364–376 (2018)

Partoazar, M., Hoodaji, M., Tahmourespour, A.: The effect of glyphosate application on soil microbial activities in agricultural land. Afr. J. Biotechnol. **10**(83), 19419–19424 (2011)

Ratcliff, A.W., Busse, M.D., Shestak, C.J.: Changes in microbial community structure following herbicide (glyphosate) additions to forest soils. Appl. Soil. Ecol. **34**, 114–124 (2006)

Šantrić, L., Radivojević, L., Gajić Umiljendić, J., Đurović-Pejčev, R., Sarić-Krsmanović, M.: Assessment of microbial activity and biomass in different soils exposed to nicosulfuron. Pestic. Phytomedicine (Belgrade) **29**(3), 213–219 (2014a)

Šantrić, L., Radivojević, L., Gajić Umiljendić, J., Sarić-Krsmanović, M., Đurović-Pejčev, R.: Effects of herbicides on growth and number of actinomycetes in soil and in vitro. Pestic. Phytomedicine (Belgrade) **31**(3–4), 121–128 (2014b)

Talaie, A.R., Jafaarzadeh, N., Talaie, M., Beheshti, M.: Biodegradation of aromatic compounds in crude oil by isolated microorganisms from environment. Sci. J. Zanjan Univ. Med. Sci. Health Serv. **18**(70), 68–80 (2010)

Tyurin, I.V.: Analytical procedure for a comparative study of soil humus. Trudy pochvovedenii Instituta Dokuchaeva **38**, 5 (1951)

Wang, M.C., Gong, M., Zang, H.B., Hua, X.M., Yao, J., Pang, Y.J., Yang, Y.H.: Effect of methamidophos and urea application on microbial communities in soils as determined by microbial biomass and community level physiological profiles. J. Environ. Sci. Health B **41**, 399–413 (2006)

Weaver, M.A., Krutz, L.J., Zablotowicz, R.M., Reddy, K.N.: Effects of glyphosate on soil microbial communities and its mineralization in a Mississippi soil. Pest Manage. Sci. **63**, 388–393 (2007)

Yamada, T., Kremer, R., Castro, P., Wood, B.: Glyphosate interactions with physiology, nutrition, and diseases of plants: threat to agricultural sustainability? Eur. J. Agron. **31**, 111–113 (2009)

Yu, X.M., Yu, T., Yin, G.H., Dong, Q.L., An, M., Wang, H.R., Ai, C.X.: Glyphosate biodegradation and potential soil bioremediation by Bacillus subtilis strain Bs-15. Genet. Mol. Res. **14**(4), 14717–14730 (2015)

Zabaloy, M.C., Garland, J.L., Gomez, M.A.: A integrated approach to evaluate impacts of the herbicides glyphosate, 2-4-D and metsulfurom-methyt on soil microbial communities in the pampas region. Argentina J. Appl. Soil Ecol. **40**, 1–12 (2008)

Zhao, C., Xie, H., Zhang, J., Xu, J., Liang, S.: Spatial distribution of organochlorine pesticides (OCPs) and effect of soil characters: a case study of a pesticide producing factory. Chemosphere **90**, 2381–2387 (2013)

Determination of Yield and Agronomic Characteristics of Some Soybean Genotypes Suitable for Double Crop Agriculture in Mediterranean Climate Conditions

Aliye Yildirim(✉), Hakan Yildiz, Emre Ilker, and Sidika Ekren

Faculty of Agriculture, Department of Field Crops, Ege University, Izmir, Turkey
aliyee.yildirim@gmail.com

Abstract. As a source of oil, protein, biodiesel, etc., soybean (Glycine max L. Merr.) is the fourth widely grown crop in the world. In Mediterranean climate conditions, soybean can be produced as a double crop. This study was conducted at the experimental area of Department of Field Crops, Faculty of Agriculture, Ege University in Turkey (2016) to determine yield and important agronomic traits of some soybean varieties and lines under double crop conditions. Ten advanced soybean lines and four registered varieties (KASM-03, KASM-02, KANA, KAMA, BDUS-04, BATEM 207, BATEM 223, BATEM 306, BATEM 317, BDSA 05 and ARISOY, ATAEM-7, BRAVO, NOVA) were used as a plant material. Field trial was conducted in a randomized complete block design (RCBD) with four replications. Results from the analysis of variance for all traits (grain yield, plant height, first pod height, pods per plant, days for maturity, days for flowering, 1000-seed weight) indicated large variations among the genotypes. Although the lines KASM-02 (3519.5 kg ha^{-1}), BATEM 207 (3456.7 kg ha^{-1}) and BATEM 223 (3414.2 kg ha^{-1}) provided the best result for grain yield, according to earliness, ARISOY, ATAEM-7 and BRAVO registered varieties were showed better performance rather than advanced lines used in the study. As a result, genotypes with high grain yields but with late maturing times are tested under the main product conditions.

Keywords: Soybean · Double cropping · Yield · Earliness

1 Introduction

Soybean is the common name for Glycine max (Merrill), a paleo-amphidiploid grain legume (2n = 4x = 40), part of the genus Glycine Willdenow, family Leguminosae, subfamily Papilionoideae, and tribe Phaseolae. The Glycine genus has its evolutionary origins in Australia and Asia, and was first named by Carl von Linne (also known as Linnaeus; Genera Plantarum, 1737) based on the Greek word 'glykys', meaning 'sweet' (done incorrectly as it referred then to the sweet tubers of Glycine apios L. which is now correctly classified as Apios americana) [4]. Soybean is an ancient Chinese grain and now a top oilseed crop, with world annual production reaching more

M. Brka et al. (Eds.): AgriConf 2019, IFMBE Proceedings 78, pp. 44–50, 2020.
https://doi.org/10.1007/978-3-030-40049-1_5

than 300 million metric tons [12]. Soybeans are known as 'yellow jewel' or 'miracle beans,' because of their unique chemical composition. By average, on a dry matter basis, mature soybeans contain about 20% oil, 40% protein, and 5% ash and the remaining 35% is total carbohydrate. With this composition, the soybean has the highest protein content between cereal and other legume species. After the peanut soybean has the second highest oil content among all food legumes. There are many reasons for the rapid worldwide growth in soybean production. Soybeans have many possible uses, including as human food, animal feed, and industrial materials.

As a current oilseed report total world soybean production 355.39 million metric tons. Brazil (123 million metric tons), the United States (112.95 million metric tons) and Argentina (53 million metric tons) meet most of the world soybean production. Total soybean area as a that report is 126.67 million hectares in the world [2]. According to 2017 data, 10714671 tonnes of soybean production was made in the European continent and the most soybean producing countries are Ukraine (3899370 tonnes) and Russia (3621344 tonnes). Turkey's soybean production is based on the same statistical data is quite low (140000 tonnes) [3].

The second crop agriculture is a very important form of production for the regions near Mediterranean. It has been demonstrated by various researchers that double crop soybean farming can be done successfully in the Mediterranean coastline [3, 5, 8, 9]. In the current research, the study was planned to determine promising soybean genotypes, which have high yielding potential and suitable for double cropping under Mediterranean climate conditions.

2 Materials and Method

Ten advanced soybean lines (Table 1) and four registered soybean varieties (from III and IV maturity group) were planted after wheat as a double crop on 23 June 2016. The trial area is located western part of Turkey under Mediterranean climate conditions (Latitude 38°28′ and Longitude 27°13′). Soil properties of this area are heavy soil structure with clay-silt soil at 0–20 cm depth and clay-loamy structure at 20–40 cm depth [6]. The Field experiment was conducted in a randomized complete block design (RCBD) with four replications. 5 m long parcels were planted by hand with 70 cm row and 45 plants per m^2. 200 kg ha^{-1} of DAP (diammonium phosphate) (36 kg ha^{-1} N, 92 kg ha^{-1} P) fertilizers were applied before planting. Sprinkler irrigation was performed 6 times starting with sowing time for the soybeans. The climatically data during the 2016 growing season in experimental area presented at Table 2 [1]. The study was aimed to find out the high yielding and earliness genotypes for double crop. The parameters grain yield, plant height, first pod height, pods per plant, days for flowering, days for maturity, 100-seed weight parameters were observed and measured. Analysis of variance was calculated for each traits and LSD (least significant difference) test was applied to compare the differences [15].

Table 1. Registered varieties and advanced lines used in the reseacrh.

Advanced lines (F9)	Pedigree
BATEM 207	Ataem-6 x A-3935
BDUS-04	Umut 2002 x Sprite 87
BATEM 223	J-357 x 9392
KAMA	Macon x Apollo
BATEM 306	Ataem-6 x ETAE-8
KANA	NE 3297 x AP 2292
BATEM 317	Prota x Ap- 2292
KASM-02	Sprite 87 x Macon
BDSA 05	Sprite 87 x Apollo
KASM-03	Sprite 87 x Macon
Registered varieties	ARISOY, NOVA, BRAVO, ATAEM 7

Table 2. The weather data during the 2016 growing season in experimental area [1].

Months	Average temp. (°C)	Relative humudity (%)	Precipitation (mm)
June	27.5	47.9	2.8
July	29.3	44.5	0.0
August	28.9	51.0	0.4
September	24.7	50.1	8.6
October	19.4	57.7	0.5

The advanced lines used in the study were hybridized by different research institutes in Turkey. BATEM 207, BATEM 223, BATEM 306, BATEM 317 advanced lines were hybridized by Batı Akdeniz Agricultural Research Institute; KAMA, KANA, KASM-02 and KASM-03 advanced lines were hybridized by Karadeniz Agricultural Research Institute; BDUS-04 and BDSA-05 advanced lines were hybridized by Bahri Dagdas International Agricultural Research Institute.

3 Results and Discussions

According to the results of variance analysis, differences were found to be significant in terms of all traits. The main values of the traits tested are presented in Tables 3 and 4. The soybean belongs to the family Leguminosae. The cultivated form, *Glycine max* (L.) Merrill, grows annually. Its plant is bushy, with a height between 0.2 and 2.0 m, and produces a fruit having three to five hairy pods in a cluster [12]. In this study the highest plant height was obtained from BATEM 207 advanced line with 82.6 cm (Table 2). The average plant height obtained in this study was similar to other studies conducted under similar conditions [7, 16, 17].

Table 3. Mean plant height (cm), first pod height (cm), days for flowering (day) and days for maturity values of the genotypes and varieties and the LSD grouping

Genotypes	Plant height (cm)		First pod height (cm)	Days for flowering (DFF)	Days for maturity (DFM)
BATEM 207	82.60	a	11.05 efg	35.50 c	108.50 c
BATEM 223	76.60	ab	11.65 cde	36.75 b	109.75 bc
BATEM 306	69.70	bc	16.35 a	39.25 a	110.25 b
BATEM 317	65.20	cd	14.45 b	39.25 a	112.00 a
BDSA 05	64.70	cd	9.05 hij	37.00 b	109.75 bc
BDUS 04	65.65	cd	11.40 def	36.75 b	108.25 cd
KAMA	56.90	d	9.90 fghi	33.5 def	108.75 bc
KANA	75.45	ab	13.20 bc	34.5 cd	109.00 bc
KASM 02	61.95	cd	8.35 ij	33.25 ef	106.75 de
KASM 03	56.90	d	7.50 j	32.75 f	105.25 efg
ARISOY	61.90	cd	10.50 efgh	33.00 ef	104.25 g
ATAEM 7	76.70	ab	13.10 bcd	32.75 f	104.50 fg
BRAVO	64.20	cd	8.95 hij	33.00 ef	104.50 fg
NOVA	62.85	cd	9.40 ghj	34.00 de	106.00 ef
Mean	67.03		11,06	35,08	107.67
LSD	8,87		1,703	1,25	1,58

The soybean plant is photoperiod-sensitive and makes the transition from vegetative to flowering stages in direct response to day length. Most soybean varieties begin flowering soon after the day length begins to shorten. Because of this feature, latitude plays a major role in the adoption of soybeans to various geographic regions [12]. The value of days for flowering and days for maturity are important parameters determining earliness. In this study days for flowering ranged from 32.75 days (ATAEM 7 and KASM 03) to 39.25 days (BATEM 306 and BATEM 317). The harvesting time of the plants grown as double crop is very important. In this study, the registered varieties were showed better performance than the advanced lines in terms of days for maturity (Table 2). The earliest days for maturity observed from ARISOY registered variety. On the other hand, there are only one-day difference between the ARISOY and the advanced line KASM 03 and two days with KASM 02.

First pod height is an agronomic trait for the mechanical harvesting of soybeans with combines. The seed loss could be minimized, if the FPH is higher than the height of the cutter bar in combines. Hence, developing soybeans with high FPH has become one of important breeding goals in current crop improvement programs [11]. In this research, the first pod height values varied between 7.5 cm–16.35 cm and the highest first pod height obtained from BATEM 306 advanced line. The mean value of first pod height (11.6 cm) obtained in this research was similar to the data obtained from other studies conducted under similar conditions [3, 8, 17].

Table 4. Mean grain yield (kg ha^{-1}), pods number plant^{-1} and 100 seed weight (g) values of the genotypes and varieties and the LSD grouping

Genotypes	Grain yield (kg ha^{-1})	Pods number plant^{-1}	1000 seed weight (g)
BATEM 207	3456.75 a	57.85 ab	165,45 cdef
BATEM 223	3414.25 a	55.77 abc	141,00 f
BATEM 306	3354.00 ab	45.65 def	165,70 cdef
BATEM 317	3036.25 bcde	45.07 def	142,98 ef
BDSA 05	2913.50 def	39.10 fgh	182,03 abc
BDUS 04	2662.00 fg	34.57 h	201,95 a
KAMA	2980.00 cdef	41.85 efgh	204,73 a
KANA	3248.00 abc	52.65 bcd	175,15 bcd
KASM 02	3519.50 a	58.75 ab	182,98 abc
KASM 03	3230.75 abcd	62.55 a	167,70 cde
ARISOY	2971.25 cdef	36.37 gh	155,83 def
ATAEM 7	2420.75 g	34.47 h	194,68 ab
BRAVO	2854.25 ef	43.60 efg	160,50 cdef
NOVA	3284.00 abc	48.00 cde	167,15 cde
Mean	3096,08	46,87	171,99
LSD	31,82	8,40	25,61

Pods number per plant plays important role for estimating grain yield before the harvesting and it contribute to yield. The pods number per plant values ranged from 34.47 pods/plant to 62.55 pods/plant in this study. The highest pod number per plant obtained from KASM 03 advanced line. The calculated mean pods number value of these 14 genotypes close to obtained data of other studies conducted similar conditions [3, 8, 15].

Seed-size traits are controlled by multiple genes in soybean, and play an important role in determining seed yield, quality and appearance [6]. 1000 seed weight parameter is one of the important yield component in soybean. In this study 100 seed weight values varied between 141 g to 204,73 g and the average 1000 seed weight was 171,99 g.

BDSA 04 and KAMA advanced lines showed the best performance in terms of 1000 seed weight. Similar results were reported by Ilker [9], Ilker et al. [10], Onat et al. [15] and Bakal et al. [4].

According to recent reports getting from USDA, World soybean yield was determined as 2.81 metric tons ha^{-1} and the highest soybean yield obtained from Turkey (3.70 metric tons ha^{-1}). In this study the grain yield per hectare ranged between 2420.45 kg ha^{-1} and 3456.75 kg ha^{-1}. The highest grain yield obtained from BATEM 207 and BATEM 223 advanced lines and KASM 02. The lines has also high pod number per plant values and 100 seed weight values which components determining yield. In this study the mean grain yield value (3096 kg ha^{-1}) showed lower performance than some other research reported by Gulluoglu et al. [6] (3238.1 kg ha^{-1}) and Ilker [9] (3594 kg ha^{-1}).

Many complex traits important in the development of cultivars can be difficult to measure or are environmentally sensitive, and thus it is often difficult to make progress in breeding programs [13]. This is especially true for traits with low heritability such as soybean yield. For these traits, extensive testing is required over multiple locations and years prior to release of the improved variety. According to our findings based on previous years and this study, it is possible to grow soybean second crop in the Mediterranean climate zone. In addition, it was determined that the obtained grain yield values were well above the world average (2.8 ton/ha, 2017) [3]. Currently, genotypes with high grain yields but with late maturing times are tested under the main product conditions.

References

1. Anonymous: The Meteorological Data for Bornova. Olive Research Institute Meteorological Station (2016)
2. Anonymous: Oilseeds: World Markets and Trade. Foreign Agricultural Service/USDA. Office of Global Analysis, June 2019
3. Anonymous: FAOSTAT (2019). http://www.fao.org/faostat/en/#data/QC, http://www.fao.org/faostat/en/#data/QC
4. Bakal, H., Arioglu, H., Gulluoglu, L., Kurt, C., Onat, B.: The determination of some important argonomical and quality properties of Soybean (Glycine max (L.) Merr) varieties in double cropped condations. J. Field Crops Central Res. Inst. **25**(special issue-2), 125–130 (2016)
5. Greesshof, P.: Soybean (Glycine max L). Brenner's Encyclopedia of Genetics, 2nd edn., vol. 6, pp. 494–497 (2013). https://doi.org/10.1016/b978-0-12-374984-0.00657-4
6. Gulluoglu, L., Bakal, H., Arioglu, H.: The effects of twin-row planting pattern and plant population on seed yield and yield components of soybean at late double-cropped planting in Cukurova Region. Turk. J. Field Crops **21**, 59–65 (2016)
7. Hu, Z., Zhang, H., Kan, G., Ma, D., Zhang, D., Shi, G., Hong, D., Zhang, G., Yu, D.: Determination of the genetic architecture of seed size and shape via linkage and association analysis in soybean (Glycine max L. Merr.), Genetica (2013). https://doi.org/10.1007/s10709-013-9723-8
8. Ilker, E.: Correlation and path coefficient analyses in sweet corn. Turk. J. Field Crops **16**(2), 105–107 (2011)
9. Ilker, E.: Performances of Soybean (Glycine Max (L.) Merr.) advanced lines grown in second cropping under mediterranean climatical conditions of western Turkey. Turk. J. Field Crops **22**(1), 104–107 (2017)
10. Ilker, E., Kocaturk, M., Kadiroglu, A., Altinbas, M., Yildirim, A., Ozturk, G., Yildiz, H.: Stability analyses for double cropping in Soybean ((Glycine max L.) Merrill). Turk. J. Field Crops **23**(2), 80–84 (2018). https://doi.org/10.17557/tjfc.467434
11. Ilker, E., Kocaturk, M., Kadiroglu, A., Yildirim, A., Ozturk, G., Yildiz, H., Koken, I.: Adaptation abilities and quality parameters of selected Soybean lines under double cropping in the mediterranean region. Turk. J. Field Crops **23**(1), 49–55 (2018). https://doi.org/10.17557/tjfc.421584
12. Kang, B.K., Kim, H.T., Choi, M.S., Koo, S.C., Seo, J.H., Kim, H.S., Shin, S.O., Yun, H.T., Oh, I.S., Kulkarni, K.P., Lee, J.D.: Genetic and environmental variation of first pod height in

Soybean (Glycine max (L.) Merr.). Plant Breed. Biotech. **51**, 36–44 (2017). https://doi.org/10.9787/PBB.2017.5.1.36
13. Liu, K.: Soybean: Overview. Agricultural Research Service, US Department of Agriculture, Aberdeen, ID, USA Elsevier Ltd. (2016)
14. Lee, M.: DNA markers and plant breeding programs. Adv. Agron. **35**, 265–344 (1995)
15. Onat, B., Bakal, H., Gulluoglu, L., Arioglu, H.: The effects of high temperature at the growing period on yield and yield components of Soybean (Glycine max (L.) Merr) varieties. Turk. J. Field Crops **22**(2), 178–186 (2017)
16. Steel, R.G.D., Torrie, J.H.: Principles and Procedures of Statistics, 2nd edn. Mc. Graw-Hill Book Company Inc., New York (1980)
17. Yildirim, A.: Determination of yield and important agronomic traits of some Soybean varieties and genotypes grown second crop condition under the Aegean Region. MSc thesis. Graduate School of Natural and Applied Sciences, Ege University, Izmir (2017). (in Turkish)
18. Yildirim, A., Ilker, E.: Determination of yield and important agronomic traits and quality traits of some Soybean varieties and lines grown under second crop condition in the Aegean Region. J. Adnan Menderes Univ. Agric. Fac. **15**(2), 1–8 (2018). https://doi.org/10.25308/aduziraat.393970

Physical and Chemical Characteristics of the Peach Varieties in the Herzegovina Area

Lamija Aliman[1(✉)], Jasmina Aliman[2], Sezai Ercisli[3], Azra Skender[4], Aida Šukalić[2], and Aleksandra Šupljeglav Jukić[2]

[1] Faculty of Pharmacy, University of Sarajevo, Sarajevo, Bosnia and Herzegovina
lamija.aliman@hotmail.com
[2] Agromediterranean Faculty, University of Mostar, Mostar, Bosnia and Herzegovina
[3] Atatürk University Agricultural Faculty, Erzurum, Turkey
[4] Biotechnical Faculty, University of Bihać, Bihać, Bosnia and Herzegovina

Abstract. The aim of this study was to examine the impact of different agro ecological conditions on the physical (weight, height, width, thickness) and chemical (the soluble solids content (%), total sugar content (%), total acid content (%) and maturity index) characteristics of five peach varieties in Herzegovina. The fruits for analysis were taken of *Maycrest*, *Springcrest*, *Spring Lady*, *Rich Lady* and *Elegant Lady* varieties from the area of Mostar, Stolac and Čapljina for three years (2009–2011). The results of research show that the locality influenced the examined physical parameters of the fruits, but had no effect on the fruit/stone ratio (randman). The highest values of the weight, height, width and thickness of the fruit and stone were recorded in Rich Lady and Elegant Lady peaches in the Dubrava plateau (Stolac).

The results show statistically significant differences in the chemical composition of peach fruit depending on variety and locality. The highest average soluble solid content and total sugars were measured in the *Elegant Lady* in Mostar, while the lowest content of these compounds was found in the Maycrest and Rich Lady.

The highest total acid content was found in Maycrest variety in Stolac and the lowest content in the *Elegant Lady* variety (Mostar). The largest sweetnes index had the variety the *Elegant Lady* (Mostar), and the lowest variety of the *Maycrest* in Stolac.

It can be concluded that the investigated varieties in all three localities had an excellent quality of fruit in terms of physical and chemical characteristics and can be recommended to farmers for wide cultivation in Herzegovina.

Keywords: Peach · Physical and chemical characteristics · Fruit

1 Introduction

Bosnia and Herzegovina has extremely favourable agro-ecological conditions for peach cultivation, but these opportunities have not been sufficiently utilized. The total production of 5,472 t in 2017 is far behind for some European countries whose total

M. Brka et al. (Eds.): AgriConf 2019, IFMBE Proceedings 78, pp. 51–58, 2020.
https://doi.org/10.1007/978-3-030-40049-1_6

production is 4,527,414 t. The largest peach producers in Europe are Spain (1,799,685 t), Italy (1,250,721 t) and Greece (938,000 t) [1]. The quality of the fruit is of fundamental importance for the acceptance of varieties of peach and nectarine by consumers, due to the great competition on the market [2–4]. Kramer and Twigg [5] define quality as a combination of physical and chemical characteristics that provide an attractive product to the consumer.

The previous studying conducted on peach [6–10] detected the connection between some pomological characteristics which are related to quality of fruits.

Physical and chemical characteristics of peach fruit was depended genetical basis and rootstock, climatic and soil conditions, such as agrotechnical measures.

The weight of fruit is major quantitative inherited factor determining yield, quality of fruit and the consumer acceptability [6]. Time of ripening significantly correlates with the weight of fruit in a way that varieties of early ripening, in general, have less fruits comparing to late peach varieties [6, 11, 12]. Jerie and Chalmers [13] concluded that it is not that only chemical compound is important for quality of fruit, but the balance of different components that are product of metabolic activities which may be altered during the growth and ripening of fruit, as well.

The relation between soluble solids and total acidity is a very important parameter in determination of fruit quality. Crisosto et al. [14] reported that in case of varieties with compound of total acids >0,9%, and soluble solids <12,0 °Brix, consumer acceptability was under the influence of interaction between total acids and soluble solids, and not just soluble solids.

The aim of this research was establishing to what extent are physical and chemical characteristics of fruit and stone the result of activity of different agroecological conditions of microlocalities, and to what extent this considered to be a variety characteristic.

2 Materials and Methods

The study was conducted on six years old peach orchard at three localities in Herzegovina region: Mostar (elevation 45 m above sea level), Čapljina (elevation 10 m) and Stolac (elevation 350 m) during tree vegetative seasons (2009–2011). In this experiment, five commercial varieties (Maycrest, Springcrest, Spring Lady, Rich Lady and Elegant Lady) were used as plant material. All experimental trees were planted in a north–south orientation with 4 m × 2.5 m spacing and trained to a slender spindle system. Rootstock for examined varieties is GF 677.

All the necessary agronomic and plant protection measures were kept for the entire experiment for all peach cultivars. Experimental design was a single tree plot complete randomized design with five individual trees as a replicate for every cultivar. Three replicates were used. At harvest time, fruit were randomly picked from marked trees at the commercial maturity stage and were placed into polyethylene bags and transported to laboratory. Fifty fruits were used to determine physical characteristics of the fruit (weight, height, width and thickness) and stone (weight) as well as chemical compositions of the fruit (soluble solid content, total sugars, total acids and sweetness index).

Weights of the fruit and stone of each peach cultivar were measured with a digital balance (Kern 440, Kern and Sohn, Germany) of 0.01 g sensitivity. Height, width and thickness of the fruit and stone were measured with a hand calliper Meba with precision of 0.1 mm. The soluble solids content was determined with the manual *Abbé* refractometer (Euromex, Holland) for each cultivar individually and the values were expressed in the percentage (%). The total sugars content was determined volumetrically according to the *Luff-Schoorl* method and expressed in the percentage (%). The determination of titratable acidity was performed by titration with 0.1 N NaOH with the phenolphthalein indicator until colour change (pH 8.1), and the result was expressed as the percentage of malic acid. The sweetness index was then calculated as the SSC:TA ratio.

All obtained data were analysed statistically (ANOVA) and differences among the means were determined for significance at $P < 0.05$ by SAS statistical software.

3 Result and Discussion

3.1 Physical Characteristics of the Peach Fruit and Stone

The results of the physical characteristics fruit of examined peach varieties are presented in Table 1.

Table 1. Physical characteristics of the fruit in the examined peach varieties

	Locality	Weight (g)	Height (mm)	Width (mm)	Thickness (mm)	Stone weight (g)	Flesh/stone ratio) (%)
Maycrest	Mostar	118.69 ± 13.04b	58.37 ± 3,00c	60.61 ± 3.04c	62.65 ± 3.13b	7.18 ± 0.93b	93.97c ± 0.26c
	Stolac	131.19 ± 17.21b	61.89 ± 3.32b	63.26 ± 3.30b	65.96 ± 3.83b	8.62 ± 1.51a	93.51 ± 1.18c
	Čapljina	114.86 ± 11.69b	57.79 ± 2.95b	59.67 ± 2.73b	61.62 ± 2.44b	6.85 ± 91b	94.07 ± 0.31c
	Average	**121.58 ± 8.54b**	**59.35 ± 2.22b**	**61.18 ± 1.86b**	**63.41 ± 2.27b**	**7.55 ± 0.23a**	**93.85 ± 0.40c**
Springcrest	Mostar	113.33 ± 14.45b	56.93 ± 2.61c	58.91 ± 2.75b	60.28 ± 3.01b	6.45 ± 0.98b	94.38 ± 0.28b
	Stolac	123.55 ± 15.37b	58.49 ± 2.90c	60.85 ± 2.79b	62.41 ± 2.95b	7.14 ± 1.06b	94.25 ± 0.25b
	Čapljina	112.83 ± 12.48b	56.00 ± 2.62b	58.19 ± 2.99b	59.58 ± 3.20b	6.24 ± 0,94b	94.43 ± 0.29b
	Average	**116.57 ± 6.05b**	**57.14 ± 1.26c**	**59.32 ± 1.38c**	**60.76 ± 1.47c**	**6.61 ± 0.64b**	**94.35 ± 0.09b**
Spring Lady	Mostar	123.87 ± 12.76b	59.37 ± 3.11b	61.33 ± 2.93c	63.19 ± 3.11b	7.71 ± 1.06a	93.74 ± 0.42c
	Stolac	126.70 ± 11.28b	59.24 ± 3.49c	61.30 ± 3.46b	63.25 ± 3.46b	7.78 ± 1.24a	94.02 ± 0.54bc
	Čapljina	119.82 ± 11.91b	59.03 ± 3.42b	60.87 ± 3.35b	62.26 ± 3.29b	7.73 ± 1.10a	93.58 ± 0.44cc
	Average	**123.46 ± 3.46b**	**59.21 ± 0.17b**	**61.17 ± 0.26b**	**62.90 ± 0.56b**	**7.74 ± 0.04a**	**93.78 ± 0.22c**
Rich Lady	Mostar	176.01 ± 23.52a	65.01 ± 3.11a	67.07 ± 3.06a	69.09 ± 3.00a	8.13 ± 1.17a	95.41 ± 0.17a
	Stolac	180.88 ± 18.86a	66.13 ± 3.06a	68.33 ± 3.11a	70.34 ± 3.06a	8.24 ± 1.36a	95.45 ± 0.30a
	Čapljina	175.78 ± 22.61a	64.85 ± 3.29a	66.90 ± 3.26a	68.91 ± 3.22a	7.90 ± 1.03a	95.51 ± 0.24a
	Average	**177.64 ± 2.88a**	**65.33 ± 0.70a**	**67.43 ± 0.78a**	**69.45 ± 0.78a**	**8.09 ± 0.17a**	**95.46 ± 0.05a**
Elegant Lady	Mostar	177.90 ± 16.50a	64.93 ± 2.12a	66.99 ± 2.23a	66.07 ± 2.42a	7.77 ± 0.96a	95.62 ± 0.17a
	Stolac	184.68 ± 15.08a	66.63 ± 2.88a	68.71 ± 2.68a	71.29 ± 2.90a	8.45 ± 1.35a	95.45 ± 0.41a
	Čapljina	175.66 ± 16.18a	64.74 ± 1.75a	66.38 ± 2.05a	68.46 ± 2.19a	7.56 ± 0.96a	95.71 ± 0.22a
	Average	**179.41 ± 4.70a**	**65.43 ± 1.04a**	**67.36 ± 1.21a**	**69.61 ± 2.61a**	**7.93 ± 0.47a**	**95.60 ± 0.13a**
Variety × year		*	*	*	*	*	*

Mean separation within columns by Tukey-Kramer test ($P \leq 0.05$). In each column, values with the same letter are not significantly different. * indicate significant influence at $p < 0.05$.

Based on the obtained values presented in Table 1, it can be concluded that the highest fruit weight (184.68 g), fruit height (66.63 mm), fruit width (68.71 mm), thickness (71.29 mm) and stone weight (8.45 g) were recorded in Elegant lady in Stolac. There were no recorded statistical significant differences between this variety and Rich Lady in the mentioned features. The smallest fruit weight (112.83 g), fruit height (56.00 mm), fruit width (58.19 mm), thickness (59.58 mm) and stone weight (6.24 g) were recorded in Springcrest in Čapljina.

Approximately values of the physical characteristics of the fruit were recorded in the varieties Maycrest and Spring Lady in the all localities. Similar results of fruit weight in the variety Elegant Lady were reported by Nenadović-Mratinić et al. [15] in Serbia (180.29 g), but significant higher than results Odalović [16] in Montenegro (117.18 g).

The obtained results of the average fruit weight Maycrest (121.58 g), Springcrest (116.57 g) and Spring Lady (123.46 g) in the all localities were higher than results Odalović in Montenegro (2002) (Maycrest 87.68 g and Springcrest 88.48 g), as well as results Pejkić et al., [17] for variety Maycrest (90.70 g) in Serbia and Gariglio et al. [18] (Maycrest 93.30 g and Spring Lady 72.80 g) in Argentina. Taheri and Hajnajari [19] were obtained higher results of the fruit weight variety Springcrest (144 g) in Iran. Similar results of the fruit weight Rich Lady in our study (177.64 g) were reported by Hudina [20] in (179.5 g) in Slovenia.

The results of the fruit height, width and thickness of Maycrest (59.35, 61.18 and 63.41 mm) were higher of the results Odalović [16] (49.17, 51.63 and 56.38 mm). Odalović [16] obtained similar values in variety Springcrest and Elegant Lady in Serbia.

The obtained results of physical characteristics all varieties confirm the allegations that early varieties have smaller fruits than late [6, 11, 12]. The highest randman (mesocarp/stone ratio) were Rich lady and Elegant Lady and the lowest in the fruit of Maycrest and Spring Lady. Locality has not statistical influence of randman.

The obtained results are in the accordance with results Nenadović-Mratinić et al. [15] and they do not deviate significantly from the standard (96%).

3.2 Chemical Characteristics of the Peach Fruit

The SSC, TA and SSC/TA ratio are chemical parameters that better permit to evaluate the fruit quality perception from consumer [21].

The results of the chemical analyses showed variations between the varieties (Table 2).

Based on the data presented in the table above, it is evident that average SSC ranged from 11.22% (Maycrest) to 12.01% (Elegant Lady). Even though differences in soluble solid content between varieties Springcrest (11.59%), Spring Lady (11.83%) and Rich Lady (11.54%) were found, differences among these varieties were not significant. The influence of locality was recorded, so the highest value of this parameter was evidenced in locality Stolac, while the lowest was found in Čapljina. It was a tendency, that late ripening varieties had higher SSC than earliest, as previously reported [6]. Similar results in the fruit of Rich Lady reported other authors [22, 23] as well as Elegant Lady [15, 16].

Table 2. Chemical characteristics of the fruit in the examined peach varieties

Variety	Locality	Soluble solid SSC (%)	Total sugars TS (%)	Total acids TA (%)	SSC/TA Ratio
Maycrest	Mostar	11.13 ± 0.92b	9.44 ± 0.50ab	0.64 ± 0.09b	17.39 ± 1.58c
	Stolac	11.62 ± 0.76b	9.76 ± 0.34b	0.77 ± 0.05a	15.08 ± 1.65d
	Čapljina	10.92 ± 0.74b	9.25 ± 0.27b	0.71 ± 0.15a	15.75 ± 2.44d
	Average	**11.22 ± 0.36b**	**9.48 ± 0.26b**	**0.71 ± 0.07a**	**16.11 ± 1.19cd**
Springcrest	Mostar	11.94 ± 0.79ab	10.05 ± 0.80a	0.70 ± 0.11a	17.37 ± 1.71c
	Stolac	11.63 ± 0.63b	9.98 ± 0.53ab	0.58 ± 0.09c	20.68 ± 3.31a
	Čapljina	11.19 ± 0.31b	9.61 ± 0.36ab	0.66 ± 0.06b	17.11 ± 1.96c
	Average	**11.59 ± 0.38ab**	**9.88 ± 0.24a**	**0.64 ± 0.06b**	**18.39 ± 1.99b**
Spring Lady	Mostar	11.84 ± 0.61ab	9.99 ± 0.44a	0.60 ± 0.13b	20.47 ± 4.67a
	Stolac	11.94 ± 0.30ab	9.99 ± 0.35ab	0.58 ± 0.05c	20.6 ± 1.99a
	Čapljina	11.19 ± 0.41b	9.74 ± 0.39a	0.61 ± 0.09b	19.7 ± 3.14b
	Average	**11.83 ± 0.41ab**	**9.91 ± 0.14a**	**0.60 ± 0.02c**	**20.26 ± 0.49b**
Rich Lady	Mostar	11.35 ± 0.41ab	9.14 ± 0.24b	0.72 ± 0.05a	15.69 ± 1.45cd
	Stolac	11.81 ± 0.30ab	9.82 ± 0.22ab	0.65 ± 0.03b	18.1 ± 0.73bc
	Čapljina	11.47 ± 0.79ab	9.64 ± 1.09ab	0.74 ± 0.09a	15.58 ± 2.52cd
	Average	**11.54 ± 0.24ab**	**9.53 ± 0.35b**	**0.71 ± 0.05a**	**16.46 ± 1.42cd**
Elegant Lady	Mostar	12.22 ± 0.68a	9.87 ± 0.43ab	0.54 ± 0.04b	22.68 ± 1.64a
	Stolac	11.99 ± 0.67a	10.25 ± 0.34a	0.59 ± 0.04c	20.32 ± 2.46b
	Čapljina	11.81 ± 0.73a	9.79 ± 0.43a	0.62 ± 0.06b	19.09 ± 1.15b
	Average	**12.01 ± 0.21a**	**9.97 ± 0.25a**	**0.58 ± 0.04c**	**20.70 ± 1.82a**
Cultivar × year		*	*	*	*

Mean separation within columns by Tukey-Kramer test ($P \leq 0.05$). In each column, values with the same letter are not significantly different. * indicate significant influence at $p < 0.05$.

Odalović [16] obtained lower value SSC in the fruit *Maycrest* (9.61%) and Springcrest (10.00%).

Sugars represent a fundamental component of fruit edible quality, predominantly conferring sweetness, one of the main attributes influencing the degree of consumer satisfaction for peaches [24, 25]. Sugar accumulation in fruit is a complex quantitative characteristics, affected by environmental conditions and controlled by multiple genes that interact with the environment and crop management [26]. SSC and acidity are determined by several factors such as cultivar [27–30] and rootstock [31].

High content of sugar and, to a lesser extent, high acid content, the consumer is a good attribute of fruit quality [2].

In our research, the highest total sugars content was obtained in Elegant Lady in Stolac (10.25%), and the lowest in Spring Lady in Mostar (9.14%). In all the varieties the similar values of total sugars were recorded.

Regarding TA, significant differences were also found among varieties because it is a variety depended characteristic. The range was from 0.54% (Elegant Lady in Mostar) to 0.77% malic acid (Maycrest in Stolac).

The contents of malic acid in the peach fruit ranging from 3.5 to 6.0 g/kg [32], while Wills et al. [33] obtained slightly higher contents, up to 8.1 g/kg.

Hilaire [34] reported that the upper limit of normal acidity in peach 0.9%. In our study all cultivars had total acids level lower than 0.8%.

Our results show SSC/TA ratio (sweetness index) ranged from 15.08 (Maycrest in Stolac) to 22.48 (Rich Lady in Mostar) depending on their SSC and TA. The higher SSC/TA ratio is usually preferred [35, 36]. Crisosto et al. (1997) reported a closer relationship of RI with eating quality than TA or SSC individually, and the higher RI considers the sweeter fruits. The above data coincides with other authors [21, 37].

From the obtained indicators, it can be concluded that the best sugar and acid ratio has Elegant Lady (the variety of the latest ripening) among all the tested varieties at all localities. It was also shown that the relationship was dependent on variety and/or localities.

4 Conclusions

In this study, the physical and chemical characteristics of the fruit five peach varieties in Herzegovina were determined during a three-year period. The physical characteristics of fruits and stone were conditioned by genetic predisposition, so that much larger fruits were measured in the late-ripening variety (Rich Lady and Elegant Lady), but also by microlocality, so all varieties had the largest fruits in locality Stolac. The varieties of Maycrest, Springcrest and Spring Lady are characterized with smaller fruits, but their maturity was earlier, so they achieve a higher price on the market.

Analysis of the chemical characteristics showed that the varieties which have late maturity are characterized by a higher content of soluble solids, total sugars and ripening index, but less total acid content of fruits compared to early varieties. Microlocality was shown significant influence on chemical composition of the fruit, so the fruits in Stolac had the best sugar and acid ratio.

Based on the research carried out, it can be concluded that the investigated varieties have largely been adapted to the specific agro-climatic conditions of SubMediterranean Herzegovina and meet the modern market demands by their quantitative and qualitative characteristics.

Conflict of Interest. The authors have no conflict of interest to this paper.

References

1. Food and Agricultural Organization of the United Nations (FAO): Statistical Database Agriculture. Available at: http://www.fao.org/faostat/en/#data/QC. Accessed 14 May 2019
2. Crisosto, C.H., Crisosto, G.M.: Relationship between ripe soluble solids concentration (RSSC) and consumer acceptance of high and low acid melting flesh peach and nectarine (*P. persica* (L.) Batsch) cultivars. Postharvest Biol. Technol. **38**, 239–246 (2005)
3. Crisosto, C.H., Crisosto, G.M., Neri, F.: Understanding tree fruit quality based on consumer acceptance. Acta Hort. **712**, 183–189 (2006)
4. Iglesias, I., Echeverría, G.: Differential effect of cultivar and harvest date on nectarine colour, quality and consumer acceptance. Sci. Hortic. **120**, 41–50 (2009). https://doi.org/10.1016/j.scienta.2008.09.011

5. Kramer, A., Twigg, B.A.: Fundamentals of Quality Control for the Food Industry, Avi Publishing, Westport, CT, p. 512 (1966)
6. Dirlewanger, E., Moing, A., Rothan, C., Svanella, L., Pronier, V., Guye, A., Plomion, C., Monet, R.: Mapping QTLs controlling fruit quality in peach (P. persica (L.) Batsch). Theor. Appl. Genet. **98**, 18–31 (1999)
7. Byrne, D.H., Nikolic, A.N., Burns, E.E.: Variability in sugar, acids, firmness and color characteristics of 12 peach genotypes. J. Am. Soc. Hort. Sci. **116**, 1004–1006. (1991). https://doi.org/10.21273/JASHS.116.6.1004
8. Genard, M., Souty, M., Holmes, S., Reich, M., Breuils, L.: Correlations among quality parameters of peach fruit. J. Sci. Food Agric. **66**, 241–245 (1994). https://doi.org/10.1002/jsfa.2740660220
9. Esti, M., Messia, M.C., Sinesio, F., Nicotra, A., Conte, L., Notte, E., Palleschi, G.: Quality evaluation of peaches and nectarines by electrochemical and multivariate analyses: relationships between analytical measurements and sensory attributes. Food Chem. **60**, 659–666 (1997). https://doi.org/10.1016/S0308-8146(96)00306-8
10. Génard, M., Reich, M., Lobit, P., Besset, J.: Correlations between sugar and acid content and peach growth. J. Hortic. Sci. Biotechnol. **74**, 772–776 (1999). https://doi.org/10.1080/14620316.1999.11511187
11. López, G., De Jong, T.M.: Spring temperatures have a major effect on early stages of peach fruit growth. J. Hortic. Sci. Biotechnol. **82**, 507–512 (2007)
12. Ruiz, D., Egea, J.: Phenotypic diversity and relationships of fruit quality traits in apricot (Prunus armeniaca L.) germplasm. Euphytica **163**, 143–158 (2008)
13. Jerie, P.H., Chalmers, D.J.: Ethylene as a growth hormone in peach fruit. Aust. J. Plant Phisiol. **3**, 429–434 (1976). https://doi.org/10.1071/PP9760429
14. Crisosto, C.H., Garner, D., Andris, H.L., Day, K.R.: Controlled delayed cooling extends peach market life. Hortic. Technol. **2004**(14), 99–104 (2004)
15. Nenadović-Mratinić, E., Milatović, D., Đurović, D.: Biološko-proizvodne osobine sorti breskve srednje poznog do poznog vremena zrenja. Zbornik naučnih radova **9**, 169–175 (2003)
16. Odalović, A.: Pomološke osobine novih sorti breskve (Prunus persica (L.) Batsch) u rejonu Ćemovskog polja, Doktorska disertacija, Novi Sad (2002)
17. Pejkić, B., Milatović, D., Vulić, T., Đurović, D.: Utjecaj prstenovanja na osobine ploda ranih sorti breskve. Zbornik naučnih radova **8**(2), 21–26 (2002)
18. Gariglio, N.F., Mendow, M., Weber, M.E., Favaro, M.A., González-Rossia, D.E., Pilatti, R. A.: Phenology and reproductive traits of peaches and nectarines in Central-east Argentina. Sci. Agric. (Piracicaba, Braz.) **66**(6), 757–763 (2009)
19. Taheri, G., Hajnajari, H.: Evaluation of 22 peach cultivars in Moghan (Iran) climatic conditions, Международная дистанционная научная конференция « Интенсификация плодоводства Беларуси: Традиции, достижения, перспективы » Минский р-н, Беларусь (2010)
20. Hudina, M.: Growing experience with some peach and nectarine cultivars. In: International Conference of "Perspectives in European Fruit Growing" on Horticulture Faculty in Lednice, pp. 58–61, Lednice (2006)
21. Di Vaio, C., Graziani, G., Marra, L., Cascone, A., Ritieni, A.: Antioxidant capacities, carotenoids and polyphenols evaluation of fresh and refrigerated peach and nectarine cultivars from Italy. Eur. Food Res. Technol. **227**, 1225–1231 (2008)
22. Gil, M.I., Tomas-Barberan, F.A., Hess-Pierce, B., Kader, A.A.: Antioxidant capacities, phenolic compounds, carotenoids, and vitamin C contents of nectarine, peach, and plum cultivars from California. J. Agric. Food Chem. **50**, 4976–4982 (2002). https://doi.org/10.1021/jf020136b

23. Herreros, A.: Agronomic evaluation of six new peach and nectarine rootstocks for Chile. Proyecto FDI-CORFO 00C7AT – 12. "Introducción, Caracterización y Evaluación Productiva de Nuevos Portainjertos de Duraznero y Nectarino en Chile (2001–2004)
24. Orazem, P., Stampar, F., Hudina, M.: Quality analysis of Redhaven peach fruit grafted on 11 rootstocks of different genetic origin in a replant soil. Food Chem. **124**, 1691–1698 (2011)
25. Delgado, C., Crisosto, G.M., Heymann, H., Crisosto, C.H.: Determining the primary drivers of liking to predict consumers' acceptance of fresh nectarines and peaches. J. Food Sci. **78**, 605–614 (2013)
26. Cirilli, M., Bassi, D., Ciacciulli, A.: Sugars in peach fruit: a breeding perspective. Hortic. Res. **3**, 15067 (2016). https://doi.org/10.1038/hortres.2015.67
27. Crisosto, C.H., Mitchell, F.G., Johnson, R.S.: Factors in fresh market stone fruit quality. Postharvest News Inf. **6**, 17–21 (1995)
28. Frecon, J.L., Belding, R., Lokaj, G.: Evaluation of white-fleshed peach and nectarine varieties in New Jersey. Acta Hort. **592**, 467–478 (2002)
29. Liverani, A., Giovannini, D., Brandi, F.: Increasing fruit quality of peaches and nectarines: the main goals of ISF-FO (Italy). Acta Hort. **592**, 507–514 (2002)
30. Byrne, D.: Breeding peaches and nectarines for mild-winter climate areas: state of the art and future directions. In: Marra, F., Sottile, F. (eds.) Proceedings of the First Mediterranean Peach Symposium Agrigento, Italy, 10 September, pp. 102–109 (2003)
31. Reighard, G.L.: Current directions of peach rootstock programs worldwide. Acta Hort. **592**, 421–428 (2002)
32. Wang, T., Gonzales, A.R., Gbur, E.E., Aselage, J.M.: Organic acid changes during ripening of processing peaches. J. Food Sci. **58**(3), 631–632 (1993)
33. Wills, R.B.H., Scriven, F.M., Greenfield, H.: Nutrient composition of stone fruit (Prunus spp. cultivars: apricot, cherry, nectarine, peach and plum). J. Sci. Food Agric. **34**(12), 1383–1389 (1983)
34. Hilaire, C.: The peach industry in France: state of art, research and development. In: Proceeding of the First Mediterranean Peach Symposium, Agrigento, Italy (2003)
35. Robertson, J.A., Meredith, F.I., Scorza, R.: Physical, chemical and sensory evaluation of high and low quality peaches. Acta Hort. **254**, 155–159 (1989)
36. Bassi, D., Selli, R.: Evaluation of fruit quality in peach and apricot. Adv. Hortic. Sci. **4**, 107–112 (1990)
37. Abidi, W., Jiménez, S., Ángeles, M.M., Gogorcena, Y.: Evaluation of antioxidant compounds and total sugar content in a nectarine (Prunus persica (L.) Batsch) Progeny. Int. J. Mol. Sci. **12**, 6919–6935 (2011). https://doi.org/10.3390/ijms12106919

Chemical Composition and Total Phenols Content of Tartary Buckwheat (*Fagopyrum tataricum* Gaertn) Grown in Different Vegetation Seasons

Teofil Gavrić[1(✉)], Drena Gadžo[1], Josip Jurković[1], Mirha Đikić[1], Dženan Hadžić[1], Blažo Lalević[2], and Saud Hamidović[1]

[1] Faculty of Agriculture and Food Sciences, University of Sarajevo, Zmaja od Bosne 8, Sarajevo, Bosnia and Herzegovina
t.gavric@ppf.unsa.ba

[2] Faculty of Agriculture, University of Belgrade, Nemanjina 6, 11080 Belgrade-Zemun, Belgrad, Serbia

Abstract. Tartary buckwheat (*Fagopyrum tataricum* Gaertn) is an annual plant that is classified as a pseudocereal with high nutritional values. It is mainly cultivated in Asia, where grain and other parts of plants are used in traditional diet. In recent years, tartary buckwheat is grown on smaller areas and mostly in a mixture with common buckwheat in Bosnia and Herzegovina. The aim of this research was to determine the chemical composition and total phenols content of tartary buckwheat grown in different vegetation seasons. Field experiments were conducted in Donje Selo, near Ilijaš during 2011, 2012 and 2013 vegetation seasons. The results show that weather in some vegetation seasons have statistically significant effect on contents of protein and starch in kernel. The largest contents of protein were found in vegetation season with a lack of precipitation (2012 year), while the largest content of starch was recorded in years with relatively higher precipitation levels (2011 and 2013 years). Results show that total phenols content depended on the plant organ and vegetation season. The highest content was found in the flower (63.63 mg GAE g^{-1}), while the smallest content was found in the kernel (5.03 mg GAE g^{-1}).

Keywords: Tartary buckwheat · Chemical composition · Protein · Starch · Total phenols

1 Introduction

Buckwheat is an annual plant which belongs to botanical family Polygonaceae. Although, there are 23 species of buckwheat in the world [1] but widely known and grown- are common buckwheat (*Fagopyrum esculentum* Moench) and tartary buckwheat (*Fagopyrum tataricum* Gaertn) [2]. Common buckwheat spreads to all continents around the world [3] and it is produced on 90% of the areas that are sown with buckwheat [4]. Tartary buckwheat is grown on smaller areas mainly in China, India, Bhutan, Nepal, Luxembourg, Slovenia [3, 4] and Bosnia and Herzegovina [5].

M. Brka et al. (Eds.): AgriConf 2019, IFMBE Proceedings 78, pp. 59–68, 2020.
https://doi.org/10.1007/978-3-030-40049-1_7

Morphological and growing technology of both varieties of buckwheat are similar, but there are certain differences. Tartary buckwheat compared to common buckwheat has some desirable characteristics such as frost resistance, self-pollinating ability and higher kernel yield [6].

As an annual crop, buckwheat is often called pseudocereal, because itś morphology is not similar to cereals. Buckwheat is nutritionally valuable food for people. Buckwheat's flour is gluten free [3] and has a good amino acid composition [7] which is particularly rich in arginine and lysine [8]. Compared to other cereals´s flour, buckwheat flour has a higher fiber content [3, 8]. Also, buckwheat and its products are rich in minerals such as Fe, Zn, Mn, Se and Cu [9]. In recent years, buckwheat became even more popular in human nutrition due to its positive impact on human health. Namely, many studies have shown that consumption of buckwheat products affects the treatment and prevention of many diseases like diabetes [10–13], elevated blood pressure [14], cardiovascular diseases [15], cancer [15, 16]. The healing properties of buckwheat are derived from the antioxidant contents that has the ability to protect the human body against oxidative stress caused by free radicals. Buckwheat is the species of pseudo-cereals which have the highest antioxidant activity [17]. Antioxidant activity of buckwheat is the result of high phenolic compounds such as rutin, quarcetin and proantho-cyanidines [11, 18]. There are total phenols in all plant organs (leaves, flowers, kernel, stem, root) [19] and they can be used as a high-quality source of antioxidants [20].

Chemical composition of buckwheat' kernel and content of total phenols in some buckwheat organs depend on variety of buckwheat [5, 21] and agroecological conditions during cultivation [22]. Significant impact on changing chemical composition of kernel and content of total phenols may have weather conditions [23, 24] mainly due to variations in rainfall [5, 25] and temperature [26] during vegetation season. Regarding the previous facts, the main focus of this research was to determine the chemical composition and total phenols content of tartary buckwheat grown in different vegetation seasons.

2 Materials and Methods

2.1 Field Experiment

A field experiment was set up at the village Donje Selo, near Ilijaš. Tartary buckwheat (Fagopyrum tataricum Gaertn) originates from Luxembourg was used in this research. It was grown in three different vegetation seasons (2011, 2012 and 2013). Sowing of buckwheat was done manually (300 seeds per m^2). Buckwheat was sown on May 25, 2011, June 2, 2012 and May 19, 2013. The plots in the experiment were arranged in a randomized block design in four replications. Each plot area was 4.8 m^2 (3.10 m long and 1.55 m wide) with 1 m wide space between two neighboring plots. The distances between rows in plot was 15 cm, respectively each plot area consisted of 10 rows. Soil type was calcocambisol. Low dose fertilizer (10 kg N, 40 kg P_2O_5 and 60 kg K_2O per hectare) have been applied in spring into a pre sowing cultivation. During the vegetation, weeds were removed manually. Meteorological data from meteorological station Sokolac were used for the analysis of weather conditions [27].

2.2 Materials for Analysis

The contents of protein, starch, cellulose, lipid and ash were determined in kernel. The content of total phenols was determined in leaves, flowers and kernels. Kernel samples were collected at the ripeness stage (August 23, 2011, August 20, 2012 and August 14, 2013). Leaves and flower samples were collected at flowering stage (July 1, 2011, July 4, 2012 and June 23, 2013). Collected plant samples were dried in a dark room at room temperature for 30 days. After drying, samples were milled to 1 mm.

2.3 Determination of Chemical Composition of Kernel

Nitrogen contents in kernel was determined according to kjeldahl method described by ISO standard method [28]. Distillation of nitrogen was done by Microkjeldahl apparatus (Foss Kjeltec 2200). After determination, protein contents in kernel was calculated by multiplying nitrogen contents in kernel with a constant factor of 6.25. Starch was determined by Ewers method according ISO method [29]. Total lipids in kernel were determined by instructions described to ISO standard method [30], contents of ash were also determined by the ISO standard method [31]. The contents of cellulose in kernel was determined according to Kürschner-Hanack method [32].

2.4 Preparation of Extracts for Analysis

The extract was prepared as follows: Approximately 1 g of each dried and milled plant material was added to 100 mL volumetric flask. Volume of flask was made up to mark with 60% of ethanol and mixed. After extraction the whole volume was filtered through Blue Ribbon and used for analysis.

2.5 Determination of Total Phenols Content

The total phenols content in the extracts of samples of tartary buckwheat was determined by modified Folin–Ciocalteu [20]. Briefly, 1 mL of extract, blank or standard was added to 10 mL volumetric flask. To the volumetric flask, 0.5 mL Folin-Ciocalteu reagent was added and the mixture was allowed to react. After 5 min, 1.5 mL of 20% Na_2CO_3 was added and after that volume of flask was made up to mark with distilled H_2O. After 120 min standing at room temperature, absorbance was determined at 765 nm using UV-VIS spectrophotometer (Ultrospec 2100 pro). The total phenols content was expressed as mg gallic acid equivalents (GAE).

2.6 Statistical Methods

The results of the research were statistically analyzed by analyzing variance in SPSS 22 software program. The average values of determined data were tested using Test Support at $p = 0.05$.

3 Results and Discussions

3.1 Analysis of Weather Conditions

Monthly precipitation amounts and average air temperatures for the period of research is presented in Fig. 1. Analysis of weather conditions were made on data for the weather station in Sokolac: about 15 km air-distance from field experiment [27]. Based on data presented in Fig. 1 it can be concluded that the weather conditions during the researched period were more unfavorable compared to the climate reference period (1961–1990). Namely, dry and warm weather was recorded during the researched period (2011–2013). Particularly unfavorable weather conditions were recorded in 2012, when during vegetation period, the air temperature was higher by 4.4 (July) compared to the climate reference period (1961–1990). In the same period, there was also a lack of precipitation when the minimum amount of precipitation was recorded in August (0.6 mm). The weather conditions in 2011 and 2013 were more favorable compared to 2012, but it was more unfavorable compared to the climate reference period (1961–1990). Buckwheat is the plant that achieves the best yield and quality of kernels in humid regions [26, 33]. Such weather conditions were not recorded during researched vegetation season, particularly in 2012 year and based on this it can be concluded that weather conditions were not the most favorable for buckwheat production.

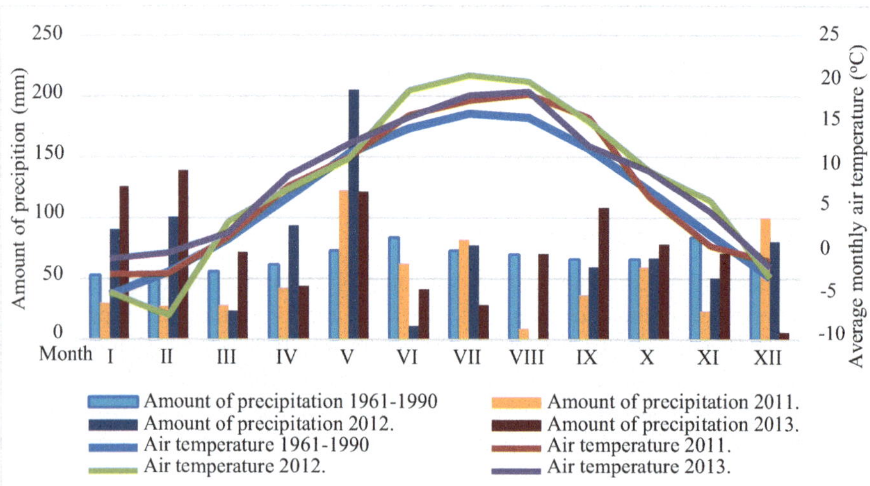

Fig. 1. Average monthly temperature and amount of rainfall

3.2 Chemical Composition of Kernel

Buckwheat kernel contains a variety of components and the main nutrients are proteins, starch, cellulose, lipids, micro, macroelements and phenols. The ratio of some nutrients to the kernel is variable and depends on the variety of buckwheat [8, 34] and environmental factors [8].

Table 1. Chemical compose of tartary buckwheat's kernel, %

Vegetation season	Protein	Starch	Cellulose	Fat	Ash
2011	10.59^{b}	60.20^{a}	13.44^{ns}	2.26^{ns}	2.21^{ns}
2012	13.40^{a}	55.71^{b}	14.41^{ns}	1.92^{ns}	2.17^{ns}
2013	11.55^{ab}	57.69^{ab}	13.65^{ns}	2.00^{ns}	2.37^{ns}
Average	11.86	57.86	13.83	2.06	2.25

a and b - Statistically significant differences at 0.05 level
ns - Statistically not significant differences at 0.05 level

In this study, it was found that some nutrients in kernel were significantly different in the years of growing (Table 1), for example in the weather conditions during vegetation season. The average contents of protein statistically significantly depended on the year of research and ranged from minimal of 10.01% (2011) to maximal of 12.40% (2012). The largest content of protein in kernel was detected (12.40%) during the 2012 year. High content of protein in kernel was caused by high temperature and drought.

Namely, during this vegetation, premature maturation kernel of occurred which reduced starch content. This fact was confirmed by correlation analysis (Table 2). Namely, date in table show a strong negative correlation relationship between crude protein and starch ($r = -0.890$).

On the basis of data for starch contents presented in Table 1, it can be concluded that its contents also depended on the year of growing. Statistically significant lower content of starch in the kernel (55.71%) was found in buckwheat grown in 2012. Lower content of starch in kernel in this vegetation season was caused by lack of precipitation (Fig. 1) and premature maturation occurred in kernel. This observation can be confirmed by the fact that in the years with a larger amount of precipitation (2011 and 2013) during vegetation season, it found the highest starch contents in kernel (60.20% and 57.69%).

Observations on changing contents of protein and starch in the kernel can be confirmed by research of many authors [35–38]. They found that lack of rain in some years led to low starch contents and high protein contents in kernel.

Based on the result in the Table 1, it can be concluded that the content of cellulose, fat and ash in kernel were not variable properties under the influence of the weather conditions in some vegetation seasons. Gavrić et al. [26] also found that rainfall and average monthly temperatures do not have a significant impact on the content of these compositions.

Table 2. The correlation between researched traits in kernel

	Ash	Cellulose	Starch	Fat	Protein
Ash	1	0.412	−0.035	−0.266	−0.139
Cellulose	0.412	1	−0.568	−0.213	0.516
Starch	−0.035	−0.568	1	0.298	−0.890**
Fat	−0.266	−0.213	0.298	1	−0.368
Protein	−0.139	0.516	−0.890**	−0.368	1

** Correlation is significant at the 0.01 level.

3.3 The Total Phenols Content

As can be seen from Table 3 contents of total phenols in tartary buckwheat depended on the year of growing and anatomic organs of tartary buckwheat. The influence of year of research on total contents were statistically significant. The highest average content of total phenols in all plant parts, especially in the kernel was found in 2012 year. If it is known that phenols in the plant have a protective role [11, 39–42] it can be concluded that the plant grown in 2012 year synthesized an additional amount of phenol to protect its self from unfavorable conditions. Unfavorable conditions for buckwheat in 2012 were caused by lack of precipitation and high average temperature (Fig. 1). This observation was confirmed by many authors [5, 43, 44] who found that high temperatures and dry weather have positive effect on total phenols content.

Table 3. Content of total phenols in tartary buckwheat, mg GAE g^{-1}

Vegetation season	Leaves	Flowers	Kernel
2011	52.46^{ns}	63.03^{ab}	1.09^{b}
2012	37.30^{ns}	76.64^{a}	12.52^{a}
2013	41.44^{ns}	51.23^{b}	1.49^{b}
Average	43.73	63.63	5.03

a and b - Statistically significant differences at 0.05 level
ns - Statistically not significant differences at 0.05 level

Comparison of total phenols in different organs (Table 3), it can be concluded that the flowers contained significantly higher amounts of total phenols. In flowers (63.63 mg GAE g^{-1}) there were found in average 8.7-times higher concentrations of total phenol than kernel (5.03 mg GAE g^{-1}). These results are in consistency with results of many authors [19, 20, 45, 46] who also found higher phenols content in flowers and leaves then in kernel. However, date about total phenols contents in some plant organs were very different. Differences were probably caused by different varieties of buckwheat, growing technology, weather conditions and the method of extraction and determination of total phenols.

Based on data in Table 3 and the research data of many authors [10, 47, 48] it can be concluded that buckwheat, especially its flowers and leaves are rich in phenolic compounds. This fact is interesting, because tartary buckwheat can be used as a potent source of phenolic compounds and antioxidants that have a positive effect on human health.

4 Conclusions

Results in this research showed that chemical composition and total phenols content of tartary buckwheat depended on the year of growing i.e. weather conditions in some vegetation seasons. Protein and starch contents in the kernel were in a negative correlation and their contents depended on the year of growing. The largest protein content was found in years with a lack of precipitation, while the largest content of the starch was recorded in years with relativity higher precipitation levels. Weather conditions in some vegetation seasons also had a significant effect on the content of total phenols. Higher concentrations of phenols were recorded in vegetation season with a lack of rainfall. The highest content of total phenols was found in the flower (63.63 mg GAE g^{-1}), while the smallest content was found in the kernel (5.03 mg GAE g^{-1}).

References

1. Rana, J., Singh, M., Chauhan, R., Chahota, R., Sharma, T., Yadav, R., Archak, S.: Genetic resources of buckwheat in India. In: Molecular Breeding and Nutritional Aspects of Buckwheat, pp. 109–135 (2016). https://doi.org/10.1016/B978-0-12-803692-1.00009-2
2. Kara, N.: Yield and mineral nutrition content of buckwheat (*Fagopyrum esculentum* Moench): the effect of harvest times. Suleyman Demirel Univ. J. Agric. (Turk.) **9**(1), 85–94 (2014)
3. Gadžo, D., Đikić, M., Oručević-Žuljević, S., Gavrić, T., Grahić, J.: Buckwheat production in mountainous regions – experiences and future challenges. In: Simpozij unapređenje poljoprivrede, šumarstva i vodoprivrede u kraškim, brdskim i planinskim područjima - racionalno korištenje i zaštita, Posebna izdanja ANUBiH Sarajevo, BiH, vol. 25, pp. 51–60 (2016)
4. Sharma, T., Jana, S.: Species relationships in Fagopyrum revealed by PCR-based DNA fingerprinting. Theor. Appl. Genet. **105**, 306–312 (2002)
5. Gavrić, T., Gadžo, D., Đikić, M., Bezdrob, M., Čadro, S., Bašić, F.: Effects of plant density on the yield and total phenolic contents of tartary buckwheat. In: IX International Scientific Agricultural Symposium "Agrosym 2018", Jahorina, pp. 97–102 (2018)
6. Tsuji, K., Ohnishi, O.T.: Origin of cultivated Tatary buckwheat (Fagopyrum tataricum Gaertn.) revealed by RAPD analyses. Genet. Resour. Crop Evol. **47**, 431–438 (2000)
7. Selimović, A., Miličević, D., Jašić, M., Selimović, A., Đurđica, Đ., Pešić, T.: The effect of baking temperature and buckwheat flour addition on the selected properties of wheat bread. Croatian J. Food Sci. Technol. **6**(1), 43–50 (2014)
8. Christa, K., Soral-Śmietana, M.: Buckwheat grains and buckwheat products – nutritional and prophylactic value of their components – a review. Czech J. Food Sci. **26**(3), 153–162 (2008)

9. Unal, H., Izli, G., Izli, N., Asik, B.B.: Comparison of some physical and chemical characteristics of buckwheat (*Fagopyrum esculentum* Moench) grains. CyTA – J. Food **15**(2), 257–265 (2017)
10. Mann, S., Gupta, D., Gupta, R.K.: Evaluation of nutritional and antioxidant potential of Indian buckwheat grains. Indian J. Tradit. Knowl. **11**(1), 40–44 (2012)
11. Kreft, I., Fabjan, N., Germ, M.: Rutin in buckwheat - protection of plants and its importance for the production of functional food. Fagopyrum **20**, 7–11 (2003)
12. Kawa, J.M., Taylor, C.G., Przybylski, R.: Buckwheat concentrate reduces serum glucose in streptozotocin-diabetic rats. J. Agric. Food Chem. **51**(25), 7287–7291 (2003)
13. Li, J., Gong, F., Li, F.: Hypoglycemic and hypolipidemic effects of flavonoids from tatary buckwheat in type 2 diabetic rats. Biomed. Res. **27**(1), 132–137 (2016)
14. Choio, Y.S., Lee, B.H., Shin, T.H., Lee, H.H.: Hypotensive effect of dietary buckwheat in SHR. In: Proceedings of the 8th International Symposium on Buckwheat, Chunchon, Korea (2001)
15. Tomotake, H., Shimaoka, I., Kayashita, J., Nakajoh, M., Kato, N.: Buckwheat protein suppress plasma cholesterol more strongly then soy protein isolate in rats by enhancing fecal excretion of steroids. In: Proceedings of the 8th International Symposium on Buckwheat, Chunchon, Korea, pp. 587–594 (2001)
16. Park, B.D., Park, C.H.: Cytotoxic activities of tartary buckwheat against human cancer cells. In: Proceedings of the 9th International Symposium on Buckwheat, Prague, Czech Republic, pp. 665–668 (2004)
17. Kerienė, I., Mankevičienė, A., Bliznikas, S., Jablonskytė-Raščė, D., Maikštėnienė, S., Česnulevičienė, R.: Biologically active phenolic compounds in buckwheat, oats and winter spelt wheat. Zemdirbyste-Agric. **102**(3), 289–296 (2015)
18. Kreft, I., Germ, M.: Organically grown buckwheat as a healthy food and a source of natural antioxidants. Agronomski Glas. **4**, 397–406 (2008)
19. Park, B.J., Park, J.I., Chang, K.J., Park, C.H.: Comparison in rutin content in seed and plant of tartary buckwheat (*Fagopyrum tartaricum*). In: Proceedings of the 9th International Symposium on Buckwheat, Prague, Czech Republic, pp. 626–629 (2004)
20. Bystrická, J., Vollmannová, A., Margitanová, E., Čičová, I.: Dynamics of polyphenolic formation in different plant parts and different growth phases of selected buckwheat cultivars. Acta Agriculturae Slov. **95**(3), 225–229 (2010)
21. Gavrić, T., Gadžo, D.: Yield and chemical composition of seeds of common buckwheat and tartary buckwheat by various cultivation technologies. Works Fac. Agric. Univ. Sarajevo **61**(1), 7–17 (2011)
22. Strakšas, A., Kučinskas, V., Vaiciukevičius, E.: The investigation and comparison of buckwheat harvesting technologies in Lithuania. Agric. Eng. Res. Pap. **43**(1), 44–59 (2011)
23. Rahimić, A., Gadžo, D.: The influence of the location and sowing term on the yield and qualitative properties of buckwheat. Works Fac. Agric. Food Sci. Univ. Sarajevo **62**(1), 15–28 (2012)
24. Mikulajová, A., Takácsová, M., Rapta, P., Brindzová, L., Zalibera, M., Németh, K.: Total phenolic contents and antioxidant capacities of cereal and pseudocereal genotypes. J. Food Nutr. Res. **46**, 150–157 (2007)
25. Vojtíšková, P., Kmentová, K., Kubáň, V., Kráčmar, S.: Chemical composition of buckwheat plant (*Fagopyrum esculentum*) and selected buckwheat products. J. Microbiol. Biotechnol. Food Sci. **1**, 1011–1019 (2012)
26. Gavrić, T., Čadro, S., Gadžo, D., Dikić, M., Bezdrob, M., Jovović, Z., Jurković, J., Hamidović, S.: Influence of meteorological parameters on the yield and chemical composition of common buckwheat (*Fagopyrum esculentum* Moench). Agric. For. **64**(4), 113–120 (2018)

27. Republički hidrometeorološki zavod RS (2015). https://rhmzrs.com
28. ISO 5983-2: International Organization for Standardization, animal feeding stuffs – determination nitrogen content and calculation of crude protein content (2005)
29. ISO 6493: International Organization for Standardization, animal feeding stuffs - Determination of starch content - Polarimetric method (2000)
30. ISO 6492: International Organization for Standardization, animal feeding stuffs - determination of fat content (1999)
31. ISO 5984: International Organization for Standardization, animal feeding stuffs – determination of ash (2002)
32. Kulić, G., Radojičić, V.: Analysis of cellulose content in stalks and leaves of large leaf tobacco. J. Agric. Sci. **56**(3), 207–215 (2011)
33. Milić, V., Govedarica, B., Šilj, M., Berjan, S., Jovović, Z.: Effect of agro-ecological conditions on grain yield in some genotypes of buckwheat. In: IV International Scientific Agricultural Symposium "Agrosym 2013", Jahorina, BiH, pp. 191–195 (2013)
34. Bonafaccia, G., Fabjan, N.: Nutritional comparison of tartary buckwheat with common buckwheat and minor cereals. Zbornik Biotehniłke fakultete Univerze v Ljubljani **81**(2), 349–355 (2003). Slovenia
35. Sial, M.A., Arain, M.A., Naqavi, S.K., Dahoti, M.M., Nizamani, N.A.: Yield and quality parameters of wheat genotypes as affected by sowing dated and high temperature stress. Pak. J. Bot. **37**(3), 575–584 (2005)
36. Pierre, C.S., Petersona, J., Rossa, A., Ohma, J., Verhoevena, M., Larsona, M., Hoefera, B.: White wheat grain quality changes with genotype, nitrogen fertilization and drought stress. Agron. J. **100**, 414–420 (2008)
37. Erekul, O., Köhn, W.: Effect of weather and soil conditions on yield components and bread-making quality of winter wheat (*Triticum aestivum* L.) and winter triticale (*Triticosecale* Wittm.) varieties in north-east Germany. J. Agron. Crop Sci. **192**, 452–464 (2006)
38. Vafa, P., Naseri, R., Moradi, M.: The effect of drought stress on grain yield, yield components and protein content of durum wheat cultivars in Ilam Province. Iran Int. J. Biol. Vet. Agric. Food Eng. **8**(6), 619–624 (2014)
39. Lim, J.H., Park, K.J., Kim, B.K., Jeong, J.W., Kim, H.J.: Effect of salinity stress on phenolic compounds and carotenoids in buckwheat (*Fagopyrum esculentum* M.) sprout. Food Chem. **135**, 1065–1070 (2012)
40. Germ, M.: Environmental factors stimulate synthesis of protective substances in buckwheat. In: Proceedings of the 9th International Symposium on Buckwheat, Prague, Czech Republic, pp. 55–60 (2004)
41. Guo, X., Ma, Y., Parry, J., Gao, J., Yu, L., Wang, M.: Phenolics content and antioxidant activity of tartary buckwheat from different locations. Molecules **16**(12), 9850–9867 (2011)
42. Gavrić, T., Jurković, J., Hamidović, S., Haseljić, S., Lalević, B., Čorbo, A., Bezdrob, M.: Yield and contents of some bioactive components of basil (*Ocimum basilicum* L.) depending on time of cutting. Stud. Univ. Vasile Goldis Ser. Stiintele Vietii (Life Sci. Ser.) **28**(4), 192–197 (2018)
43. Kreft, S., Janeš, D., Kreft, I.: The content of fagopyrin and polyphenols in common and tartary buckwheat sprouts. Acta pharm. **63**, 1–8 (2013)
44. Lumingkewas, A., Koesmaryono, Y., Aziz Impron, S., Sugimoto, H.: The influence of temperature to rutin concentration of buckwheat grains in humid tropic. IJSBAR **20**(1), 1–9 (2015)
45. Kalinová, J., Dadakova, E.: Varietal differences of rutin in common buckwheat (*Fagopyrum esculentum* Moench) determined by micellar electrokinetic capillary chromatography. In: Proceedings of the 9th International Symposium on Buckwheat, Prague, Czech Republic, pp. 719–722 (2004)

46. Bonafaccia, G., Maccati, F., Galli, V.: Dietary fiber and phenolic compounds in common and tartary buckwheat. In: Proceedings of the 11th International Symposium on Buckwheat, South Korea (2009)
47. Zielińska, D., Turemko, M., Kwiatkowski, J., Zieliński, H.: Evaluation of flavonoid contents and antioxidant capacity of the aerial parts of common and tartary buckwheat plants. Molecules **17**(8), 9668–9682 (2012)
48. Seo, J., Lee, D., Valan Arasu, M., Wu, Q., Suzuki, T., Yoon, Y., Lee, S., Park, S., Kim, S.: Quantitative differentiation of phenolic compounds in different varieties of buckwheat cultivars from China, Japan and Korea. J. Agric. Chem. Environ. **2**(4), 109–116 (2013)

Root Pathogens Occurring on Pea (*Pisum sativum*) and Faba Bean (*Vicia faba*) in Germany

Adnan Šišić[1(✉)], Jelena Baćanović-Šišić[2], Harald Schmidt[3], and Maria R. Finckh[1]

[1] Department of Ecological Plant Protection, Kassel University, Witzenhausen, Germany
{adnan_sisic,mfinckh}@uni-kassel.de
[2] Section of Organic Plant Breeding and Agrobiodiversity, Kassel University, Witzenhausen, Germany
bacanovic@uni-kassel.de
[3] Foundation Ecology & Agriculture (SOEL), Ahrweiler, Germany
schmidt@soel.de

Abstract. An extensive survey was undertaken in ten German States in three consecutive years (2016–2018) in order to determine the incidence of root rot on pea (*Pisum sativum*) and faba bean (*Vicia faba*) and identify pathogens involved. In total, 185 pea and faba bean fields were sampled (102 conventional and 83 organically managed) within the demonstration network on peas and beans in the framework of the German Protein Crop Strategy. A total of 5167 *Fusarium* and 1438 *Didymella* isolates were recovered representing 14 different *Fusarium* and *4 Didymella* species. *Didymella pinodella* was the only pathogen strongly influenced by the management systems and was found predominantly in organic fields on both crops. Preliminary results further indicate increasing importance and significance of *F. redolens* and *D. pinodella* for grain legume production in Germany. However, their role and direct pathogenicity need to be further investigated.

Keywords: Grain legumes · Foot and root rot · Pathogens · *Fusarium* · *Didymella*

1 Introduction

There is a great interest in Germany and Europe to convert to more diversified and sustainable cropping systems aiming to restore soil ecosystem functions by promoting internal on-farm regulatory processes while reducing external inputs [1]. Suitable cropping systems need to rely on soil biological processes and management strategies that promote and enhance positive soil biology. Plants actively influence soil processes through root exudates that promote the growth of specific soil microorganisms. In turn, these microorganisms help in plant nutrition by making mineral nutrients available and reducing the need for external N inputs. In addition, they may compete with or directly

M. Brka et al. (Eds.): AgriConf 2019, IFMBE Proceedings 78, pp. 69–75, 2020.
https://doi.org/10.1007/978-3-030-40049-1_8

antagonize plant pathogens in the soil and thus they are an important component of natural soil disease suppressiveness [2, 3].

It is of great agronomical importance, and in organic farming in particular, to integrate more legumes into crop rotations. Largely seen in the context of fixed-N contribution to the system, legumes perform a range of ecosystem services beyond biological N fixation. When planned properly in terms of adequate diversity and appropriate density, legume based species mixtures and crop rotations can enhance physical, chemical and biological characteristics of soils. Increase of legumes in rotations, particularly in cooler areas, requires an integrated approach that should include higher frequencies of legume cash crops. In Germany, pea (*Pisum sativum*) and faba bean (*Vicia faba*) are of particular interest that in addition to important ecological services, provide valuable sources of proteins for human and animal nutrition, representing cash crops capable of partially replacing soybean imports. However, despite this potential, their production in Germany and many other European countries is in long-term decline. Since pea and faba bean share important soil-borne pathogens, farmers are often forced to grow these species in wide rotations that usually only fit into five or more year rotation plans [4].

As the functional integrity of leguminous roots is a basic prerequisite for biological N fixation and ecological services provision, root pathogens have profound effects on these processes with direct negative consequences on weed suppression and on nutrient availability in subsequent main crops. Numerous *Fusarium* and several *Didymella* spp. play a key role in the decline of the area planted to the crops. In Germany, *D. pinodella*, *D. pinodes*, *F. redolens* and *F. avenaceum* have recently been reported as the primary causal agents of the disease [5, 6]. These results are in contrast with studies from France, Sweden, USA, Canada and Australia where *D. pinodella* was not considered as a major part of the complex, and *Aphanomyces euteiches* and several different *Fusarium* spp. were identified as the main contributing factors to the decline of the area planted to the crops [7]. In addition to the complex ecology and difficulties to precisely determine combinations of soil and environmental conditions that influence individual densities of particular species, prevalence of the single pathogens may vary greatly depending on location, climate, and agricultural practice [8].

The main objectives of the current study were to assess root health status and to characterize diversity, frequency and possible underlying patterns of host or regional preference of *Fusarium* and *Didymella* spp. associated with roots of pea and faba bean in Germany. The survey included a 3-year study period (2016–2018) conducted within the demonstration network on peas and beans in the framework of the German Protein Crop Strategy. The network provided the opportunity to scientifically investigate root health status and pathogen complex in pea and faba bean roots in arable farming systems covering a north-south German transect and to discuss existing research findings with regard to their practical relevance. Furthermore, as the farming system (organic versus conventional) may impact *Fusarium* and *Didymella* infestation and favor different pathogenic species in the roots, we investigated possible effects of some of the abiotic soil properties on the occurrence of these species. Data presented in this paper provide a preliminary overview with respect to a limited number of the total parameters assessed as sampling is continuing in 2019.

2 Materials and Methods

In the period between 2016 and 2018, 185 fields of the pea/bean Demonstration Network were sampled and included 102 conventional and 83 organically managed fields. The fields were located in 10 German states, namely Mecklenburg Vorpommern, Lower Saxony, Saxony-Anhalt, Saxony, Thuringia, Hesse, North Rhein-Westphalia, Rhineland-Palatinate, Baden Württemberg and Bavaria. From each field, 18–20 root samples of either winter or summer pea or faba bean were collected at two different locations and the severity of root symptoms, the frequency of relevant pathogenic fungi was determined as previously described [9, 10]. Briefly, following sample collection roots were washed from adhering soil and individual plants were evaluated for root rot disease severity using a visual 0–8 score scale based on external and internal root tissue discoloration levels according to Pflughöft et al. 2012 [11]. Subsequently, the plant samples were surface sterilized with 3% sodium hypochlorite for 10 s, rinsed in distilled water and air dried. Each plant was cut in 3 pieces (ca. 1 cm in length) placed on appropriate media for incubation at 20 °C under 12 h cycles of UV light and dark. After 1–2 weeks of incubation, pure cultures were generated either through single spore transfers or hyphal tipping, and each isolate was examined microscopically and identified to the species level based on cultural appearance (colony color and pigmentation) and microscopic examination of conidiogenous cells [12, 13]. Morphological identity of a subset of randomly selected isolates was confirmed by phylogenetic placement using partial sequences of the translation elongation factor 1 alpha and, if necessary, the second largest subunit of RNA polymerase II (rpb2) [14–16].

3 Results

A total of 1992 pea and 1698 faba bean plants were analyzed over the sampling period. Foot and root rot symptom severity were at low to moderate level and similar for pea and faba bean across years in both management systems (data not shown). Overall, 5167 *Fusarium* and 1438 *Didymella* isolates were recovered and these represented 14 different *Fusarium* and 4 different *Didymella* species. The number of recovered *Fusarium* isolates was similar in conventional and organic fields in each year, whereas *Didymella* was strongly influenced by the management systems and was found predominantly in organic fields on both crops (Figs. 1 and 2). *Fusarium redolens*, *F. oxysporum*, *F. solani*, *F. avenaceum*, *F. tricinctum*, *F. culmorum*, *F. equiseti* and *D. pinodella* were the most frequently isolated pathogens from pea and faba bean roots in Germany (Table 1).

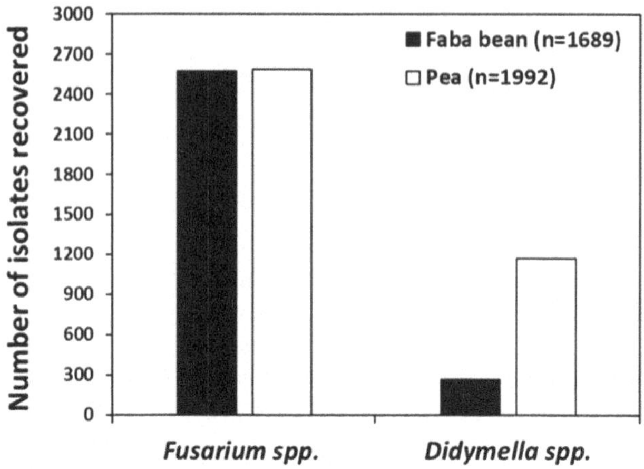

Fig. 1. Number of isolates of *Fusarium* spp. and *Didymella* spp. recovered from roots of peas and faba beans in the period 2016–2018.

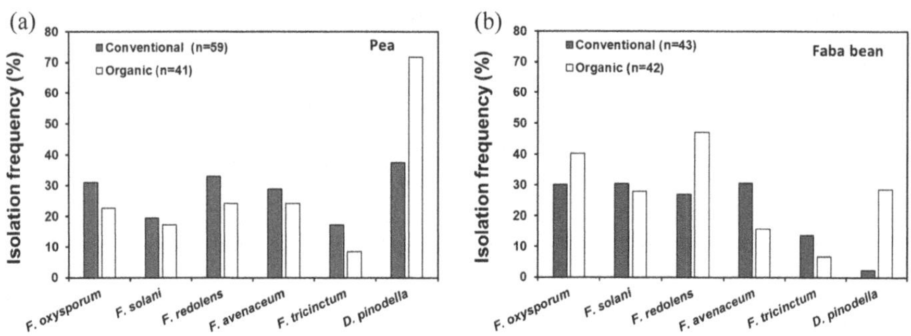

Fig. 2. Isolation frequency (%) of six the most common pathogens isolated from roots on pea and faba bean grown under conventional and organic farming system.

Table 1. Frequency (%) of isolation of eight the most commonly found pathogens on the roots of pea and faba bean in the period 2016–2018

Year	Plant species	N[1]	n[2]	*D. pin*	*F. ave*	*F. oxy*	*F. red*	*F. sol*	*F. tric*	*F. equi*	*F. culm*
2016	Pea	28	558	50.9	26.7	28.0	17.1	33.1	42.5	8.4	2.3
	Faba bean	26	520	8.3	28.3	26.0	30.6	34.6	29.2	10.2	0.4
2017	Pea	39	778	55.5	18.7	39.0	17.3	25.3	24.7	3.8	4.0
	Faba bean	29	577	21.1	25.4	29.0	20.0	31.7	22.1	5.9	0.2
2018	Pea	33	656	47.0	43.8	33.0	21.3	25.7	16.4	3.8	0.9
	Faba bean	30	592	16.0	55.5	30.0	37.2	38.8	19.3	2.7	0.0

[1]Number of fields; [2]Number of roots evaluated; *D. pin* = *Didymella pinodella; F. ave = F. avenaceum; F. oxy = Fusarium oxysporum; F. red = F- redolens; F. sol = F. solani; F. tric = F. tricinctum; F. equi = F. equiseti; F. culm = F. culmorum.*

In faba bean, *Fusarium redolens* and *D. pinodella* were the species that showed both, the higher prevalence (the percentage of the total number of fields from which the pathogens were isolated) and a higher root colonization rates in organic fields compared to conventional fields. In contrast, *F. avenaceum*, *F. tricinctum* and *F. culmorum* were more prevalent and colonized more frequently roots in conventional farms. However, although the two latter species were found in 63% conventional and (*F. culmorum*) 40 to 48% (*F. tricinctum*) organic farms, both species were recovered in low frequencies from the roots in both management systems. The species *F. oxysporum, F. solani* were common in both organic and conventional systems (data not shown). Soil abiotic properties did not correlate strongly with frequencies of any of the most frequently isolated pathogens, except for moderate negative correlations with pH for *F. oxysporum*.

To further examine if combination of abiotic soil properties played a role in the ability of a single species to colonize roots of faba bean, hierarchical clustering on principle component (HCPC) analysis was performed. Using the HPCP, the most influential combination of soil properties were found to be soil organic matter content (SOM) and pH. The subsequent HPCP analysis grouped the faba bean fields into 3 separate clusters; the dimension 1 explained 50.1% of the variance and separated the fields mainly based on their differences in pH (e.g. the clusters 1 and 2), whereas the dimension 2 contained the remaining 49.9% of the variance and separated the fields based on their differences in SOM content. Main effects on isolation frequencies of *Fusarium* species between organic and conventional management system were observed in soil clusters 1 and 2, whereas in soil cluster 3 which was characterized by particularly high mean SOM content ($\geq$ 4% on average) there was no difference in isolation frequencies of the single species. *Fusarium redolens* was more frequently present in roots collected from organic fields in both, cluster 1 and 2 (SOM content of around 2.2% for both clusters, and pH of around 6.2 and 7.0 in organic and conventional fields for cluster 1 and 2, respectively). The species root colonization rates, however were considerably higher in cluster 2 (higher pH) in both management systems. In contrast, *F. avenaceum* was more frequent in the conventional fields in cluster 1 and 2, with the species showing particularly low frequencies in organic fields under higher pH content e.g. cluster 2. The highest isolation frequencies of *D. pinodella* were observed in soil cluster 3 (high SOM content), moderate in soil cluster 1 (moderate average SOM and pH 6.2) while in cluster 2 (moderate average SOM and pH of 7.0) frequencies of isolation for the species were particularly low. The species *F. oxysporum* and *F. solani* mean isolation frequencies were similar between the systems regardless of the cluster, whereas *F. tricintum, F. culmorum* and *F. equiseti* showed some variability but were generally recovered at lower frequencies in each of the soil clusters in both growing systems (data not shown).

For pea, preliminary data suggest that different tendencies of the single species are visible and isolation frequencies are strongly influenced by weather conditions. For example, *F. oxysporum* was found to be more frequent on spring cultivars compared to winter cultivars in both, organic and conventional systems, whereas *D. pinodella* was found at particularly high frequencies in organic winter cultivars. *Fusarium solani* and *F. redolens* showed similar tendencies to colonize less frequently roots of organically managed winter pea compared to conventional winter pea and overall was lower on spring pea cultivars in both management systems. The analysis of the effects of soil abiotic properties on frequencies of individual species is ongoing.

4 Discussion

Overall, there was no effect of the growing system, conventional vs organic, on the spectrum of pathogens isolated. However, the prevalence, i.e. frequency of isolation, differed among pathogens and crops. The most frequently isolated fungal species from pea and faba bean roots in Germany were *D. pinodella, F. redolens, F. oxysporum, F. solani, F. avenaceum, F. tricinctum, F. culmorum* and *F. equiseti*. Interestingly, *Didymella* was strongly affected by the management systems and was found predominantly in organic fields on both, pea and faba bean. However, the number of conventional fields with winter pea cultivars sampled was low and no robust conclusions about differences between organically and conventionally grown winter peas can be made. More samples, weather data, and crop rotation details among others will be required in order to potentially determine the main driving forces leading to the occurrence of specific pathogens in given fields.

In conclusion, our preliminary results indicate increasing importance and significance of *F. redolens* and *D. pinodella* for grain legume production in Germany. The dominance of *F. redolens* in the pathogen complex causing root rot of legumes in Germany is relatively recent, and its epidemiology, impact on yield and aggressiveness is not known and needs to be further investigated. Among species isolated from root samples, *D. lethalis* was for the first time identified in Germany [6], including several other *Didymella* spp. that may represent one or several novel species in the genus. For breeders, the clear identification and role of important pathogens in the complex that is causing disease is an indispensable prerequisite for making progress in resistance breeding, since resistances can only be identified if they can be clearly assigned to pathogens.

Acknowledgments. This work was carried out within the framework of the research projects PATHO-ID (2814EPS40) and APSOLU (2814EPS035), funded by the Federal Ministry of Food and Agriculture within the framework of the BMEL protein plant strategy, in cooperation with the Demonstration network Pea/Bean.

Competing Interests. The authors declare that they have no competing interests.

References

1. Basch, G., González-Sánchez, E.J., Gómez McPherson, H., Kassam, A.: Opportunities for conservation agriculture in the EU common agricultural policy 2014–2020. In: Proceedings of the 5th World Congress on Conservation Agriculture and 3rd International Farming System Design Conference, pp. 316–317. WCCA/FSD Local Organizing Commitee (2011)
2. Schlatter, D., Kinkel, L., Thomashow, L., Weller, D., Paulitz, T.: Disease suppressive soils: new insights from the soil microbiome. Phytopathology **107**, 1284–1297 (2017)
3. van Bruggen, A.H.C., et al.: One health - cycling of diverse microbial communities as a connecting force for soil, plant, animal, human and ecosystem health. Sci. Total Environ. **664**, 927–937 (2019)

4. Finckh, M.R., Yli-Mattila, T., Nykänen, A., Kurki, P., Hannukkala, A.: Organic temperate legume disease management. In: Finckh, M.R., van Bruggen, A.H.C., Tamm, L. (eds.) Plant Diseases and Their Management in Organic Agriculture, pp. 293–311. APS Press (2015)
5. Baćanović-Šišić, J., Šišić, A., Schmidt, J.H., Finckh, M.R.: Identification and characterization of pathogens associated with root rot of winter peas grown under organic management in Germany. Eur. J. Plant Pathol. **151**, 745–755 (2018)
6. Šišić, A., Baćanović-Šišić, J., Schmidt, H., Finckh, M.R.: First report of *Didymella lethalis* associated with roots of pea, subterranean clover and winter vetch in Germany, Switzerland and Italy. Plant Disease (2018). https://doi.org/10.1094/pdis-05-18-0837-pdn
7. Wille, L., Messmer, M.M., Studer, B., Hohmann, P.: Insights to plant–microbe interactions provide opportunities to improve resistance breeding against root diseases in grain legumes. Plant, Cell Environ. **42**, 20–40 (2019)
8. Summerell, B.A., Laurence, M.H., Liew, E.C.Y., Leslie, J.F.: Biogeography and phylogeography of *Fusarium*: a review. Fungal Divers. **44**, 3–13 (2010)
9. Šišić, A., et al.: Roots of symptom-free leguminous cover crop and living mulch species harbor diverse *Fusarium* communities that show highly variable aggressiveness on pea (*Pisum sativum*). PLoS ONE **13**, e0191969 (2018)
10. Šišić, A., et al.: The 'forma specialis' issue in *Fusarium*: a case study in *Fusarium solani* f. sp. *pisi*. Sci. Rep. **8**, 1252 (2018)
11. Pflughöft, O., Merker, C., von Tiedemann, A., Schäfer, B.C.: Zur Verbreitung und Bedeutung von Pilzkrankheiten in Körnerfuttererbsen (*Pisum sativum* L.) in Deutschland. Gesunde Pflanzen **64**, 39–48 (2012)
12. Leslie, J.F., Summerell, B.A.: The *Fusarium* laboratory manual. Blackwell Publishing, Hoboken (2006)
13. Boerema, G.H., de Gruyter, J., Noordeloos, M.E., Hamers, M.E.C.: *Phoma* Identification Manual: Differentiation of Specific and Infra-specific Taxa in Culture. CABI Publishing, Wallingford (2004)
14. O'Donnell, K., Kistler, H.C., Cigelnik, E., Ploetz, R.C.: Multiple evolutionary origins of the fungus causing Panama disease of banana: concordant evidence from nuclear and mitochondrial gene genealogies. Proc. Natl. Acad. Sci. **95**, 2044–2049 (1998)
15. Liu, Y.J., Whelen, S., Hall, B.D.: Phylogenetic relationships among ascomycetes: evidence from an RNA polymerse II subunit. Mol. Biol. Evol. **16**, 1799–1808 (1999)
16. Sung, G.H., Sung, J.M., Hywel-Jones, N.L., Spatafora, J.W.: A multi-gene phylogeny of *Clavicipitaceae* (*Ascomycota*, Fungi): identification of localized incongruence using a combinational bootstrap approach. Mol. Phylogenet. Evol. **44**, 1204–1223 (2007)

Influence of Different *Plum Pox Virus* Strains on Chemical Composition of 'Čačanska Lepotica' Plum Fruit Cultivar

Nemanja Miletić[1(✉)], Darko Jevremović[2], Olga Mitrović[2], Olivera Gvozdenović, Marko Pajić[1], and Svetlana Paunović[2]

[1] Faculty of Agronomy, University of Kragujevac, Cara Dušana 34, 32000 Čačak, Serbia
n.m.miletic@gmail.com
[2] Fruit Research Institute, Kralja Petra I 9, 32000 Čačak, Serbia

Abstract. *Plum pox virus* (PPV) is economically the most important virus of stone fruits. In sensitive plum (*Prunus domestica* L.) cultivars, PPV causes foliar symptoms, fruits deformation, changes in the chemical composition of fruits and premature fruit drop.

Fruits of 'Čačanska lepotica' plum cultivar were collected from an experimental orchard in locality Čačak. Samples were collected from 15 trees (5 infected with PPV-D, 5 infected with PPV-Rec and 5 from PPV-free trees). Trees for sampling were selected after the RT-PCR analysis of the trees from the plantation using PPV strain-specific primers. The fruits were picked at five different terms at seven-days interval during the entire maturation stage.

All tested samples, 10 PPV-infected and 5 uninfected, were further chemically analyzed. The analyses included determination of soluble solid content, titratable acidity and total phenolic content.

The aim of this work was to reveal the possible changes in chemical composition in the fruits of 'Čačanska lepotica' cultivar, grown at the same locality, caused by the two different PPV strains.

Keywords: PPV · Plum · RT-PCR · Phenolics · Antioxidant capacity

1 Introduction

European plum (*Prunus domestica* L.) is economically important fruit crop in Serbia. According to FAOSTAT (Food and Agriculture Organization Corporate Statistical Database), Serbia is one of the leading world producer, with about 460.000 t produced in 2016. On the other hand, plum production per unit is low. One of the limiting factors for successful plum fruit production is *Plum pox virus* (PPV), the causal agent of Sharka disease. In the susceptible cultivars ('Požegača') PPV reduces yield (up to 100%) and fruit quality. Tolerant cultivars may be infected with PPV and express leaf symptoms of the disease, but the fruits are without any visible damage, or they occur very rarely (Jevremovic 2013). One of the main PPV-tolerant plum cultivar that is grown in Serbia and Central and East Europe is 'Čačanska Lepotica'. It was bred at the Fruit Research Institute Čačak, Serbia, as a result from the cross of 'Wangenheims

M. Brka et al. (Eds.): AgriConf 2019, IFMBE Proceedings 78, pp. 76–81, 2020.
https://doi.org/10.1007/978-3-030-40049-1_9

Frühzwetsche' and 'Požegača' in 1961, named and released in 1975. Fruits are of attractive dark blue color, coated by intensive fatty layer, with green-yellowish mesocarp and harmonious sweet-acid flavor (Lukić et al. 2016).

The objective of this research was to examine the influence of two different strains of *Plum pox virus*, PPV-Rec and PPV-D, on chemical composition of fresh plum 'Čačanska Lepotica' fruits, by comparing the results of the analyzed fruits from PPVinfected and healthy trees.

2 Materials and Methods

2.1 Collecting Samples

Plum fruits of 'Čačanska Lepotica' were collected in 2017 from an experimental orchard established in 2008 in the village Ostra (43°54'54.56"N, 20°30'07.31"E), Serbia. The orchard included 400 plum trees, with 4 × 3.5 m tree spacing. Samples were collected from 15 trees. Visual assessment of leaves from the selected trees showed that five trees were without any leaf symptoms, while ten trees showed strong sharka-like symptoms.

The fruits were picked at five different terms at seven-days interval during maturation stage, starting from a fully green and immature fruits as the initial harvest stage, and terminating four weeks later with fruits being soft and overripened (Table 1).

The sampling addressed the following parameters: position of the three in the orchard, position of the branch on the tree, position of fruit on the branch, and sun exposure to assess representative sample. Fifty samples of fruits, with no mechanical injures, were collected. Each subsequent sampling was conducted on the same threes. After harvesting, fruits were immediately transported to the laboratory and stored at −80 °C until the analyses. The whole edible part of the fruit was used. Plum fruits were carefully cut in halves and pits were removed. Mesocarp and exocarp were frozen by pouring into liquid nitrogen and homogenized, using a stainless steel blender.

Table 1. Selected harvest stages for sample collection of fruits of cv 'Čačanska Lepotica'.

Denomination	Abbreviation	Maturity stage	Harvesting date
Harvesting stage 1	HS1	very firm, green, sour	22.06.2017
Harvesting stage 2	HS2	firm, locally light blue, sour	30.06.2017
Harvesting stage 3	HS3	Partly soft, light blue, slightly sweet	07.07.2017
Harvesting stage 4	HS4	soft, fully blue, balanced ac- id/sweet ratio	14.07.2017
Harvesting stage 5	HS5	very soft, fully blue, very sweet	21.07.2017

2.2 IC-RT-PCR Analysis

From the establishment of the orchard in 2008, all 400 trees were annually visually inspected and tested on the PPV presence using immunocapture reverse transcription polymerase chain reaction (IC-RT-PCR). Each sample for the analysis consisted of the 24 randomly collected leaf samples around the canopy of one tree. IC was done with polyclonal antibodies (Fruit Research Institute, Čačak) and RT step was performed with random hexamers using Maxima Reverse Transcriptase (Thermo Fisher Scientific, USA). In PCR analysis, each sample was tested in four PCR reactions using PPV-D and PPV-M specific primers. To amplify the fragments of 467 bp located in the C-ter Nib–Nter-CP coding region primer pairs P4/P3M and P4/P3D were used (Candresse et al. 1998). To amplify the fragments of 880 bp and 468 bp that are located in the CI coding region, sets of primers CIP-M/CIP-MR and CIP-D/CIP-DR were used, respectively (Kamenova et al. 2011). All PCR reactions were carried out in T Personal thermal cycler (Biometra, Germany). PCR products were analyzed in 1.5% agarose gel, stained with ethidium-bromide and visualized with a Gel Doc EZ System (Biorad, USA). The presence of the amplified fragment of the expected size with each primer pair was considered as positive reaction. Strain typing of the isolates was done according to the PCR results with each primer pair, as described in detail by Jevremović (2013).

2.3 Determination of Soluble Solid Content, Dry Matter Content and Titratable Acidity

The soluble solid content of the fruits was determined on a manual refractometer (Milwaukee MR 200, USA). The dry matter content was determined by drying at 105 °C until constant mass. Titratable acidity was determined by neutralization of fruit extract with 0.1 N NaOH to pH 8.2, using phenolphthalein as indicator. Acidity was expressed as mg malic acid/100 g fresh weight. Data are reported as means for at least three replications.

2.4 Determination of Total Phenolic Content

The total phenolic content was determined using a modified Folin-Ciocalteu colorimetric method (Singleton et al. 1999; Liu et al. 2002) and the results were expressed as milligrams of gallic acid equivalents per 100 g fresh weight (mg GAE/100 g FW). Data are reported as means for at least three replications.

3 Results and Discussion

3.1 PPV Strain Detection and Characterization

Out of 400 tested plum trees in spring 2017, IC-RT-PCR revealed that 175 trees were infected with PPV.PPV-Rec strain was detected in 160 trees, and PPV-D strain in 15 trees. Based on these results, we selected 10 systematically infected trees (5 per each strain) and 5 PPV-free trees for the main part of the study (Table 2).

Table 2. IC-RT-PCR typing results of the trees selected for further chemical analysis.

No.	Tree position[a]	PCR analysis results				Strain
		N-ter CP		CI		
		P4-P3M	P4-P3D	CIP-M	CIP-D	
1	1/2	-[b]	+[c]	–	+	D
2	3/1	–	+	–	+	D
3	3/3	–	+	–	+	D
4	3/18	–	+	–	+	D
5	3/20	–	+	–	+	D
6	4/29	+	–	–	+	Rec
7	4/28	+	–	–	+	Rec
8	5/29	+	–	–	+	Rec
9	5/28	+	–	–	+	Rec
10	7/28	+	–	–	+	Rec
11	14/4	–	–	–	–	/
12	14/5	–	–	–	–	/
13	15/4	–	–	–	–	/
14	15/5	–	–	–	–	/
15	15/6	–	–	–	–	/

[a]row/tree; [b]negative reaction; [c]positive reaction

All infected trees expressed typical Sharka symptomson leaves: chlorotic light green to yellow rings, patterns and vein clearing. No differences on the type or intensity of the symptoms were observed between PPV-Rec and PPV-D infected trees.

3.2 Physicochemical Properties of Fresh Plum Fruits

Fruit ripening is associated with important biochemical changes that modify color, texture, taste and other quality traits. Soluble solid content was determined for all infected and uninfected samples and the results are presented in Table 3. It is obvious that soluble solid content increased during ripening in all tested fruits. Such trend was expected since the content of sugars, as the main soluble component, increases by ripening. The presence of PPV strains, both PPV-D and PPV-Rec, had no influence on soluble solid content if compared to the uninfected fruit samples. In contrary, Usenik et al. (2015) detected that PPV infection caused earlier ripening, as reflected via soluble solid content, of fresh plums cv 'Brkinska cespa'.

Titratable acidity was followed during ripening and the results are presented in Table 3. In all fruit samples acidity decreased with ripening, and no influence of the *Plum pox virus* presence on acidity, regardless the virus strain, was detected. A decrease of acidity over ripening in all samples was expected, due to the oxidation of organic acids in the later stages of maturity. Again, Usenik et al. (2015) obtained that early ripening of the plum fruits cv 'Brkinska cespa' was caused by PPV infection, as seen via rapider decrease in acid content.

Table 3. Changes of soluble solid content and titratable acidity during fruit ripening in infected (PPV-D and PPV-Rec) and uninfected (Neg) plum cv 'Čačanska Lepotica'.

	Virus strain	HS1	HS2	HS3	HS4	HS5
Soluble solid content	PPV-D	7.64 ± 0.75	11.04 ± 0.92	13.26 ± 2.47	15.14 ± 2.15	16.08 ± 1.72
	PPV-Rec	7.26 ± 0.47	10.55 ± 1.59	12.31 ± 0.70	14.95 ± 0.72	16.90 ± 1.00
	Neg	7.75 ± 0.29	11.01 ± 1.00	11.35 ± 0.98	16.55 ± 3.00	15.00 ± 0.58
Titratable acidity	PPV-D	4.39 ± 0.70	3.04 ± 1.03	2.01 ± 0.55	1.53 ± 0.41	1.07 ± 0.24
	PPV-Rec	4.17 ± 0.84	2.89 ± 0.77	2.39 ± 0.24	1.33 ± 0.42	1.15 ± 0.21
	Neg	5.60 ± 1.59	2.09 ± 0.32	2.57 ± 0.32	1.97 ± 0.40	1.56 ± 0.18

3.3 Total Phenolic Content and Antioxidant Capacity

Maturity stage dramatically affects the total phenolic content in the fruit. The changes in total phenolic content during maturation are presented in Fig. 1. Content of polyphenolics in infected samples increased with on-tree fruit ripening. On the other hand, in negative samples polyphenolics content increased during HS1, HS2 and HS3, while subsequently decreased.

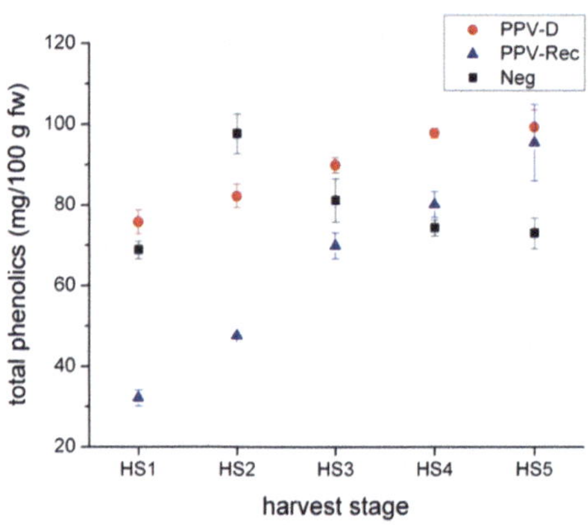

Fig. 1. Changes of total phenolic content during maturity period in uninfected (Neg) and infected (PPV-D and PPV-Rec) plum fruits cv 'Čačanska Lepotica'.

In here, an influence of the virus presence in fruit on chemical composition is quite evident. Considering the last harvesting stage, HS5, as an optimal technological maturity level, average contents of polyphenols in negative sample and infected samples PPV-D and PPV-Neg were 73.00, 99.49 and 95.61 mg/100 g fw, respectively. Taking into account the last two harvesting stages, polyphenolic content in PPV-D and PPVRec plum samples were 1.34- and 1.19-fold increased, compared to the negative samples. PPV infection significantly changes the composition of phenolic compounds (Usenik et al. 2015; Sochor et al. 2015). Apart from the visual changes caused by *Plum pox virus*, easily notified on fruits, leaves and stones, and being quite specific for each cultivar, it is well known that the presence of the virus is significant stress factor, which prompts a production of free radicals, as reactive oxygen forms (Mittler 2002; Lamb and Dixon 1997). Free radicals are mainly being formed as products of the following processes: peroxidation of membrane lipids, oxidation of proteins, inhibition of enzymes and destruction of nucleic acids. An increased production of polyphenolics, as main antioxidants in plants, in infected samples is a natural defensive plant's reaction.

4 Conclusion

The aim of this study was to correlate the presence of certain strain of PPV with physiochemical properties and polyphenolic profile in fresh plum fruits cv 'Čačanska Lepotica'. It was revealed that presence of PPV, both strains PPV-D and PPV-Rec, had no influence on soluble solid content and titratable acidity. On the other hand, polyphenolic profile of the fresh plums was highly influenced by the viral infection. Virus infection, as significant stress factor, causes a high production of free radicals. The plant neutralizes these radicals by intensive formation of polyphenolics. Therefore the content of polyphenolics in infected plums cv 'Čačanska Lepotica' is higher, if compared to the negative samples, especially in the last two harvesting stages.

References

Candresse, T., Cambra, M., Dallot, S., Lanneau, M., Asensio, M., Gorris, M.T., Revers, F., Macquaire, G., Olmos, A., Boscia, D., Quiot, J.B., Dunez, J.: Comparison of monoclonal antibodies and polymerase chain reaction assays for the typing of isolates belonging to the D and M serotypes of Plum pox potyvirus. Phytopathology **88**, 198–204 (1998)

Jevremović, D.: Distribution of PPV-D and PPV-Rec strains of Plum pox virus in Serbia and the dynamics of their spread in plum orchard. (PhD thesis), University of Belgrade, Belgrade, Serbia (2013)

Kamenova, I., Dallot, S., Bozkova, V., Milusheva, S.: First report of the Plum pox virus recombinant strain on peach in Bulgaria. Plant Dis. **95**, 1320–1320 (2011)

Lamb, C., Dixon, R.A.: The oxidative burst in plant disease resistance. Annu. Rev. Plant Physiol. Plant Mol. Biol. **48**, 251–275 (1997)

Liu, M., Li, X.Q., Weber, C., Lee, C.Y., Brown, J. Liu, R.H.: Antioxidant and antiproliferative activities of raspberries. J. Agric. Food Chem. **50**(10), 2926–2930 (2002)

Lukić, M., Pešaković, M., Marić, S., Glišić, I., Milošević, N., Radičević, S., Leposavić, A., Đorđević, M., Miletić, R., Karaklajić-Stajić, Ž., Tomić, J., Paunović, S.M., Milinković, M., Ružić, Đ., Vujović, T., Jevremović, D., Paunović, S.A., Popović, B., Mitrović, O., Kandić, M.: Sorte voćaka stvorene u Institutu za voćarstvo, Čačak (1946–2016), pp. 1–182. Fruit Research Institute, Čačak (2016)

Mittler, R.: Oxidative stress, antioxidants and stress tolerance. Trends Plant Sci. **7**, 405–410 (2002)

Singleton, V.L., Orthofer, R., Lamuela-Raventos, R.M.: Analysis of total phenols and other oxidation substrates and antioxidants by means of Folin-Ciocalteu reagent. Methods Enzym. **299**, 152–178 (1999)

Sochor, J., Krska, B., Polak, J., Jurikova, T.: The influence of viral infections on antioxidant levels in the genetically modified plum variety "Honeysweet" (*Prunus domestica* L.). Potravinarstvo **9**, 195–200 (2015)

Usenik, V., Kastelec, D., Stampar, F., Marn, M.V.: Effect of Plum pox virus on chemical composition and fruit quality of plum. J. Agric. Food Chem. **63**, 51–60 (2015)

Molecular and Biological Characterization of *Black Raspberry Necrosis Virus* on Red Raspberry in Serbia

Darko Jevremović(✉), Aleksandar Leposavić, and Svetlana A. Paunović

Fruit Research Institute, Kralja Petra I 9, 32000 Čačak, Serbia
darkoj@ftn.kg.ac.rs

Abstract. *Black raspberry necrosis virus* (BRNV) is considered to be economically important virus infecting raspberries. In black raspberry (*Rubus occidentalis* L.) BRNV causes cane tip necrosis, but infected red raspberry (*Rubus idaeus* L.) cultivars in general do not express foliar symptoms.

From 2014–2018, a number of 105 samples from 9 red raspberry cultivars ('Amira', 'Fertödi Zamatos', 'Glen Ample', 'Meeker', 'San Rafael', 'Polana', 'Polka', 'Tulameen' and 'Willamette') were collected from different localities in Serbia. Samples were tested on the BRNV presence by reverse transcription-polymerase chain reaction (RT-PCR) using specific primers targeting 417 bp nucleotide (nt) fragment of the RNA-dependent RNA polymerase (RdRp) region of RNA 1 of the BRNV genome. BRNV was detected in 19 samples in cultivars 'Fertödi Zamatos', 'Polana', 'Tulameen' and 'Willamette'.

A 417 nt sequence of the RdRp was determined for 5 isolates. The nt sequences of these isolates were 94.2–99.7% identical. When compared to available sequences of BRNV isolates from North America and Europe, Serbian isolates showed significant divergence (80.0–82.2% of nt identity).

A portion of the BRNV infected plants of cultivar 'Fertödi Zamatos' were further tested on indicator *R. occidentalis* by inarch bottle grafting. Two months after inoculation cane tip begun to curl down and became necrotic on all inoculated plants. Non-grafted control plants were symptomless.

Keywords: BRNV · Red raspberry · RT-PCR · Bottle grafting · Diversity

1 Introduction

Red raspberry and other species from the *Rubus* genus may be infected with more than 30 viruses and virus-like agents (Martin et al. 2013). *Black raspberry necrosis virus* (BRNV) was first described in North America in the 1950s as aphid-transmitted agent inducing latent or mild symptoms in infected red raspberry (*Rubus idaeus* L.) cultivars, but severe tip necrosis in black raspberry (*Rubus occidentalis* L.) (Martin et al. 2013). BRNV is considered economically important virus of raspberries, decreases yield up to 30% in sensitive cultivars in initial cropping years. Additionally, sensitive cultivars produce shorter and thinner canes and smaller fruits (Stace-Smith and Jones 1987). BRNV is common virus in mixed infections with *Rubus yellow net virus* (RYNV),

M. Brka et al. (Eds.): AgriConf 2019, IFMBE Proceedings 78, pp. 82–87, 2020.
https://doi.org/10.1007/978-3-030-40049-1_10

Raspberry bushy dwarf virus (RBDV) and *Raspberry leaf mottle virus* (RLMV) and it was consistently associated with the raspberry mosaic disease in North America. The natural hosts of BRNV are wild and cultivated *Rubus* species. Most of red raspberry cultivars infected with BRNV do not express any foliar symptoms, but in some chlorotic spots and venial chlorotic mottle may appear. First phylogenetic analysis indicated that BRNV belongs to the genus *Sadwavirus* (Halgren et al. 2007). According to the latest data, BRNV is unassigned species in the family *Secoviridae* (Thompson et al. 2017). In North America, large raspberry aphid (*Amphorophora agathonica* Hottes) and green peach aphid (*Myzus persicae* Sulzer) are described as its vectors. BRNV is not seed-borne in red raspberries.

The aim of our study was to characterize BRNV in Serbia using molecular and biological methods.

2 Materials and Methods

2.1 Plant Material

Material for this study was collected during a large survey conducted from 2014 to 2018 in red raspberry orchards in Serbia. A total of 105 symptomatic and asymptomatic samples from 9 raspberry cultivars ('Amira', 'Fertödi Zamatos', 'Glen Ample', 'Meeker', 'San Rafael', 'Polana', 'Polka', 'Tulameen' and 'Willamette') were collected from commercial orchards and Fruit Research Institute berry fruits collection orchard (Table 1). Collected samples were stored at 4 °C (up to 2 days) or at −80 °C (longer storage).

Table 1. Detection of BRNV using RT-PCR in red raspberry samples from Serbia.

Cultivar	Number of analyzed samples	Number of BNRV positive samples	%
Amira	2	0	–
Fertödi Zamatos	16	14	87.5
Glen Ample	5	0	–
Meeker	20	0	–
San Rafael	2	0	–
Polana	6	1	16.6
Polka	8	0	–
Tulameen	6	1	16.6
Willamette	40	3	7.5
Total	105	19	18.1

2.2 TNA Extraction

Total nucleic acids (TNA) were extracted with a modified CTAB protocol (Li et al. 2008). Fresh or frozen leaf tissue (200 mg) was grinded in extraction bags with 2 ml of 2% Cetyltrimethylammonium-bromide (CTAB) buffer. One ml of the extract was incubated at 65 °C for 15 min and centrifuged at 10,400 rpm for 5 min. Obtained supernatant (approx. 650 µl) was transferred to a new 1.5 ml tube and vortexed with the equal volume of (24:1) chloroform/isoamyl alcohol. The mixture was centrifuged at 12,800 rpm for 10 min. Five hundred µl of the supernatant was transferred to a new tube with 350 µl ice-cold isopropanol, mixed and centrifuged at 12,800 rpm for 10 min. The aqueous phase was decanted and TNA pellet was washed with 1 ml cold 70% ethanol by centrifugation at 12,800 rpm for 5 min, dried at RT and dissolved in 100 µl Tris-EDTA.

2.3 Reverse Transcription and Polymerase Chain Reaction (RT-PCR)

Samples were analyzed for the BRNV presence by reverse transcription-polymerase chain reaction (RT-PCR) using specific primers. RT-PCR reaction was performed in two steps.

Reverse transcription (RT) step was done with random hexamers and Maxima Reverse Transcriptase (Thermo Scientific, USA). In the second step, obtained cDNA was used as a template in a PCR reaction with BRNV 1-F/BRNV 1-R primer set targeting 417 bp nucleotide (nt) fragment of the RNA-dependent RNA polymerase (RdRp) region of RNA 1 of the BRNV genome (Halgren et al. 2007). PCR reaction was carried out in TPersonal thermal cycler (Biometra, Germany) using Taq DNA Polymerase (Thermo Scientific, USA).

PCR products were analyzed by electrophoresis in 1.5% agarose gel, stained with ethidium bromide and visualized with the Gel Doc EZ System (Biorad, USA). The presence of a fragment of the expected size was considered as a positive reaction.

2.4 Sequence Analysis

PCR products of five selected isolates were custom sequenced by Macrogen Europe (The Netherlands). Phylogenetic relationships were reconstructed by a Maximum Likelihood (ML) tree with Kimura 2-parameter model of nucleotide substitution with gamma-distributed rate variation across sites (K2 + G) using MEGA7 software (Kumar et al. 2016).

2.5 Biological Test

Three BRNV infected plants of cultivar 'Fertödi Zamatos' were further tested on indicator *R. occidentalis* by inarch bottle grafting. Vegetatively propagated potted plants of *R. occidentalis* were bottle-grafted with the shoots of 'Fertödi Zamatos' (three plants per each tested plant, 9 plants in total). Three non-grafted *R. occidentalis* plants represented negative control. Test plants were observed for the symptom development in the next 6 months. Additionally, three months after grafting all test plants were tested on the BRNV presence by RT-PCR.

3 Results

3.1 Molecular Characterization

Collected leaf samples of nine red raspberry cultivars were analyzed on the BRNV presence. RT-PCR results are presented in Table 1. BNRV was detected in 19 samples (18.1%) in cultivars 'Fertödi Zamatos', 'Willamette', 'Polana' and 'Tulameen'. The highest incidence of BRNV was detected in cultivar 'Fertödi Zamatos' (87.5%), followed by 'Polana' and 'Tulameen' (16.6% each), and 'Willamette' (7.5%). BRNV was detected from both asymptomatic and samples showing venial chlorotic mottle (Fig. 1).

Fig. 1. Venial chlorotic mottle in a leaf of the red raspberry 'Fertödi Zamatos' infected with *Black raspberry necrosis virus*

A 417 nt sequence of the RdRp was determined for 5 selected BRNV isolates. Obtained nucleotide sequences were deposited in the GenBank sequence database (https://www.ncbi.nlm.nih.gov/) and assigned the accession numbers MK992735 – MK992739. Serbian BRNV isolates share 94.2–99.7% nucleotide (nt) identity. Four isolates from raspberry 'Fertödi Zamatos' (RS-BNRV-85, RS-BNRV-118, RS-BNRV-124 and RS-BNRV-126) showed very high nt identity (99.0–99.7%). When compared to available sequences of BRNV isolates from North America and Europe, Serbian isolates showed significant divergence (80.0–82.2% of nt identity). Phylogenetic analysis of RNA-dependent RNA polymerase in RNA1 sequences of five Serbian and four isolates from the United States, Great Britain and Finland confirmed the existence of four clusters of isolates (Fig. 2). All Serbian isolates were grouped within the same cluster.

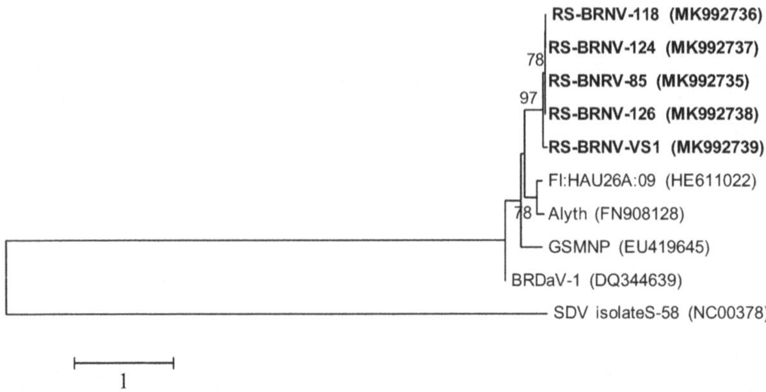

Fig. 2. Phylogenetic analysis of partial RdRp sequences of nine BRNV isolates. Bootstrapped Maximum likelihood (ML) tree using Kimura 2-parameter method. Only bootstrap values of >70% calculated from 1,000 replications are shown at nodes. Serbian isolates are presented in bold. The NCBI accession numbers are in parentheses. The tree was rooted with *Satsuma dwarf virus* (SDV) as an outgroup.

3.2 Biological Characterization

Two months after graft inoculation first symptoms developed on tested *R. occidentalis* plants. At first, cane tip begun to curl down and started to wilt, and within a week become necrotic. Symptoms developed on all grafted test plants. Non-grafted control plants were symptomless.

All test plants were tested by RT-PCR on BRNV presence. BRNV was confirmed in all 9 grafted test plants, and not in non-grafted control plants.

4 Discussion

Black raspberry necrosis virus was identified in Serbia during 1990s. Based on the differential reaction of *R. occidentalis,* Dulić-Marković and Ranković (1997) identified BRNV in cultivars 'Gradina', 'Malling Promise' and 'Zeva'. Today, these cultivars are not important for raspberry production in Serbia, and they are only present in collection orchards and gardens. In the present study the BRNV presence was confirmed using RT-PCR assay in four other cultivars that are grown in commercial raspberry orchards in Serbia: 'Fertödi Zamatos', 'Willamette', 'Polana' and 'Tulameen'. 'Willamette' is the main cultivar grown in Serbia with 90% share in production. This study showed that only 7.5% of the analyzed samples of this cultivar were infected with BRNV. On the other side, the highest percentage (87.5%) of BRNV was detected in samples of cultivar 'Fertödi Zamatos'. This Hungarian cultivar is slowly spreading in the orchards in some localities in Western Serbia. RT-PCR analysis confirmed the presence of BRNV in both asymptomatic and in symptomatic plants. All of the tested leaf samples of cultivar 'Fertödi Zamatos' showing venial chlorotic mottle were found to be infected with BRNV. These symptomatic samples were also tested on the presence of other

Rubus viruses (*Rubus yellow net virus, Raspberry bushy dwarf virus, Raspberry leaf mottle virus, Rubus vein chlorosis virus* and *Raspberry leaf blotch emaravirus*) and none of these viruses was detected. The presence of the symptoms may be associated with the interaction of BRNV and other viruses infecting *Rubus* species and require further studies.

Sequence analysis showed significant divergence among Serbian isolates (94.2–99.7%). Sequences of isolates from 'Fertödi Zamatos' were almost identical (4 nucleotides divergence), but quite divergent from the isolate RS-BNRV-VS1 from cultivar 'Polana'. When compared with available BRNV sequences of isolates from other countries our isolates showed 80–82.2% nucleotide identity. Significant divergence is evidenced among the isolates from different countries. Further study is needed to better access into the sequence variability of Serbian isolates. Larger portions of RNA1 and RNA2 of BRNV genome of our isolates should be sequenced and compared with sequences from other countries to reconstruct their phylogenetic relationships.

Biological tests confirmed that selected BRNV isolates from cultivar 'Fertödi Zamatos' are graft transmissible and able to induce tip necrosis on *R. occidentalis* test plants. Tip necrosis is used as a diagnostic test for BRNV detection during pathogen testing of *Rubus* plants in recommended certification schemes.

This study demonstrates the results of the first analysis of *Black raspberry necrosis virus* in Serbia combining biological and molecular tests. Further analysis of the obtained isolates and their possible interactions with other viruses is underway.

Acknowledgement. We acknowledge the financial support from the Ministry of Education, Science and Technological Development of the Republic of Serbia through the projects TR-31064 and TR-31093.

Compliance with Ethical Standards

Conflict of Interest. The authors declare that they have no conflict of interest.

References

Martin, R.R., MacFarlane, S., Sabanadzovic, S., et al.: Viruses and virus diseases of *Rubus*. Plant Dis. **97**, 168–182 (2013)

Stace-Smith, R., Jones, A.: Black raspberry necrosis. In: Converse, R.H. (ed.) Virus Diseases of Small Fruits. US Government Printing Office, Washington (1987)

Halgren, A., Tzanetakis, I.E., Martin, R.R.: Identification, characterization, and detection of Black raspberry necrosis virus. Phytopathology **97**, 44–50 (2007)

Thompson, J.R., Dasgupta, I., Fuchs, M., et al.: ICTV virus taxonomy profile: secoviridae. J. Gen. Virol. **98**, 529–531 (2017)

Dulić-Marković, I., Ranković, M.: Virusne bolesti maline u Jugoslaviji. Jugoslovensko voćarstvo **31**, 253–257 (1997)

Li, R., Mock, R., Huang, Q., et al.: A reliable and inexpensive method of nucleic acid extraction for the PCR-based detection of diverse plant pathogens. J. Virol. Methods **154**, 48–55 (2008)

Kumar, S., Stecher, G., Tamura, K.: MEGA7: molecular evolutionary genetics analysis version 7.0 for bigger datasets. Mol. Biol. Evol. **33**, 1870–1874 (2016)

National Center for Biotechnology Information (NCBI) (1988) National Library of Medicine (US), Bethesda. https://www.ncbi.nlm.nih.gov. Accessed 30 May 2019

Determination of the Barley Diseases in Bingöl Province of Turkey

Işıl Saraç Sivrikaya[1], Aziz Karakaya[2(✉)], and Arzu Çelik Oğuz[2]

[1] Faculty of Agriculture, Department of Plant Protection, Bingöl University, Bingöl, Turkey
[2] Faculty of Agriculture, Department of Plant Protection, Ankara University, Dışkapı, Ankara, Turkey
karakaya@agri.ankara.edu.tr

Abstract. Five barley fields in Genç, Adaklı and Central district of Bingöl province of Turkey were investigated for the presence of barley diseases in 2018. Spot form of net blotch caused by *Drechslera teres* f. *maculata,* net form of net blotch caused by *Drechslera teres* f. *teres*, scald caused by *Rhynchosporium commune*, powdery mildew caused by *Erysiphe graminis* f. sp. *hordei*, barley stripe caused by *Drechslera graminea* and covered smut caused by *Ustilago hordei* were found among the fields inspected. Prevalence and severity of these diseases showed variation.

Keywords: Barley diseases · Bingöl · Turkey

1 Introduction

Barley (*Hordeum vulgare* L.) is an important crop both in Turkey and in the world. Barley crop has high feed value. In addition to its use as a feed crop, it is also used in malt industry and in human consumption. Barley grows well in a wide range of environmental conditions including sea level and high altitudes. It is one of the most sowed plants among the cool climate cereals. Barley is especially suitable for growing in temperate regions of the world (Mathre 1982; Kün 1996; Geçit 2016). Turkey is one of the important gene centers of barley. In Turkey, barley ranks second after wheat in terms of cultivation and production (Mızrak and Yalvaç 2001). Barley is cultivated throughout Turkey including Bingöl province.

Diseases affecting barley crop reduce the yield and quality of barley. A number of diseases affecting barley crops have been reported (Mathre 1982). Among these diseases, scald caused by *Rhynchosporium commune* (previously known as *Rhynchosporium secalis*) (Zaffarano et al. 2011), both forms of net blotch caused by *Drechslera teres* f. *teres* and *D. teres* f. *maculata*, leaf rust caused by *Puccinia hordei*, powdery mildew caused by *Erysiphe graminis* f. sp. *hordei* and barley stripe caused by *Drechslera graminea* were common in Turkey (Çelik and Karakaya 2015; Özdemir et al. 2017; İlgen et al. 2017; Ertürk et al. 2018). In this study, diseases occurring on barley crops grown in Bingöl province of Turkey were determined.

M. Brka et al. (Eds.): AgriConf 2019, IFMBE Proceedings 78, pp. 88–91, 2020.
https://doi.org/10.1007/978-3-030-40049-1_11

2 Materials and Methods

Two, 2 and 1 barley fields were inspected for the presence of diseases at Central district and Adaklı and Genç districts of Bingöl province of Turkey during 2018 (Table 1). The planting areas, surveyed areas, numbers of inspected fields and percentages of inspected areas were presented in Table 1. For disease identification, Mathre (1982), Zaffarano et al. (2011) and Zillinsky (1983) were used. For assessing the severity of diseases, a scale developed by Saari and Prescott was used (Saari and Prescott 1975). Disease occurring and no disease occurring fields were considered together for calculation of mean disease prevalences.

Table 1. Districts of Bingöl province of Turkey that barley disease survey was carried out in 2018. Barley planting areas, surveyed areas, number of inspected fields and percentages of inspected areas were presented.

Districts	Planting area (ha) (Tüik 2017)	Surveyed area (ha)	Number of fields inspected	Percentage of inspected area
Central	119.7	6	2	5.01
Genç	55.9	3	1	5.36
Adaklı	105.6	3	2	2.84

3 Results and Discussion

Spot form of net blotch caused by *Drechslera teres* f. *maculata*, net form of net blotch caused by *Drechslera teres* f. *teres*, scald caused by *Rhynchosporium commune*, powdery mildew caused by *Erysiphe graminis* f. sp. *hordei*, barley stripe caused by *Drechslera graminea* and covered smut caused by *Ustilago hordei* were found among the fields inspected (Table 2). Spot form of net blotch was found in two and one fields in Adaklı and Central districts, respectively. The mean prevalence of this disease was 27.5% in Adaklı district and 5% in one Central district field. The severity value of this disease in all fields was 5. Net form of net blotch was found only in one field in Central district. In this field, the prevalence of this disease was 7.5% and severity value was 7. Scald was found in two fields in Central district and in one field in Genç district. The mean prevalence of this disease was 50% in two Central district fields and 10% in one Genç district field. The severity values of this disease in two Central district fields were 5 and 9. The severity values of this disease in Genç district field was low (3). Powdery mildew, covered smut and barley stripe diseases were found in one Central district field. The prevalence of powdery mildew was 5% and severity value was 7. The prevalence of covered smut and barley stripe was 10% and 1%, respectively. Scald and spot form of net blotch were the most common diseases. Karakaya et al. (2014) found that these two diseases were common in Turkey. Spot form of net blotch, net form of net blotch, scald, powdery mildew, covered smut and barley stripe were also reported from various regions of Turkey (Karakaya et al. 2016; Çelik and Karakaya 2015; Karakaya et al. 2016; Özdemir et al. 2017; İlgen et al. 2017; Ertürk et al. 2018).

Table 2. Barley diseases occurring in Bingöl province of Turkey during a survey in 2018, number of fields inspected, number of fields with disease, prevalence of diseases and their severity values. For determination of disease severity, a scale developed by (Saari and Prescott 1975) was used.

District	Number of fields inspected	*Drechslera teres* f. *maculata*			*Drechslera teres* f. *teres*			*Rhynchosporium commune*		
		Number of fields with disease	Prevalence (%)	Severity	Number of fields with disease	Prevalence (%)	Severity	Number of fields with disease	Prevalence(%)	Severity
Central	2	1	5%	5	1	7.5%	7	2	50%	5-9
Genç	1	-	-	-	-	-	-	1	10%	3
Adaklı	2	2	27.5%	5	-	-	-	-	-	-

District	Number of fields inspected	*Erysiphe graminis* f. sp. *hordei*			*Ustilago hordei*			*Drechslera graminea*		
		Number of fields with disease	Prevalence (%)	Severity	Number of fields with disease	Prevalence (%)	Severity	Number of fields with disease	Prevalence (%)	Severity
Central	2	1	5%	7	1	10%	-	1	1%	-
Genç	1	-	-	-	-	-	-	-	-	-
Adaklı	2	-	-	-	-	-	-	-	-	-

4 Conclusions

This study summarized the disease situation of barley crops grown in Bingöl province of Turkey. Both forms of net blotch, scald, powdery mildew, barley stripe and covered smut were present in Bingöl barley fields. Prevalence and severity of these diseases showed variation. Scald and spot form of net blotch were the most common diseases. Strategies controlling diseases occurring at Bingöl barley fields should be implemented. Growing disease resistant barley cultivars could be the preferred option because of environmental and economical concerns.

References

Çelik, E., Karakaya, A.: Eskişehir ili arpa ekim alanlarında görülen fungal yaprak ve başak hastalıklarının görülme sıklıklarının ve yoğunluklarının belirlenmesi. Bitki Koruma Bülteni **55**(2), 157–170 (2015)

Ertürk, H., Karakaya, A., Oğuz, A.Ç.: Leaf diseases occurring on barley plants in Bala district of Ankara province Turkey. Ecol. Life Sci. **13**(4), 204–207 (2018)

Geçit, H.H.: Serin İklim Tahılları (Buğday, Arpa, Yulaf, Triticale). Yayın No: 1640 Ankara Üniversitesi Ziraat Fakültesi Yayınları, Ankara (2016)

İlgen, M.Z., Karakaya, A., Oğuz, A.Ç.: Leaf diseases occurring on barley and wheat fields in Çubuk district of Ankara, Turkey. Works Fac. Agric. Food Sci. Univ. Sarajevo **67**(2), 210–215 (2017)

Karakaya, A., Mert, Z., Oğuz, A.Ç., Azamparsa, M.R., Çelik, E., Akan, K., Çetin, L.: Current status of scald and net blotch diseases of barley in Turkey. In: IWBLD–1st International Workshop on Barley Leaf Diseases, Abstract Book, Salsomaggiore Terme, Italy, 3–6 June 2014, p. 31 (2014)

Karakaya, A., Mert, Z., Oğuz, A.Ç., Çetin, L.: Distribution of barley stripe disease in Central Anatolia, Turkey. Selcuk J. Agric. Food Sci. **30**(2), 59–61 (2016)

Kün, E.: Tahıllar-I Serin iklim tahılları. Üçüncü Baskı. Yayın No: 1451. Ankara Üniversitesi Ziraat Fakültesi, Ankara (1996)

Mathre, D.E.: Compendium of Barley Diseases. APS Press, Minnesota (1982)

Mızrak, G., Yalvaç, K.: Ülkemizde buğday ve arpa tarımının bugünkü durumu ve geleceği. Ziraat Mühendisliği **332**, 7–15 (2001)

Özdemir, H.Y., Karakaya, A., Oğuz, A.Ç.: Kırıkkale ilinde buğday ve arpa ekim alanlarında görülen fungal yaprak hastalıklarının belirlenmesi. Bitki Koruma Bülteni **57**(2), 89–112 (2017)

Saari, E.E., Prescott, J.M.: A scale for appraising the foliar intensity of wheat diseases. Plant Dis. Rep. **59**, 377–380 (1975)

Tüik (2017). https://biruni.tuik.gov.tr/medas/?kn=92&locale=tr. Accessed 28 June 2018

Zaffarano, P.L., McDonald, B.A., Linde, C.C.: Two new species of *Rhynchosporium*. Mycologia **103**(1), 195–202 (2011)

Zillinsky, F.J.: Common diseases of small grain cereals a guide to identification. International Maize and Wheat Improvement Center, Mexico (1983)

Current Situation of Rice Blast Disease in Edirne, Turkey

Melis Seidi[1], Aziz Karakaya[2](✉), Kemal Akın[1], and Arzu Çelik Oğuz[2]

[1] Trakya Agricultural Research Institute, Edirne, Turkey
[2] Faculty of Agriculture, Department of Plant Protection, Ankara University, Dışkapı, Ankara, Turkey
karakaya@agri.ankara.edu.tr

Abstract. Rice (*Oryza sativa*) is one of most important cereal crops in Turkey. More intensive rice production areas in Turkey are located in Thrace region. Rice blast disease caused by the fungus *Pyricularia oryzae* is one of the most important pathogens that affect the yield and quality of rice plants. In this study, a total of 22 rice fields were surveyed in Keşan, Meriç, Uzunköprü, İpsala, Enez and Havsa districts of Edirne province of Turkey in 2018 for the presence of rice blast disease. The disease severity and prevalence values of rice blast disease in these districts were recorded. The highest disease severity was observed in Havsa district (48.40%), followed by Uzunköprü (27.68%), Meriç (26.20%), Keşan (19.23%), İpsala (19.12%) and Enez (6.40%) districts. The highest prevalence of the disease was observed in Havsa district (100%), followed by Meriç (71.43%), Uzunköprü (69.15%), Keşan (62.73%), İpsala (33.66) and Enez (13.00%) districts. It appears that rice blast is common in Edirne province of Turkey. Precautions should be taken against this important disease.

Keywords: Rice blast disease · Rice · *Pyricularia oryzae* · *Magnaporthe oryzae* · Edirne · Turkey

1 Introduction

Rice (*Oryza sativa* L.) crop which belongs to *Gramineae* family is a warm season cereal grown in water. In Turkey, rice is grown mainly in Thrace and Black Sea regions. Seventy percent of the rice production in Turkey occurs in Thrace-Marmara regions. Twenty-five percent of the Turkish rice is produced in Black Sea region and the rest is produced in other regions (Sürek 2002). Rice blast disease caused by the fungus *Pyricularia oryzae* Cavara (teleomorph: *Magnaporthe oryzae* B.C. Couch) is an important rice disease in the world (Bonman 1992; Ou 1985). The characteristics of lesions on the leaf blades vary with the environment and host resistance. When lesions first appear, they are often white to gray-green with darker green borders. Older lesions are generally whitish to gray with necrotic borders (Bonman 1992). Their shape show variation but lesions are characteristically diamond-shaped. This disease lowers the quality and quantity of rice plants and it is considered as one of the biggest threats to rice production (Ou 1985). Rice blast disease causes significant crop losses every year

M. Brka et al. (Eds.): AgriConf 2019, IFMBE Proceedings 78, pp. 92–97, 2020.
https://doi.org/10.1007/978-3-030-40049-1_12

and it is estimated that rice destroyed due this disease could feed more than 60 million people. *P. oryzae* is known to occur in 85 countries worldwide (Scardaci et al. 1997). In India, 60% to 90% crop losses due to this disease have been reported (Rangaswamy and Subramanian 1957; Sing 1987). This disease also caused significant crop losses in China (Shen and Lin 1996). In Turkey, the first studies related to rice diseases were carried out by Bremer and Özkan (1946) and Göbelez (1953). They concluded that rice blast disease was the most important disease in rice cultivation areas. Göbelez (1953) reported yield losses ranging between 25%–75% in rice fields of Black Sea region of Turkey. In Mediterranean region of Turkey, 90% yield loss has been reported (Tekinel et al. 1980). In Edirne province of Turkey, in 1995 and 1997, 20% and 15% yield losses due to rice blast disease occurred. Also in 2002, in İpsala region of Edirne province yield losses due to rice blast disease was observed (Sürek 2007). This study was carried out in order to determine rice blast disease situation in Edirne province of Turkey.

2 Materials and Methods

In 2018, a survey was conducted in Meriç, İpsala, Enez, Uzunköprü, Keşan and Havsa districts of Edirne province of Turkey for determination of rice blast disease ratio, disease severity and prevalence. For disease identification, Ou (1985) and Bonman (1992) were used. Sixty to eighty-five days old rice plants were examined. In survey, systematic sampling method was used (Aktaş 2001). Three, two, five, nine, two and one rice fields in Keşan, Meriç, Uzunköprü, İpsala, Enez and Havsa districts were examined, respectively. Samples were taken at every 20 km. Fields were surveyed diagonally or a zigzag pattern was followed. Area of each surveyed field was also recorded (Table 1).

Table 1. The sizes of fields and number of places sampled (Aktaş 2001).

Field size	Number of places sampled
Up to 10 decares	>5 different places
11–100 decares	>10 different places
101–500 decares	>15 different places
>501 decares	>20 different places

At each location, 100 plants were examined. Plants were counted as diseased and disease free and the diseased plant ratio was calculated by dividing the diseased plant numbers to total plants counted (Aktaş 2001).

$$\text{Disease ratio} = \frac{\text{Plant numbers with the disease}}{\text{Total plant numbers counted}} \times 100$$

Disease severity for each field was calculated using Townsend-Heuberger formula (Townsend and Heuberger 1943).

$$\text{Disease severity} = \frac{\sum \text{infection frequencies} \times \text{number of plants at each class}}{\text{Total of observed plants} \times \text{highest value of the evaluation scale}} \times 100$$

The prevalence of the disease was calculated using the following formula (Bora and Karaca 1970).

$$\text{Prevalence of the disease} = \frac{\sum \text{Disease ratio in field} \times \text{area of the field (decares)}}{\text{Total area examined (decares)}} \times 100$$

Disease assessment was accomplished using a 0–5 scale using diseased leaves (Table 2) (Anonymous 2017).

Table 2. Leaf disease evaluation scale for rice blast (Anonymous 2017)

Scale value	Symptoms on the plants
0	Immune, no disease
1	Very small, brown, slightly elongated necrotic gray lesions, 1–2 mm, covering less than 1% leaf area, resistant
2	Typical scald lesions, 1–2 cm in length, with brown edges, covering 2-10% leaf area, moderately resistant
3	Typical blight lesions covering 11–25% leaf area, moderately susceptible
4	Typical blight lesions covering 26–50% leaf area, susceptible
5	Typical blight lesions covering more than 50% leaf area, plant is dead, very susceptible

3 Results and Discussion

In 2018, a total of 2360 decares of rice fields in Keşan, Meriç, Uzunköprü, İpsala, Enez and Havsa districts of Edirne province of Turkey were examined for the presence of rice blast disease. The disease was common among the fields inspected (Fig. 1). Numbers of rice fields inspected, total area examined, mean disease ratio (%), mean disease severity (%) and mean disease prevalence (%) were presented in Table 3.

Rice blast disease was found in all districts, however, their prevalence and severity values showed variation. The rice blast severity values ranged between 6.40% and 48.40%. The disease prevalence values were between 13.00% and 100%. The highest and lowest severity and prevalence values were observed in Havsa and Enez districts, respectively. However, one and two fields were surveyed in these districts. For more reliable results, more fields should be surveyed in these districts.

Fig. 1. Rice blast disease symptoms on rice leaves

Table 3. Numbers of rice fields inspected, total area examined, mean disease ratio (%), mean disease severity (%) and mean disease prevalence (%) in Edirne province in 2018.

The districts of Edirne province	Numbers of fields examined	Total area examined (decares)	Mean disease ratio (%)	Mean disease severity (%)	Mean disease prevalence (%)
Keşan	3	410	55.33	19.23	62.73
Meriç	2	70	59.00	26.20	71.43
Uzunköprü	5	800	64.40	27.68	69.15
İpsala	9	730	41.33	19.12	33.66
Enez	2	250	15.00	6.40	13.00
Havsa	1	100	100	48.40	100.00
TOTAL	22	2360			

Rice blast is a serious disease of rice in Turkey and in the world (Göbelez 1953, 1956; Oran 1975; Tekinel 1982; Zeigler et al. 1994). This disease was responsible in 75% of yield loss in some rice fields of Black Sea region of Turkey. Also in Central Anatolia region of Turkey rice blast caused yield losses (Göbelez 1953, 1956). Oran (1975) reported 8.33% yield loss in Southeast Anatolia region of Turkey. Tekinel et al. (1982) investigated the rice diseases in the fields of Adana, Ankara, Erenköy and Samsun Research Institutes and concluded that rice blast was the most important disease. Rice blast disease was also reported from Çanakkale, Bursa and Balıkesir provinces of Turkey (Elmacı 2012). Ergün (2017) conducted a survey in Biga district of Çanakkale province of Turkey and found the prevalence of rice blast disease as 75.33%. In our current study we also observed this disease as common in rice fields of Edirne province of Turkey.

4 Conclusions

It appears that rice blast is common in Edirne province of Turkey. The disease was found in every district that surveyed. Rice blast disease has been reported from different regions of Turkey and significant yield losses have been reported (Göbelez 1953, 1956; Oran 1975; Sürek and Beşer 1997). Precautions should be taken against this important disease in Turkey.

References

Aktaş, H.: Önemli hububat hastalıkları ve sürvey yöntemleri. T. C. Tarım ve Köyişleri Bakanlığı, Tarımsal Araştırmalar Genel Müdürlüğü, Bitki Sağlığı Araştırmaları Daire Başkanlığı, Ankara, 80 p. (2001)

Anonymous. Çeltik entegre mücadele teknik talimatı. T.C. Gıda Tarım ve Hayvancılık Bakanlığı, Tarımsal Araştırmalar ve Politikalar Genel Müdürlüğü, Bitki Sağlığı Araştırmaları Daire Başkanlığı, Ankara, 61 p. (2017)

Bonman, M.: Blast. In: Webster, R.K., Gunnell, P.S. (eds.) Compendium of rice diseases, pp: 14–17. APS Press, The American Phytopathological Society, USA (1992)

Bora, T., Karaca, I.: Kültür bitkilerinde hastalığın ve zararın ölçülmesi. Ege Üniversitesi, Ziraat Fakültesi Yardımcı Ders Kitabı, Yayın No: 167, Ege Üniversitesi Matbaası, Bornova, Turkey, 43 p. (1970)

Bremer, H., Özkan, H.: Türkiye'de çeltik hastalıkları. Ziraat Dergisi **73–74**, 41–53 (1946)

Elmacı, A.: Güney Marmara çeltik ekim alanlarında çeltik yanıklık hastalığının (*Pyricularia oryzae* Cavara) yaygınlık ve yakalanma oranları ile hastalık şiddetinin belirlenmesi. Yüksek Lisans Tezi. Ege Üniversitesi, Fen Bilimleri Enstitüsü, Bitki Koruma Anabilim Dalı, İzmir, 55 p. (2012)

Ergün, S.B.: Biga'da çeltik yanıklığı (*Pyricularia oryzae*) hastalığının çıkışı ve yaygınlığının tespiti ile tohumluklarda bulaşıklık oranlarının saptanması. Yüksek Lisans Tezi, Çanakkale Onsekiz Mart Üniversitesi, Fen Bilimleri Enstitüsü, Bitki Koruma Anabilim Dalı, Çanakkale, 38 p. (2017)

Göbelez, M.: Karadeniz bölgesi çeltiklerinde kavrulma *Pyricularia oryzae* Bri. Cav. Tomurcuk **22**, 12–13 (1953)

Göbelez, M.: Orta Anadolu'nun bazı illerinde yetiştirilen kültür bitkilerinde to-humla geçen bakteri ve mantar hastalıklarının türleri, yayılış alanları ve bunların takribi zarar derecelerinin tespiti üzerine araştırmalar. Ankara Üniversitesi Ziraat Fakültesi Yayınları, no. 107, Ankara (1956)

Oran, Y.K.: Güneydoğu Anadolu'da çeltik yanıklığı fungusu (*Pyricularia oryzae* Bri. et Cav.) 'nun taksonomisi, bio ekolojisi, zararı ve çeltik çeşitlerinin dayanık-lılığı üzerinde araştırmalar. Bitki Koruma Bülteni Ek yayın (1), 1–49 (1975)

Ou, S.H.: Rice Diseases, 2nd edn. Commonwealth Mycological Institute, Kew (1985)

Rangaswamy, G., Subramanian, T.V.: Estimation of loss due to blast disease of rice. Sci. Cult. **23**, 192–193 (1957)

Scardaci, S.C., Webster, R.K., Greer, C.A., Hill, J.E., Williams, J.F., Mutters, R.G., Brandon, D. M., McKenzie, K.S., Oster, J.J.: Rice blast: a new disease in California. Agron. Fact Sheet Series, No. 1997-2. Department of Agronomy and Range Science, University of California, Davis (1997)

Shen, M., Lin, J.Y.: The economic impact of rice blast disease in China. In: Evenson, R.E., Herdt, R.W., Hossain, M. (eds.) Rice Research in Asia: Progress and Priorities, International Rice Research Institute (IRRI). Manila, Philippines, pp. 317–324 (1996)

Sing, N.: Incidence of rice panicule stalk blast (BI) in Manipur. Int. Rice Res. Newsl. **4**, 34–35 (1987)

Sürek, H.: Ülkemizde sorun olabilen önemli çeltik hastalıkları. In: Sürek, H. (ed.) Çeltik tarımı, Hasad Yayıncılık, İstanbul, Turkey, pp. 116–123 (2002)

Sürek, H.: Çeltik yanıklık hastalığı (*Pyricularia oryzae*). Hasad Bitkisel Üretim **23**(265), 80–86 (2007)

Sürek, H., Beşer, N.: The effects of blast disease infection on rice and milling yield. Int. Rice Res. Notes **22**(1), 25–26 (1997)

Tekinel, N., Babaoğlu, B., Yılmaz Demir, F.Y., Bilgin, O.: Türkiye'de çeltik hastalıkları üzerine araştırmalar. A. 103308 Nolu Ülkesel Proje Sonuç Raporu (1980)

Tekinel, N., Yılmazdemir, F.Y., Babaoğlu, B., Bilgin, O.: Türkiye'de çeltik hastalıkları üzerine araştırmalar. T.C. Tarım Orman ve Köyişleri Bakanlığı Zirai Mücadele ve Zirai Karantina Genel Müdürlüğü Araştırma Dairesi Başkanlığı, Zirai Mücadele Araştırma Yıllığı **17**, 124–125 (1982)

Townsend, G.K., Heuberger, J.W.: Methods for estimating losses caused by diseases in fungicide experiments. Plant Dis. Rep. **27**, 340–343 (1943)

Zeigler, R.S., Leong, S.A., Teng, P.S.: Rice Blast Disease. CAB International, Wallingford (1994)

Prevalence of *Botrytis cinerea* in Tomato Greenhouses in Antalya Province of Turkey and Phenotypic Characterization of Isolates

Esra Gül and Aziz Karakaya(✉)

Faculty of Agriculture, Department of Plant Protection, Ankara University, Dışkapı, 06110 Ankara, Turkey
esragul@ankara.edu.tr

Abstract. *Botrytis cinerea* is one of the most important fungal diseases that decrease the yield and quality of tomato plants especially under greenhouse conditions where intensive cultivation was carried out. The fungus can infect plants under field, garden, greenhouse and storage conditions. *B. cinerea* has the ability to infect all aboveground parts of the plants during the vegetation period, it is prone to fungicide resistance and has high genetic variation. These features make control of the pathogen difficult. Antalya province of Turkey is an important tomato growing area and greenhouse production of tomatoes in this province is common. In 2018, a survey was conducted in Serik, Aksu, Kumluca, Finike and Kepez districts of Antalya province. Diseased tomato plant samples were obtained from 35 greenhouses. *B. cinerea* was found in all surveyed districts of Antalya province. The disease was more common in Serik district. The percentages of *B. cinerea* isolated from diseased tomato parts were 64, 20, 35, 22 and 29 from Serik, Kumluca, Kepez, Finike and Aksu districts, respectively. Phenotypic characterization of the isolates revealed that the most common phenotype was S3.

Keywords: Antalya · Tomato · *Botrytis cinerea* · Phenotypic characterization

1 Introduction

Botrytis cinerea, a heterothallic fungus in the Ascomycetes class, can infect more than 200 plants (Choquer et al. 2007). The pathogen can directly infect the flower and leaf parts of plants. Infections in other tissues of the plants are also observed. During the cultivation period of the plants, it is difficult to control the pathogen since it infects the vegetative and generative organs (Gullino, 1992).

Conidia of pathogen can infect different hosts by spreading through the air. Airborne spread of the pathogen may affect control studies, in particular fungicide resistance management and sanitation measures (Ma and Michailides 2005).

B. cinerea can be present in field, garden, greenhouse and storage conditions. It is one of the most important fungal diseases that decrease the yield and quality of tomatoes especially under greenhouse conditions where intensive cultivation was carried out. Gray mold disease caused by *B. cinerea* was determined as 17.39% in tomato fields in Elazığ province of Turkey (Mutlu and Üstüner 2017), and it was

M. Brka et al. (Eds.): AgriConf 2019, IFMBE Proceedings 78, pp. 98–106, 2020.
https://doi.org/10.1007/978-3-030-40049-1_13

determined as 21.2% in Kumluca district of Antalya, Turkey where a significant portion of tomato cultivation was carried out in greenhouses (Duran and Özkaya 2016).

Considering the greenhouse production in the world and in Turkey, tomatoes are among the most important plant species (TUIK 2018). In Turkey, 65% of the total tomato production is carried out in Antalya province where intensive greenhouse production occurs (TUIK 2018). This study was carried out in order to determine the prevalence and phenotypic characterization of *B. cinerea* in greenhouses in Antalya province of Turkey.

2 Materials and Methods

2.1 Greenhouse Studies

In December 2018, surveys were conducted in Serik, Aksu, Kepez, Kumluca and Finike districts of Antalya, Turkey. Thirty-five greenhouses were inspected and plant samples were taken from different parts of tomato plants.

2.2 Isolation Studies

Plant parts containing infected and healthy tissues from leaves, petioles and stems of infected plants were cut with a scalpel and maintained in 1% NaOCI solution for 1 min. These samples were then dried between the sterile filter papers and transferred to Petri plates containing Potato Dextrose Agar (PDA) medium. For some samples Nutrient Agar (NA) medium was used. After isolation, 39 *B. cinerea* isolates were selected from different greenhouses for phenotypic characterization studies. *B. cinerea* isolates were identified according to colony growth, conidia and conidiophore shapes (Ellis 1971; Ellis and Waller 1974; Jones et al. 2016). The isolates were purified by taking hyphal tips and stored at +4 °C in agar slants. In some greenhouses, other diseases were also present. For identification of these diseases, Jones et al. (2016) was used.

2.3 Phenotypic Characterization Studies

Thirty-nine isolates were selected for phenotypic characterization and they were inoculated into PDA medium. Following inoculation, they were incubated in the dark for 4 days. Agar pieces, 0.6 cm in diameter, were cut from the edges of growing colonies and transferred to PDA medium. There were three replications. After incubation for 3 weeks at 20 °C in darkness, mycelial and sclerotial phenotypes based on mycelium formation, sporulation density and sclerotia production were determined according to Martinez et al. (2003), Kuzmanovska et al. (2012) and Zhang et al. (2017).

3 Results and Discussion

Isolations and greenhouse observations revealed that *Botrytis cinerea* was widespread in Antalya province. *B. cinerea* was found in all surveyed districts of Antalya province, however, the disease was more common in Serik district. The percentages of *B. cinerea* isolated from diseased tomato parts were 64, 20, 35, 22 and 29 from Serik, Kumluca, Kepez, Finike and Aksu districts, respectively (Tables 1 and 2). Under greenhouse conditions, *B. cinerea* infections were observed especially at the shoot tips of the plants. In addition, stem infections on pruned plant parts, flower and fruit infections were present (Fig. 1).

Table 1. The number of greenhouses, total number of isolates obtained, numbers of *Botrytis cinerea* isolations and the percentages of *Botrytis cinerea* isolations by districts

Districts	Number of greenhouses	Total isolation counts	Numbers of *Botrytis cinerea* isolations	Percentages of *Botrytis cinerea* isolations
Serik	10	87	56	64
Kumluca	4	5	1	20
Kepez	14	46	16	35
Finike	5	49	11	22
Aksu	2	7	2	29
Total	**35**	**194**	**86**	**44**

Table 2. Some information about the isolates selected for phenotypic characterization

Antalya Districts	Greenhouse numbers	Plant parts that isolations were carried out	Isolate numbers	Phenotypes	Number of isolates by districts
Serik	1	Leaf	Ant2	S4	20
Serik	2	Leaf	Ant5	S4	
Serik	2	Leaf	Ant6	S3	
Serik	3	Leaf	Ant8	M1	
Serik	3	Leaf stalk	Ant11	S5	
Serik	4	Leaf	Ant14	S3	
Serik	7	Leaf	Ant22	S3	
Serik	7	Leaf	Ant23	S5	
Serik	7	Leaf	Ant26	S2	
Serik	8	Leaf	Ant33	S3	
Serik	8	Stem	Ant34	S3	
Serik	8	Leaf	Ant38	S3	
Serik	9	Leaf	Ant43	S1	
Serik	9	Stem	Ant45	S3	
Serik	9	Leaf	Ant52	S3	
Serik	9	Leaf	Ant53	S4	
Serik	9	Leaf	Ant54	S4	
Serik	10	Leaf stalk	Ant55	S3	
Serik	10	Leaf stalk	Ant57	M6	
Serik	10	Leaf	Ant59	S4	
Finike	11	Leaf stalk	Ant60	S4	7
Finike	11	Stem	Ant61	S3	
Finike	11	Leaf stalk	Ant63	S4	
Finike	11	Leaf	Ant65	M1	

(*continued*)

Table 2. (*continued*)

Antalya Districts	Greenhouse numbers	Plant parts that isolations were carried out	Isolate numbers	Phenotypes	Number of isolates by districts
Finike	12	Leaf	Ant69	S5	
Finike	12	Fruit	Ant72	M1	
Finike	12	Leaf	Ant74	S1	
Kepez	13	Leaf	Ant80	S4	9
Kepez	13	Leaf	Ant81	M6	
Kepez	13	Leaf	Ant84	S4	
Kepez	14	Stem	Ant91	S3	
Kepez	14	Leaf stalk	Ant92	M1	
Kepez	15	Leaf	Ant95	M3	
Kepez	16	Leaf stalk	Ant96	S4	
Kepez	16	Leaf stalk	Ant97	S2	
Kepez	17	Leaf stalk	Ant98	S4	
Aksu	18	Leaf stalk	Ant102	S2	2
Aksu	18	Leaf stalk	Ant103	S3	
Kumluca	19	Leaf	Ant143	S3	1
Total isolates					**39**

Fig. 1. Shoot, stem and fruit infections caused by *Botrytis cinerea*

In some greenhouses, no disease was found except for the black spots observed in tomato leaves, whereas in other greenhouses gray mold (*B. cinerea*) and powdery mildew (*Leveillula taurica*) diseases were common. In Finike district *Cladosporium fulvum*, in Serik and Kepez districts *Alternaria* spp., and *Ulocladium* sp. were also isolated. In the isolation made from black superficial spots observed on tomato stems in Kepez district, bacterial growth was observed in NA medium. Although no detailed biochemical tests were carried out it is thought that this pathogen could be *Pseudomonas syringae* pv. *tomato*.

In this study, it was determined that *B. cinerea* was widespread in Antalya province according to isolation studies and greenhouse observations. *B. cinerea* was more common in greenhouses in Serik district of Antalya province in December (Table 1).

Phenotypic characterization of the isolates revealed that sclerotial phenotypes and S3 phenotype which was a sclerotial phenotype were more common in greenhouse tomatoes in Antalya (Tables 2 and 3, Fig. 2). Similar results were reported by different researchers working on different hosts (Tanović et al. 2014; Zhang et al. 2017) and tomatoes (Kuzmanovska et al. 2012). In our current study, the percentage of sclerotial isolates was 82. Among the sclerotial phenotypes, S3 was the most common phenotype with a percentage of 33. Other common phenotype following S3 phenotype was S4 with a percentage of 28. M1 phenotype was common among the mycelial phenotypes (Tables 2 and 3, Fig. 2). Except these phenotypes, M3, M6, S1, S2 and S5 phenotypes were also found. There were differences in mycelium density, mycelium color and sclerotia size in phenotypic groups. No relationship was found between the phenotypic groups and the districts where isolates were obtained.

In a study related to phenotypic and genetic characterization of *B. cinerea* isolates obtained from tomato greenhouses, 9 different phenotypes were determined. It has been stated that the newly defined S5 sclerotype may be a transition type between mycelial and sclerotial phenotypes (Kuzmanovska et al. 2012). Furthermore, it was reported that the phenotypic diversity of *B. cinerea* was more complex than those reported (Kuzmanovska et al. 2012). New M5, M6 and S6 phenotypes have been described later (Zhang et al. 2017).

Phenotypic variation in *B. cinerea* is based on differences in colony morphology, mycelium growth, sporulation density, virulence, enzyme production and fungicide resistance (Martinez et al. 2003; Leroch et al. 2011; Yourman et al. 2001; Valiuskaite et al. 2010). Phenotypic variation level in *B. cinerea* populations is controlled by genotype, environment and their interactions (Corwin et al. 2016).

Genetic diversity in the pathogen can result in phenotypic differences. For example, a relationship between the transposons and the conidia size was reported. It was found that conidia size of the *transposa* isolates were larger compared to the *vacuma* isolates and there was no difference in the size of conidia between *boty* and *flipper* isolates (Kumari et al. 2014). Moreover, it is thought that the genetic diversity in the pathogen and its phenotypic results may effect host range of the pathogen (Corwin et al. 2016).

It is suggested that transposons which may affect the genotypic and phenotypic diversity of the pathogen are also effective on the pathogenicity and fungicide resistance (Amiri et al. 2018; Törün 2018). However, there are different views related to the relationship between transposons and pathogenicity. Pei et al. (2019) reported no significant relationship among the *B. cinerea* isolates, transposon distribution, colony type, geographic region and pathogenicity. However, according to other reports, *transposa* isolates were more virulent than *vacuma* isolates and *vacuma* isolates had a higher tendency to be saprophytes (Martinez et al. 2005; Kumari et al. 2014). When the relationship between the phenotypes of the isolates and their pathogenicity was evaluated, most of non-pathogenic isolates were determined to be the mycelial phenotypes (Mirzaei et al. 2009). However, according to another study, no significant relationship was found among the pathogenicity, geographic region and colony type (Pei et al. 2019).

Table 3. Phenotypic characterization of *Botrytis cinerea* isolates (Martinez et al. 2003; Kuzmanovska et al. 2012; Zhang et al. 2017)

Phenotype	Mycelial type 'M'						Sclerotial type 'S'					
	M1	M2	M3	M4	M5	M6	S1	S2	S3	S4	S5	S6
Mycelium	Short	Aerial	Mycelial masses	Thick and woolly	Thick and woolly	Short in the center and aerial in the edge of Petri dish	Rather short	Rather short	Rather short	Rather short	Aerial	Rather short
Sporulation[1]	-	+	±	-	+	+	±	±	±	-	+	±
Sclerotia[1]	0	-	-	-	-	-	In the edge of Petri dish	Often large, in circles	Often large, placed irregularly	Numerous, small and scattered	Numerous, small and scattered	Numerous, small and agminated
Number of isolates	4	-	1	-	-	2	2	3	13	11	3	-

[1] 0: absent; -: absent or very rare; ±: more or less sporulating; +: sporulating profusely

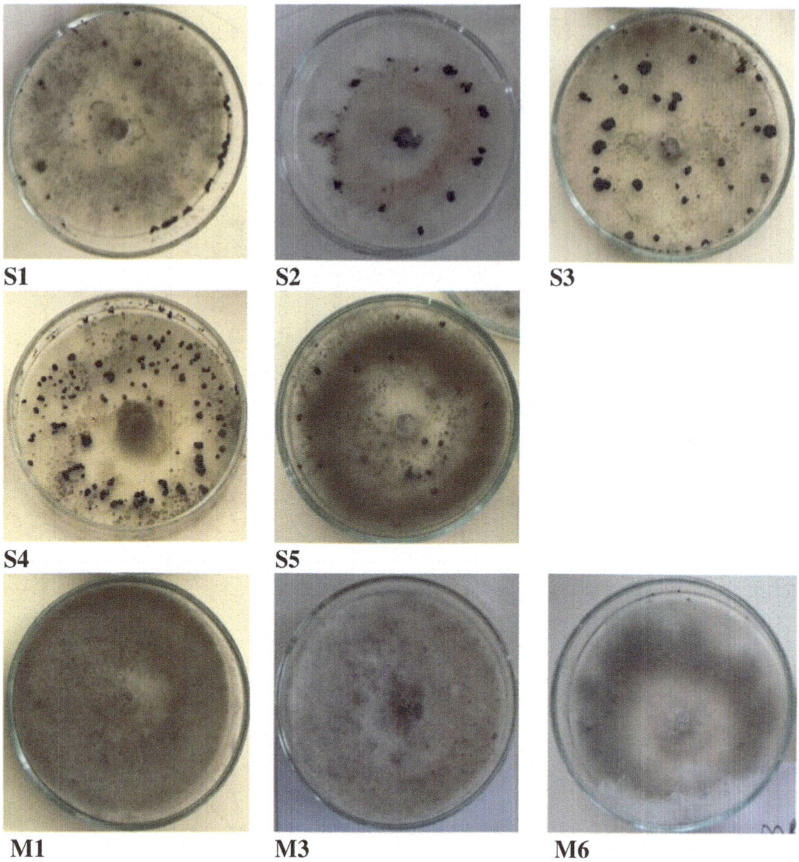

Fig. 2. Morphological types of *Botrytis cinerea* isolates determined in Antalya greenhouses

Zhang et al. (2017) also reported that the phenotype was not related to the growth rate and virulence.

Effects of the environment on the phenotype of pathogen was evaluated by various researchers. A correlation was found between the phenotype and climatic region where the isolate was obtained (Zhang et al. 2017). It was stated that isolates collected from mostly humid areas exhibited high growth rates (Mirzaei et al. 2009). In addition, a correlation was observed between the virulence and the climatic region where the isolate was obtained (Zhang et al. 2017).

The high genetic variation of *B. cinerea* is one of the reasons that makes the control of pathogen difficult (Mirzaei et al. 2009). In addition, fungicide resistance problems against different active substances (Kretschmer and Hahn 2008; Leroch et al. 2011; Konstantinou et al. 2015; Kanetis et al. 2017; Wessels et al. 2016) lead to difficulties in the control of the pathogen. Therefore, there is a need for alternative or integrated control methods. It is thought that determination of the variation in populations of this fungus may provide the development of more effective methods for the pathogen control (Mirzaei et al. 2009).

Phenotypic characterization accomplished in our study could be improved with evaluating virulence and fungicide resistance of these isolates which were macroscopically divided into sclerotial and mycelial groups. This study provided basic information which can be useful for future studies related to the variation of *B. cinerea* isolates.

4 Conclusions

Botrytis cinerea was found in all surveyed districts of Antalya province of Turkey, however, the fungus was more common in Serik district of Antalya. Sclerotial phenotypes were more common among the isolates obtained from tomato greenhouses in Antalya. Among these isolates S3 phenotype was more common. In our study, eight different phenotypes were determined. Three of these phenotypes were mycelial types and 5 of the phenotypes were sclerotial types.

References

Amiri, A., Zuniga, A.I., Peres, N.A.: Prevalence of *Botrytis* cryptic species in strawberry nursery transplants and strawberry and blueberry commercial fields in the eastern United States. Plant Dis. **102**(2), 398–404 (2018)

Choquer, M., Fournier, E., Kunz, C., Levis, C., Pradier, J.M., Simon, A., Viaud, M.: *Botrytis cinerea* virulence factors: new insights into a necrotrophic and polyphageous pathogen. FEMS Microbiol. Lett. **277**, 1–10 (2007)

Corwin, J.A., Subedy, A., Eshbaugh, R., Kliebenstein, D.J.: Expansive phenotypic landscape of *Botrytis cinerea* shows differential contribution of genetic diversity and plasticity. Mol. Plant Microbe Interact. **29**(4), 287–298 (2016)

Duran, İ., Özkaya, H.Ö.: Kumluca ilçesi sera alanlarında toprak ve yapr ak kökenli fungal hastalık etmenlerinin belirlenmesi. Süleyman Demirel Üniversitesi Fen Bilimleri Enstitüsü Dergisi **20**(1), 111–122 (2016)

Ellis, M.B.: Dematiaceous Hyphomycetes. Commonwealth Mcological Institute, Kew (1971)

Ellis, M.B., Waller, J.M.: *Sclerotinia fuckeliana* (conidial state *Botrytis cinerea*). In: Descriptions of Pathogenic Fungi and Bacteria, no. 431. Commonwealth Mycolgoical Institute, Kew (1974)

Gullino, M.L.: Chemical control of *Botrytis* spp. In: Verhoeff, K., Malathrakis, N.E., Williamson, B. (eds.) Proceedings of the 10th International *Botrytis* Symposium Recent Advances in *Botrytis* Reserarch, Heraklion, Crete, Greece, pp. 217–222. Pudoc Sci. Publisher, Wageningen (1992)

Jones, J.B., Zitter, T.A., Momol, T.M., Miller, S.A.: Compendium of Tomato Diseases and Pests, 2nd edn. APS Press, Minnesota (2016)

Kanetis, L., Christodoulou, S., Iacovides, T.: Fungicide resistance profile and genetic structure of *Botrytis cinerea* from greenhouse crops in Cyprus. Eur. J. Plant Pathol. **147**(3), 527–540 (2017)

Konstantinou, S., Veloukas, T., Leroch, M., Menexes, G., Hahn, M., Karaoglanidis, G.: Population structure, fungicide resistance profile, and sdhB mutation frequency of *Botrytis cinerea* from strawberry and greenhouse-grown tomato in Greece. Plant Dis. **99**, 240–248 (2015)

Kretschmer, M., Hahn, M.: Fungicide resistance and genetic diversity of *Botrytis cinerea* isolates from a vineyard in Germany. J. Plant Dis. Prot. **115**, 214–219 (2008)

Kumari, S., Tayal, P., Sharma, E., Kapoor, R.: Analyses of genetic and pathogenic variability among *Botrytis cinerea* isolates. Microbiol. Res. **169**(11), 862–872 (2014)

Kuzmanovska, B., Rusevski, R., Jankuloski, L., Jankulovska, M., Ivic, D., Bandzo, K.: Phenotypic and genetic characterization of *Botrytis cinerea* isolates from tomato. Genetika **44**(3), 633–647 (2012)

Leroch, M., Kretschmer, M., Hahn, M.: Fungicide resistance phenotypes of *Botrytis cinerea* isolates from commercial vineyards in South Germany. J. Phytopathol. **159**, 63–65 (2011)

Ma, Z., Michailides, T.J.: Genetic structure of *Botrytis cinerea* populations from different host plants in California. Plant Dis. **89**, 1083–1089 (2005)

Martinez, F., Blancard, D., Lecomte, P., Levis, C., Dubos, B., Fermaud, M.: Phenotypic differences between vacuma and transposa subpopulations of *Botrytis cinerea*. Eur. J. Plant Pathol. **109**, 479–488 (2003)

Martinez, F., Dubos, B., Fermaud, M.: The role of saprotrophy and virulence in the population dynamics of *Botrytis cinerea* in vineyards. Phytopathology **95**, 692–700 (2005)

Mirzaei, S., Goltapeh, E.M., Shams-Bakhsh, M., Safaie, N., Chaichi, M.: Genetic and phenotypic diversity among *Botrytis cinerea* isolates in Iran. J. Phytopathol. **157**, 474–482 (2009)

Mutlu, G., Üstüner, T.: Elazığ ili domates alanlarında fungal hastalıkların yaygınlığı ve şiddetinin saptanması. Türk Tarım ve Doğa Bilimleri Dergisi **4**(4), 416–425 (2017)

Pei, Y.G., Tao, Q.J., Zheng, X.J., Li, Y., Sun, X.F., Li, Z.F., Qi, X.B., Xu, J., Zhang, M., Chen, H.B., Chang, X.L., Tang, H.M., Sui, L.Y., Gong, G.S.: Phenotypic and genetic characterization of *Botrytis cinerea* population from kiwifruit in Sichuan Province. China. Plant Dis. **103**(4), 748–758 (2019)

Tanović, B., Hrustić, J., Mihajlović, M., Grahovac, M., Delibašić, G.: *Botrytis cinerea* in raspberry in Serbia I: morphological and molecular characterization. Pesticides Phytomed. **29**(4), 237–247 (2014)

Törün, B.: Aydın ve Mersin illerinden toplanan çileklerde *Botrytis cinerea* populasyonlarındaki transpozon sıklığı ve fungusit dirençliliği, Doktora tezi, Adnan Menderes Üniversitesi, Fen Bilimleri Enstitüsü, Biyoloji Anabilim Dalı, Aydın (2018)

TUIK Homepage. https://biruni.tuik.gov.tr/medas/?kn=92&locale=tr. Accessed 15 May 2019

Wessels, B.A., Linde, C.C., Fourie, P.H., Mostert, L.: Genetic population structure and fungicide resistance of *Botrytis cinerea* in pear orchards in the Western Cape of South Africa. Plant. Pathol. **65**, 1473–1483 (2016)

Valiuskaite, A., Surviliene, E., Baniulis, D.: Genetic diversity and pathogenicity traits of *Botrytis* spp. isolated from horticultural hosts. Žemdirbystė Agric. **97**(4), 85–90 (2010)

Zhang, Y., Xu, H., Shen, F., Li, X., Li, Y., Liu, D.: Phenotypes and virulence variability among grape gray mold isolates from grapes (*Vitis vinifera*) in China. J. Agric. Biotechnol. **25**(11), 1740–1755 (2017)

Yourman, L.F., Jeffers, S.N., Dean, R.A.: Phenotype instability in *Botrytis cinerea* in the absence of benzimidazole and dicarboximide fungicides. Phytopathology **91**, 307–315 (2001)

Animal Production

Polymorphisms of the *ovine prion protein* (PrP) Gene in the Pramenka Sheep Breed Population(s) in Bosnia and Herzegovina - Kupreski Strain

Ervin Zecevic(✉), Admir Dokso, Alma Rustempasic, and Muhamed Brka

Faculty of Agriculture and Food Sciences, University of Sarajevo, Zmaja od Bosne 8 (UNSA Campus), 71 000 Sarajevo, Bosnia and Herzegovina
e.zecevic@ppf.unsa.ba

Abstract. Transmissible spongiform encephalopathy (TSE) is a group of diseases that affects nervous system, and which leads to death. Scrapie is an infectious disease of sheep and goats that belongs to this TSE group and with always fatal outcome. The putative infectious agent is the host-encoded prion protein (PrP) and it appears that the development of scrapie in sheep is closely related to polymorphisms in the host PrP gene. The aim of this study was to investigate three well-known polymorphisms in exon 3 of the PrP gene (on chromosome 13) that have a clear and significant effect on scrapie in sheep (at codons 136, 154 and 171), but also to investigate four other polymorphisms in exon 3 (at codons 145, 185, 231 and 237) that are undefined yet, but might have some effect on disease. The study was performed on the Bosnian sheep – Pramenka breed, Kupreski strain from central Bosnia. Genomic DNA was isolated from venous blood and then a 628 bp long DNA fragment from exon 3 was amplified using the PCR method. The results from this study indicate the presence of nine different genotypes associated with susceptibility of scrapie in investigated Bosnian sheep population(s). Five haplotypes were found.

Keywords: PrP · Transmissible spongiform encephalopathy · Scrapie · Sheep

1 Introduction

Scrapie is a prion-associated transmissible spongiform encephalopathy that occurs in sheep and goats. Scrapie was found in most breeds of sheep, but there are differences between breeds, strains and individuals in susceptibility on it. It has been found that genetics (PrP gene) is responsible for disease susceptibility and pathogenesis. Sheep PrP gene consists of three exons and two introns and is located on chromosome 13 (Ryan et al. 1993), the region which is evolutionarily preserved (Lee et al. 1998). In general, the PrP gene is in the promoter coding region, but in different areas in different species. Sequence of this gene was determined in 1998 and it was found high similarity in coding and non-coding sequences in sheep and humans what indicates that the two sequences are conserved in these two species (Lee et al. 1998). The ovine PrP gene is

M. Brka et al. (Eds.): AgriConf 2019, IFMBE Proceedings 78, pp. 109–116, 2020.
https://doi.org/10.1007/978-3-030-40049-1_14

31 412 bp long. In sheep, the genetic variability of the prion protein gene coding sequences is the most important in susceptibility to scrapie. Seven polymorphic loci (codons) were described, but the most important and with established significance to expressing disease are locicodons 136 (valine – alanine), 154 (histidine – arginine) and 171 (glutamine – arginine). Allelic variants (haplotypes) found on these loci (codons) are: ARR, ARH, AHQ, ARQ and VRQ (Dawson et al. 1998). These alleles are associated with different levels of scrapie susceptibility. Five mentioned Allelic variants (haplotypes) were tested the most, and it was found that ARR haplotype (allelic variant) was always connected with partial or complete protection of disease. Based on this information, all farming selection programs have the aim to improve frequency of these alleles in population.

Table 1. PrP genotype classification, based on scrapie susceptibility (3).

Classification	Genotype	Resistance
NSP1	ARR/ARR	Highly resistant
NSP2	ARR/ARQ	Resistant
	ARR/ARH	
	ARR/AHQ	
NSP3	ARQ/ARQ	Low resistant
	ARQ/AHQ	
	AHQ/AHQ	
	ARH/ARH	
	ARQ/ARH	
NSP4	ARR/VRQ	Susceptible
NSP5	ARQ/VRQ	Highly susceptible
	VRQ/VRQ	
	VRQ/ARH	
	VRQ/AHQ	
	VRQ/ARR	

Different sheep genotypes are classified in groups related to susceptibility level (see Table 1). The highest level of resistance to scrapie is provided by ARR/ARR genotype belonging to NSP1 group, while highly susceptible genotypes are in the NSP5 group, namely: ARQ/VRQ, VRQ/VRQ, VRQ/ARH, VRQ/AHQ and ARQ/ARR.

2 Materials and Methods

Research has been conducted on a sheep breed Pramenka, Kupreski strain from Bosnia and Herzegovina. Blood samples were collected from 84 randomly chosen animals belonging to different flocks from livestock sheep farms in the Central part of Bosnia. Venous blood (approx. 3 ml) was taken from the jugular vein of each individual animals and collected in sterile 3 ml K3E BD Vacutainer tubes (BD Vacutainer

System, Plymouth, UK) containing Ethylene Diamine Tetra acetic Acid (EDTA), and then transported to the laboratory where they were stored at −300C until needed for DNA isolation. Genomic DNA was extracted from blood sample (200 μl) using Sigma's GenElute Mammalian Genomic DNA Miniprep Kit (Sigma Aldrich Chemie, GmbH, Germany) according to the manufacturer's instructions. Concentration of extracted DNA was measured by spectrophotometer (Genova, JENWAY, UK), whereas the quality of DNA was checked by gel electrophoresis (at 120 V for 45 min) on 1% TBE-agarose gel stained with ethidium bromide (EtBr).

The PrP gene fragment size of 628 bp from exon 3 was amplified using the following primers (Gmür et al. 2004):

Forward 5' – CCGCTATCCACCTCAGGGA – 3'

Reverse 5' – TTGCCCCTATCCTACTATGAGA – 3'.

Optimisation of PCR conditions by temperature gradient PCR was carried out by using iCycler thermal cycler (BIO-RAD, Germany). Thermocycling was performed by initial denaturation for 8 min at 950C, followed by 34 cycles of 1 min at 940C, 1 min at 640C annealing temperature and 1 min extension at 720C. A final extension was performed for 7 min at 720C. The amplification reactions were carried out in a total volume of 20 μl containing 4 μl of DNA, 1 μl of each primer (10 pmol/μl) and 10 μl of ready-to-use GoTaq® Green master mix (Promega), in iCycler thermal cycler (BIO - RAD, Germany).

PCR products (20 μl) were analyzed by electrophoresis (at 80 V for 30 min) on a 0.8% agarose gels stained by ethidium bromide (EtBr). The visualization was performed in a UV transilluminator and the images were captured with a video documentation system (UVITEC). Amplified fragments of the appropriate size (628 bp) were cut directly from the gel and purified by Promega Wizard® SV Gel and OCR Clean – Up System kits following the supplier's instructions. Then, amplified products were sequenced by standard Sanger dideoxynucleotide method, using the same primers as for PCR reaction. Sequencing was performed on ABI PRISM® 3100 – Avant Genetic Analyzer.

The sequences were analyzed using BioEdit software, version 7.0.9.0. (Hall 1999) while the Clustal W program, version 2.2.10. (Larkin et al. 2007) was used for multiple sequences alignment.

In this study, polymorphisms at loci (codons) 136, 154, 171, 145, 185, 231 and 237 were investigated on the Bosnian sheep – Pramenka breed, Kupreski strain from the Central Bosnia. First, the allelic variants (haplotypes) in the codons (loci) 136, 154 and 171 for the SSHL ("scrapie susceptible haplotype locus") were identified and estimated using the Bayesian approach and, then, the analysis and estimation of polymorphisms, allele frequencies and genotypes were carried out at loci (codons): 145, 185, 231 and 237. The PROC Haplotype and PROC Allele procedures, implemented in SAS/Genetics 9.1.3. software package (SAS Institute Inc. 2008), were used for all calculations.

3 Results

Nine genotypes on scrapie susceptible codons have been found (Table 2), The most frequent genotypes, with the 28.57% frequency appearance, were ARR/ARQ belonging to the risk group 2 (NSP2 – Resistant) and ARQ/ARQ belonging to the risk group 3 (NSP3 – Low Resistance). High frequency of genotype ARR/ARR was found (21.24%) which belongs to the risk group 1 (NSP1 – High Resistance). The other genotypes appeared in frequencies ARQ/AHQ (10.71%; NSP3), AHQ/AHQ (2.38%; NSP3), ARR/VRQ (2.38%; NSP4), ARQ/VRQ (2.38% NSP5) and ARR/AHQ (1,19%; NSP2).

Table 2. Genotype frequency

Genotype codon			Risk group	Frequency (%)	n
136	154	17			
AA	HH	QQ	3	2.76	2
AA	RH	QQ	3	10.71	9
AA	RH	RQ	2	1.19	1
AA	RR	QQ	3	28.57	24
AA	RR	HQ	3	2.38	2
AA	RR	RQ	2	28.57	24
AV	RR	QQ	5	2.38	2
AA	RR	RR	1	21.24	18
AV	RR	RQ	4	2.38	2

Table 3. Allelic variants (haplotypes) frequency

Allelic variants (haplotypes)	n	Frequency (%)
AHQ	14	8.33
ARQ	86	51.20
ARR	62	36.90
VRQ	4	5.36
ARH	2	1.19

The most frequent haplotype was ARQ with a frequency of 51.20%, while haplotype ARH was with the lowest appearance with a frequency of 1.19% (Table 3).

Table 4. Polymorphism and frequency of SNPs for (scrapie susceptible codons)

Codon	Allele	n	Frequency (%)
136	A_{136}GCC	39	92.86
	V_{136}GTC	3	7.14
154	H_{154}CAT	5	11.90
	R_{154}CGT	37	88.10
171	Q_{171}CAG	35	83.33
	R_{171}CGG	7	16.67

In this study, single nucleotide polymorphisms (SNPs) were found in all of three investigated and well-known scrapie susceptible loci (at codons 136, 154 and 171) of the PrP gene (see Table 4).

Looking at amino acid variations at locus (codon) 136 which is responsible for synthesis of two amino acids (alanine and valine) (Belt et al. 1995), we found that the most frequent was alanine variant with a frequency of 97.62%, while valine variant occurred with a frequency of 2.38% (see Table 4). Alanine variant at codon 136 is preferred due to the fact it reduces the risk of TSE occurrence.

Considering the locus (codon) 154, which is responsible for synthesis two amino acids - arginine and histidine (Belt et al. 1995), we found that the arginine variant is more prevalent in our Bosnian population(s) (with a frequency of 91.08%) while histidine variant was found with a frequency of 8.92% (see Table 4).

Locus (codon) 171, which is responsible for synthesis of four amino acids (arginine, glutamine, histidine and lysine) (Goldman et al. 1990; Clouscard et al. 1995 and Gombojav et al. 2003) was present in our study three variants: glutamine, arginine and histidine. Arginine synthesis at codon 171 forms non risk haplotype ARR while glutamine variant increases risk of TSE. In the Bosnian population(s) under the study, arginine is represented with a frequency of 36.90% until glutamine was with a frequency of 61.90% (Table 4). Combination of alanine variant from the codon 136 together with arginine variant of codon 171 makes scrapie risk free ARR haplotype.

Table 5. Polymorphisms at undefined loci that might have some effect on disease

Codon	Allele	Frequency (%)
145	G_{145}GGC	98.21
	S_{145}AGC	1.79
185	I_{185}ATC	100.00
	T_{185}ACC	0.00
231	R_{231}AGG	87.50
	R_{231}CGG	12.50
237	L_{237}CTC	85.12
	L_{237}CTG	14.88

Besides investigations on polymorphisms of those three well-known scrapie susceptible codons (loci), examination was also extended to the investigations on polymorphisms of four other loci (codons 145, 185, 231 and 237) for which is unknown the effect of developing scrapie in sheep (unknown and undefined levels of risk) – see Table 5. The polymorphism was found at codons 145 in the Bosnian population(s) under the investigation (GGC → AGC) with frequencies 98,21% and 1,79% respectively. The silent mutations have been recorded on codons 231 (AGG → CGG) and 237 (CTC→ CTG). In both cases, there were no amino acids changes found. The silent mutation that was found at locus (codon) 231, which is responsible for synthesis of amino acid arginine, had a frequency of variant AGG 87.50% while a variant CGG was present with a frequency of 12.50% (see Table 5). Silent mutation that was found at codon 237 (responsible for synthesis of amino acid leucine) was present with a frequency of variant CTC 85.12% and variant CTG with a frequency of 14.88% (Table 5).

4 Discussion

This is the very first study conducted on sheep populations in Bosnia and Herzegovina Kupreski strain, regarding the genetic aspects of scrapie and PrP gene sequencing. In a study, it was identified the presence of five risk groups according to NSP classification. It was determined a high frequency of the haplotype ARQ (51.20%) at scrapie susceptible codons, which is in concordance with the values obtained in other studies (Lühken et al. 2008). The ARQ haplotype could be considered as a "wild" haplotype and it is present with a high frequency in other countries as well: in Italy for the breed Gentile di Puglia is 53.23% (12), in Austria in the Tyrolean mountain sheep 65.7% (Sipos et al. 2002) and in Hungary with the breed Tsigai 53.45% (Lühken et al. 2008), while in the neighboring country Croatia is present with 67.4% in the Istrian Sheep breed and with 52.6% in the crossbreeds of Istrian sheep (Cubric-Curik et al. 2009). In our studied population(s) of Pramenka breed from Kupreski strain, the most frequent genotypes were ARQ/ARQ and ARR/ARQ. The first one is susceptible of developing scrapie - belongs to the risk group 3 (NSP3 – Low resistance). The second one belongs to the second group of NSP (NSP2 – Resistant)

Haplotype frequency of ARR (36.90%), which belongs to the risk group 1 (NSP1 – High resistance), was relatively high, despite the fact that the way of farming in Bosnia and Herzegovina (BH) is rather primitive, that is without systematic breed selection, particularly regarding to scrapie monitoring and selection. Comparing the frequency of haplotype ARR with the neighboring country Croatia, they found the presence of this ARR haplotype 22.8% in the Istrian sheep and 13.2% for hybrids (Cubric-Curik et al. 2009). In the region, a frequency of haplotypes ARR was found as 35.48% in Italian breed Gentile di Puglia, in Hungary - in the breed Tsigai is 32.76% (Lühken et al. 2008), while in Austria - the frequency of this haplotype is 25.8% in the Tyrolean mountain sheep (Sipos et al. 2002). Important fact is that the presence of the ARR alleles, as well as the ARR/ARR genotype, providing the opportunity to increase the resistance to scrapie in sheep populations of BH if better farming practice and monitored selection will take place in Bosnia and Herzegovina (BH) in order to increase the number of animals and flocks with this favorable genotype.

The most unfavorable haplotype (VRQ) with its frequency of 5.36% in our studied sheep flocks can be considered as a reasonable in comparison with other European sheep breeds, where the frequency of this haplotype ranges from 0.00% to 17.40%. For example, in Italian sheep breed Gentile di Puglia is 0.00% (Lühken et al. 2008), in Croatian Istrian Sheep is 1.1% and in crossbreeds of Istrian sheep 0.00% (Cubric-Curik et al. 2009), in Austria in the Tyrolean mountain sheep is 0.00% (Sipos et al. 2002), while the Hungarian Tsigai breed is with a higher frequency than ours – it is 10.34% (Lühken et al. 2008).

The polymorphisms at loci (codons) 145 showed amino acid change G $\rightarrow$ S, while on codon 185 were absent. The polymorphisms at loci (codons) 231 and 237 resulted in silent mutations R $\rightarrow$ R or L $\rightarrow$ L. These polymorphisms were also found in sheep populations in the neighboring Croatia (Cubric-Curik et al. 2009), but the investigations on these loci (codons) were not carried out in other countries and it is not possible further comparison of our results with others. Mentioned mutations occur at a rather low frequency and their impact on the occurrence of scrapie has not been determined yet.

It is important to continue investigations and screening of other sheep flocks in Bosnia and Herzegovina, as well as to involve larger number of loci (codons) that might have some effect on scrape susceptibility. More organized selection and breeding programs in the farms in Bosnia and Herzegovina is necessary in order to increase ARR combination of alleles that will lead to improvement and increase of scrapie resistant genotypes.

References

Ryan, A.M., Womack, J.E.: Somatic cell mapping of the bovine prion protein gene and restriction fragment length polymorphism studies in cattle and sheep. Anim. Genet. **24**, 23–26 (1993)

Lee, I.Y., Westaway, D., Smit, A.F.A.: Complete genomic sequence and analysis of the prion protein gene region from three mammalian species. Genome Res. **8**, 1022–1037 (1998)

Dawson, M., Hoinville, L.J., Hoise, B., Hunter, N.: Guidance on the use of PrP genotyping as an aid to the control of clinical scrapie. Vet. Rec. **142**, 623–625 (1998)

Gmür, A., Gaillard, C., Dolf, G.: Characterization of the prion protein gene (PRNP) region in Swiss sheep breeds. J. Anim. Breed. Genet. **121**, 216–220 (2004)

Hall, T.A.: BioEdit: a user – friendly biological sequence alignment editor and analysis program for Windows 95/98/NT. In: Nucleic Acid Symposium Series, no. 41, pp. 95–98 (1999)

Larkin, M.A., Blackshields, G., Brown, N.P., Chenna, R., McGettigan, P.A., McWilliam, H., Valentin, F., Wallace, I.M., Wilm, A., Lopez, R., Thompson, J.D., Gibson, T.J., Higgins, D. G.: Clustal W and Clustal X version 2.0. Bioinformatics **23**, 2947–2948 (2007)

SAS Institute Inc.: SAS® 9.1.3 Foundation System Requirements for AIX®. SAS Institute Inc., Cary, NC (2008)

Belt, P.B.G.M., Muileman, I.H., Schreuder, B.E.C., Ruijter, J.B., Gielkens, A.L.J., Smits, M.A.: Identification of five allelic variants of the sheep PrP gene and their association with natural scrapie. J. Gen. Virol. **76**, 509–517 (1995)

Goldmann, W., Hunter, N., Foster, J.D., Salbaum, J.M., Beyreuther, K., Hope, J.: Two alleles of a neural protein gene linked to scrapie in sheep. In: Proceedings of the National Academy of Science, U.S.A., vol. 87, pp. 2476–2480 (1990)

Clouscard, C., Beaudry, P., Elsen, J.M., Milan, D., Dussaucy, M., Bounneau, C., Schelcher, F., Chatelain, J., Launay, J.M., Laplanche, J.L.: Different allelic effects of the codons 136 and 171 of the prion protein gene in sheep with natural scrapie. Journal of General Virology **76**, 2097–2101 (1995)

Gombojav, A., Ishiguro, N., Horiuchi, M., Serjmyadag, D., Byambaa, B., Shinagawa, M.: Amino acid polymorphisms of PrP gene in Mongolian sheep. J. Vet. Med. Sci. **65**, 75–81 (2003)

Lühken, G., Lipsky, S., Peter, C., Erhardt, G.: Prion protein polymorphisms in autochthonous European sheep breeds in respect to scrapie eradication in affected flocks. Small Rumin. Res. **75**, 43–47 (2008)

Sipos, W., Kraus, M., Schmoll, F., Achmann, R., Baumgartner, W.: PrP genotyping of Austrian sheep breeds. J. Vet. Med. A **49**, 415–418 (2002)

Cubric-Curik, V., Feligini, M., Ferencakovic, M., Alen, D.A., Salajpal, K., Ambriovic-Ristov, A., Cetkovic, H., Majhen, D., Curik, I.: Sequence polymorphism of PrP exon 3 gene in Istrian and crossbred sheep. Ital. J. Anim. Sci. **8**, 86–88 (2009)

Polymorphism of PrP Gens in the Population of Chamois in the Area of Biokovo - Republic of Croatia

Ervin Zecevic(✉), Admir Dokso, Alma Rustempasic, and Muhamed Brka

Faculty of Agriculture and Food Sciences, University of Sarajevo, Zmaja od Bosne 8 (UNSA Campus), 71 000 Sarajevo, Bosnia and Herzegovina
e.zecevic@ppf.unsa.ba

Abstract. Transmissible Spongiform Encephalopathies (TSEs) are a group of contagious neurodegenerative diseases that attack neural tissue by the formation of aggregates of prion particles in it. The final outcome of these diseases is lethal. Causative agent is prion, a cellular glycoprotein which have two forms, not contagious with the α - projection, denoted PrP^{C} and infectious with the β - folded plate and labeled PrP^{Sc}. This disease is characteristic for many mammals and humans also. Research has shown that there are genetic predispositions to disease or resistance to it. In the case of goats, various codons that may be associated with disease resistance are reported in the literature, mostly for codon 146. In the study of polymorphisms in chamois, the same information as for goats was used. The genomic DNA was used in the study, which was isolated from the full blood of animals and was subjected to a polymerase chain reaction with specially designed primers. The populations of the chamois have shown that the investigated population is monomorphic when it comes to PrP gene.

Keywords: Chamois · PrP gene · Scrapie

1 Introduction

Transmissible Spongiform Encephalopathies (TSEs) are a group of diseases caused by prions. These are neurodegenerative diseases of man and domestic animals that can be transmitted from the individual to individual of the same species, as well as to individuals of different species.

The ultimate outcome of the disease is lethal. The name of disease spongiform encephalopathies comes from typical vacuoles spongy gray matter of the brain in patients specimens.

The causes of the disease are prions, small protein particles are composed of 250 amino acids. Griffith et al. (1967) first postulated that the cause of neurodegenerative diseases of protein nature. Prusiner et al. (1982) called it prion (prion = proteinaceous infectious particle PrP). There are two isoforms of protein particles. The normal isoform, labeled as PrP^{C}, is located in the cells of all vertebrates and plays a role in

M. Brka et al. (Eds.): AgriConf 2019, IFMBE Proceedings 78, pp. 117–121, 2020.
https://doi.org/10.1007/978-3-030-40049-1_15

defending the cells of infection. Abnormal isoform is denoted as PrP^{Sc}, and from normal, it differs by three-dimensional conformation.

Transmissible spongiform encephalopathies have characteristic forms for different animal species. In this group of diseases are: scrapie, bovine spongiform encephalopathy (BSE), chronic wasting disease (CWD), Creutzfeldt - Jacobs disease (CJB), kuru, transmissible encephalopathy of American vultures, spongiform encephalopathy of cats and exotic animals in zoos and a new variant of Creutzfeldt - Jacob disease (nvCJB) (Prusiner 1998).

Medical science has ignored scrapie until 1959, when American vets connect scrapie with the recently described human disease, called kuru, infectious neurological disorder that is found only in the eastern mountainous areas of Papua New Guinea. In the following period, many studies have been carried out to reveal the physical and chemical properties of the infectious agent, its distribution and titer in the tissue of infected animals. In the early 1980s, it was discovered that one protein (called PrP, or "prion protein") could be isolated from brain tissue and its association with infectivity was discovered (Prusiner 1982). The gene encoding this protein is the normal gene of the organism. At present, the role of a gene encoding prion protein (PrP) is intensively investigated in TSE pathogenesis. The obtained information and genotypes are used in breeding programs and selection of animals genetically resistant to TSE.

Prion (PrP^{C}), is conserved glycoprotein bounded to the cell wall by glycosylphosphatidylinositol bond. This is an infectious agent and does not contain nucleic acids. Permanent efforts to identify even small amounts of DNA or RNA are unsuccessful. The only component that has been identified so far is the protein. Prion is 28 kD hydrophobic glycoprotein, which is encoded by a cellular gene (PrP). The protein exists in two forms. The product found in normal brain tissue is called PrP^{C} and is completely susceptible to degradation by protease. Protein found in infected brain tissue is called PrP^{Sc} (Hunter 1998) and it is highly resistant to degradation by protease. Normal cellular PrP^{C} converts to PrP^{Sc} through the process in which the part of its α - helical coil unfoldes in β - plate. This structural transmission was followed by fundamental changes in the physical and chemical properties of PrP protein. The nucleotide sequence of amino acid found in PrP^{Sc} corresponds to that which is encoded by the PrP gene of the host mammal which was last replicated. This is in contrast to nucleic acid - containing pathogens and encodes specific characteristics of the gene. Transgenetic studies assure that PrPSc acts as a template by which PrPC is restructured into the initial PrPSc molecule in a process that is supported by another protein. More than twenty mutations of PrP genes that cause hereditary prion disease are known and significant genetic correlation has been established for five of these twenty mutations.

The prion protein having the PrPC structure under certain conditions is transformed into PrPSc. Recent studies have shown that cellular osmolites, also referred to as Chaperon, have a particular role in the restatement of PrPC into its infectious form PrPSc. The accompanying proteins support without covalent bonding, wrapping or scraping, joining or splitting macromolecular structures, but they do not appear in these structures when they carry out their normal biological functions. The accompanying proteins do not necessarily bring the stereochemical information needed to form the protein envelope. One of the most important functions of these proteins is to prevent the formation of nonfunctional units after aggregation of the polypeptide chain and associated subunits (DebBurman 1997).

1.1 Genetic Aspects of Scrapie in Chamois

So far, no significant number of polymorphisms on the PrP gene has been found in chamois, except synonym polymorphism in heterozygous form at 119th codon (gct → gcc) (Peletto et al. 2009).

Chamois as wild animals were not a common subject of the PrP genotype analysis so far, so there is currently no comprehensive frequency and distribution data available in the literature as PrP genotypes in general and risky genotypes in terms of resistance to or susceptibility to the disease.

2 Materials and Methods

Balkan Chamois belongs to the mammals, the artiodactyla, and to the bovidae. It can grow up to 80 cm in height and in length up to 130 cm, and stronger males weighing more than 40 kg. Females are slightly smaller and lighter. The natural predator is a wolf, and for youth, danger can also be an eagle.

2.1 Samples

Single samples of the chondrocyte tissue were taken from the ear shells during regular culling, transported to the laboratory and stored at −70°C until DNA was extracted. DNA extraction of all samples was performed in the Laboratory for Conservation Genetics, Institute for Livestock at the Faculty of Agriculture, University of Zagreb.

Before isolation, the samples were prepared by taking a small amount of tissue from the earbuds as itis provided in the reagent kit manufacturer's protocol used to extract DNA from the tissue. (FastDNA) MP Biomedicals.

2.2 Chain Polymerase Reaction (PCR)

Using designed primers Gmür i sur. (2004):

forward 5' – CCGCTATCCACCTCAGGGA – 3'
reverse 5' – TTGCCCCTATCCTACTATGAGA – 3'.

2.3 Nucleotide Sequencing of the PCR Product

Nucleotide sequencing of the PCR product was performed using the same primer that was made using PCR. In 13 ml of purified product with concentration of 200 ng/μl was added 3 pM/μL primers which were used for polymerase chain reaction.

Nucleotide sequence was determined on an ABI Prism 3100 machine - Avant Genetic Analyzer using the Sanger dideoxy method.

2.4 Nucleotide Sequence Analysis

The nucleotide sequence analysis obtained with the ABI PRISM® 3100 - Avant Genetic Analyzer was performed using the BioEdit version 7.0.9.0 program package. (Hall 1999). The sequences were checked for quality and for identification of heterozygous locuses. By Clustal 2.2.10. (Larkin et al. 2007) the sequences are aligned.

3 Results and Discussion

In the population of chamois, monomorphism was found on all examined locus. The entire sample had one genotype (TSIHNRRTRQQSP/TSIHNRRTRQQSP) and an haplotype TSIHNRRTRQQSP (Table 1).

Table 1. Comparison of the frequency of haplotypes in the examined populations of goats and chamois

Haplotype	Goats Bosnia		Goats Croatia		Chamois	
	n	Frequency %	n	Frequency %	n	Frequency %
TSTHNRTRQQSP	8	20,00	13	29,55	0	0,00
TSIHNRTQRQSP	1	2,50	0	0,00	0	0,00
TSIHNRTQRQSP	1	2,50	0	0,00	0	0,00
TSIHNRTRQKSP	6	15,00	0	0,00	0	0,00
TSIHNRTRQQSP	7	17,50	0	0,00	16	100,00
TSMHNRTRQQSP	1	2,50	0	0,00	0	0,00
TS^{1}IHNRTRQQPP	6	15,00	1	38,64	0	0,00
TS^{1}IHNRTRQQPH	1	2,50	0	0,00	0	0,00
TS^{1}IHNRTRRQPP	3	7,50	0	0,00	0	0,00
TS^{1}IHDRTRQQPP	1	2,50	1	2,27	0	0,00
TS^{1}MHNRTRQQPP	4	10,00	2	4,55	0	0,00
TS^{1}MHDRTRQQPP	1	2,50	0	0,00	0	0,00
TSHHNRTRQKSP	0	0,00	1	2,27	0	0,00
TSIHNRPRQQSP	0	0,00	1	2,27	0	0,00
TSIHDRTRQQSP	0	0,00	1	2,27	0	0,00
TS^{1}IHNRTRQQSP	0	0,00	1	2,27	0	0,00
TS^{1}IHDRTRQQPH	0	0,00	1	2,27	0	0,00
TS^{1}IRNRTRQQPP	0	0,00	4	9,09	0	0,00
PS^{1}IHNRTRQQPP	0	0,00	1	2,27	0	0,00
PS^{1}IHDRTRQQPP	0	0,00	1	2,27	0	0,00
TSIHNRTRQQSP	0	0,00	0	0,00	0	0,00

As the case of animals living in the wild, tracking the appearance of scrapie is hampered by the fact that the diseased animals easily fall prey to predators. No cases of scrapie at the chamois is not officially registered in the Republic of Croatia.

In the analyzed sample from the populations of chamois in the Republic of Croatia, one genotype was found in the homozygous form TSIHNRRTRQQSP/TSIHNRRTRQQSP. One haplotype was found and that: TSIHNRRTRQQSP (100,00%). According to the type of alleles, it is apparent that all belong to the so-called "wild" type as in goat populations.

From the obtained results it is apparent that the chamois are monomorphic when it comes to PrP gen.

The sample taken from the population of chamois from the territory of the Republic of Croatia showed the monomorphism of the PrP gene. One genotype and one haplo - type were found, $T_{110}S_{138}I_{142}H_{143}N_{146}R_{154}T_{194}R_{211}Q_{215}Q_{222}S_{240}P_{242}/T_{110}S_{138}I_{142}H_{143}N_{146}R_{154}T_{194}R_{211}Q_{215}Q_{222}S_{240}P_{242}$. An examination of the literature shows that other polymorphisms of the PrP gene in chamois are not registered with the scientific community.

References

Griffith, J.S.: Self replication and scrapie. Nature **215**, 1043–1044 (1967)

Hunter, N.: Scrapie, review. Mol. Biotechnol. **93**, 225–234 (1998)

Peletto, S., Perucchini, M., Acin, C., Dalgleish, P.M., Reid, W.H., Rasero, R., Sacchi, P., Stewart, P., Caramelli, M., Ferroglio, E., Bozzetta, E., Meloni, D., Orusa, R., Robetto, S., Gennero, S., Goldman, W., Acutis, P.L.: Genetic variability of the prion gene (PrP) in wild ruminants from Italy and Scotland. J. Vet. Sci. **10**, 115–120 (2009)

Prusiner, S.B.: Novel proteinaceous infectious particles cause scrapie. Science **216**, 136–144 (1982)

Prusiner, S.B.: Prions. Proc. Natl. Acad. Sci. USA **95**, 13363–13383 (1998)

Scientific opinion of the panel on biological hazards on a request from the European commission on genetic TSE resistance in goats. EFSA J. **995**, 1–25 (2009)

DebBurman, S.K., Gregory, J.R., Byron, C., Lindquist, S.: Chaperone – supervised conversion of prion protein to its protease – resistant form. Proc. Natl. Acad. Sci. USA **94**, 13938–13943 (1997)

Hall, T.A.: BioEdit: a user-friendly biological sequence alignment editor and analysis programfor Windows 95/98/NT. Nucl. Acids Symp. Ser. **41**, 95–98 (1999)

Larkin, M.A., Blackshields, G., Brown, N.P., Chenna, R., McGettigan, P.A., McWilliam, H., Valentin, F., Wallace, I.M., Wilm, A., Lopez, R., Thompson, J.D., Gibson, T.J., Higgins, D. G.: Clustal W and Clustal X version 2.0. Bioinformatics **23**, 2947–2948 (2007)

Bottleneck Analysis of Turkish and Algerian Sheep Breeds Using Microsatellite Markers

Nezih Ata[1], Abdelkader Ameur Ameur[2], and Onur Yilmaz[1(✉)]

[1] Department of Animal Science, Faculty of Agriculture, Aydin Adnan Menderes University, 090100 Aydin, Turkey
oyilmaz@adu.edu.tr

[2] Laboratory of Physiopathology and Biochemistry of Nutrition, Department of Biology, University of Tlemcen, Tlemcen, Algeria

Abstract. The present study was performed to reveal genetic diversity and bottleneck of Turkish and Algerian autochthonous sheep breeds using fifteen microsatellites marker recommended by FAO (2011). Animal material for the study was consisted of 180 head sheep raised in different location in Turkey and Algeria. A total of 349 alleles were detected from fifteen markers studied. The mean number of alleles (23.26), observed heterozygosity (0.76) and polymorphic information content (0.89) findings indicated that noticeable genetic variability in sheep population studied. Fourteen out of the sixteen microsatellite markers studied had a positive FIS value. The mean value of FIS was 0.061.The infinite allele model (IAM), two-phase mutation model (TPM) and stepwise mutation model (SMM) in the Bottleneck software were used to check genetic bottleneck. The Lshaped curve obtained from analyze indicates absence of bottleneck in studied sheep population raised in Turkey and Algeria. Consequently, it can be said that these results will help to develop conservation and breeding strategies for the sheep population.

Keywords: Genetic diversity · Genetic resource · Microsatellite

1 Introduction

Indigenous animal breeds that are very well adapted to the ecological and economic conditions of different geographies all over the world, is one of the most important cornerstones of animal breeding. The origin of domestic goats remain ambiguous and controversial issue, but archaeological findings probably indicate that they were domesticated 10 000 years ago in the Fertile Crescent region covering Israel, Turkey, Lebanon, Jordan, and Syria (Zeder and Hesse 2000; Luikart et al. 2001; Taberlet et al. 2011; Amills et al. 2017).

Algeria and Turkey has a rich genetic diversity especially small ruminants due to the originality of climate and vegetation existed. It can be said that the majority of sheep population, which serves multiple purposes, in Turkey as well as Algeria is raised under extensive conditions. The number of animal in Turkey and Algeria are approximately 31.2 and 28 million head, respectively according to statistics provided by Turkish Statistics Agency (TUIK 2018) and FAOSTAT (2017).

M. Brka et al. (Eds.): AgriConf 2019, IFMBE Proceedings 78, pp. 122–131, 2020.
https://doi.org/10.1007/978-3-030-40049-1_16

The change of consumer habits and the increasing population needs caused the domestic breeds to be threatened with extinction and leave their places to culture breeds. Indigenous breeds, which are very well adapted to the area where they are raised, are defined as animals with unique properties, resistant to adverse environmental conditions and diseases, even if their productivity is low. Extinction of native breeds leads to the loss of the original features of these animals that may be needed in the future (Cemal et al. 2013; FAO 2015; Rege and Gibson 2003; Wollny 2003). It is noteworthy that the variation which is very important for biological systems with each passing day is decreasing due to many factors.

Over the last few centuries, the dramatic increase in world populations has had a negative impact on natural habitats, causing some species and breeds to decrease or disappear. Rapid reduction in the population or occurrence of bottlenecks can have a deep impact on the effective population size and preservation of genetic diversity in farm animal as well as other animal population. For this reason, it is essential to unfold the genetic diversity of indigenous breeds for national breeding and conservation studies (Cornuet and Luikart 1996).

The aims of the present study was to describe genetic diversity and potential genetic bottleneck in sheep breeds raised in Algeria and Turkey using microsatellite markers.

2 Materials and Methods

All experimental procedures were achieved according to the Animal Care and Ethics rules in presented article.

2.1 Animal Material and DNA Extraction

Blood samples were obtained from 180 sheep breeds which consist of Barbarine (BB, n = 20), Rembi (RB, n = 20), Berbere (BR, n = 20), Taâdmit (TD, n = 20), D'men (DM, n = 20), and Awassi (AW, n = 20), Karakas (KK, n = 20), Karayaka (KY, n = 20) and Tuj (TU, n = 20) raised in different location in Algeria and Turkey.

Blood samples were collected from Vena jugularis into containing K3EDTA tubes and stored at −20°C until DNA extraction. Salting-out technique, reported by Miller et al. (1988) and Montgomery and Sise (1990), was used for DNA extraction from blood. Subsequently, the quality and quantity of the DNA sample were checked using NanoDrop 2000 (Thermo Scientific, Waltham, MA).

2.2 PCR Procedure and Microsatellites Genotyping

In the present study, fifteen microsatellite markers, recommended by FAO (2011), was used to reveal genetic diversity and bottleneck test. Microsatellites used, grouped as two multiplexes, were amplified with Touchdown PCR method reported by Hecker and Roux (1996) (Table 1).

Table 1. Thermal cycling conditions according to Touchdown PCR

Loci	MG	I. Denat.	II. Denat.	Annealing	Extension	Cycle	Final extension
OarFCB193	1	95 °C	95 °C	63–54 °C	72 °C	40	72 °C
OarFCB304		(5 min)	(40 s)	(40 s)	(60 s)		(10 min)
INRA0023							
OarCP34							
INRA0132							
D5S2							
BM1818							
BM8125	2	95 °C	95 °C	60–50 °C	72 °C	34	72 °C
McM0527		(5 min)	(40 s)	(40 s)	(60 s)		(10 min)
CSRD0247							
OarFCB128							
BM1329							
HSC							
OarJMP29							
MAF214							

Polymerase chain reaction (PCR) were implemented in a total of 20 μl volume including 0.10 μM for each forward and reverse primers, 0.20 mM dNTPs, 2.0 mM MgCl2, 1X PCR buffer, 1 U of Taq DNA polymerase (Applied Biological Materials Inc.) and ~50 ng of genomic DNA. Afterward, PCR fragments was separated with Beckman GeXP fragment analyzer (Beckman Coulter, Inc., USA). GenomeLab™ DNA Size Standard Kit 400 was used to determine fragment size belong to microsatellite markers.

2.3 Statistical Analysis

Genetic variation analysis was revealed using GenAlEx (Peakall and Smouse 2006; Peakall and Smouse 2012) and POPGENE (Yeh et al. 2012) to compute number of alleles per locus (Na), effective number of alleles (Ne), observed heterozygosity (Ho), expected heterozygosity (He), FIS value, known as inbreeding coefficient, from Wright's F-statistics (Weir and Cockerham 1984; Wright 1990) and Hardy–Weinberg equilibrium. CERVUS 3.0.3 (Kalinowski et al. 2007; Marshall et al. 1998) was used to calculate polymorphic information content (PIC) and null allele frequencies.

Bottleneck events were tested with Sign, Standardized differences and Wilcoxon sign–rank tests under the different mutation models such as Infinite Allele Model (IAM), Stepwise Mutation Model (SMM), and Two Phase Model of Mutation (TPM) model in Bottleneck software version 1.2.02 (1 000 simulation) (Piry et al. 1999).

3 Results

A total 349 alleles was observed from fifteen microsatellites used in this study. Computed genetic diversity statistics was given in Table 2.

Table 2. Genetic polymorphism parameters at 15 microsatellite loci

Locus	N	Na	Ne	Ho	He	PIC	F_{IS}*	HWE	F(Null)
OarFCB193	174	21	7.56	0.80	0.87	0.86	0.004	***	0.0330
OarFCB304	172	24	8.46	0.80	0.88	0.87	0.035	***	0.0483
INRA0123	172	22	13.08	0.78	0.92	0.92	0.084	***	0.0826
OarCP34	176	25	10.84	0.78	0.91	0.90	0.029	***	0.0744
INRA0132	171	19	9.00	0.88	0.89	0.88	-0.086	***	0.0032
D5S2	170	17	10.85	0.75	0.91	0.90	0.032	***	0.0928
BM1818	166	28	13.79	0.60	0.93	0.92	0.272	***	0.1204
BM8125	174	21	7.17	0.68	0.86	0.85	0.133	***	0.1166
MCM0527	171	18	7.97	0.74	0.87	0.86	0.088	***	0.0872
CSRD0247	175	27	15.34	0.80	0.93	0.93	0.051	***	0.0786
OarFCB128	177	22	7.19	0.76	0.86	0.85	0.044	***	0.0602
BM1329	174	24	15.81	0.72	0.94	0.93	0.132	***	0.1281
HSC	173	27	20.51	0.83	0.95	0.95	0.036	***	0.0705
OarJMP29	180	29	8.63	0.77	0.88	0.87	0.056	***	0.0765
MAF213	177	25	8.30	0.74	0.88	0.87	0.005	***	0.0841
Mean		**23**	**10.97**	**0.76**	**0.90**	**0.89**	**0.061**		

Na: Number of alleles, Ne: Effective number of alleles, PIC: Polymorphic information content, *: Wright's statistic according to Weir and Cockerham (1984), Ho: Observed heterozygosity, He: Expected heterozygosity, HWE: Significance level of Hardy-Weinberg Equilibrium, F(Null): Null allele frequency, *: $P < 0.05$, **: $P < 0.01$, ***: $P < 0.001$

The highest number of alleles and effective number of alleles were obtained from OarJMP29 (29) and HSC (20.51), respectively. When the PIC values ranging from 0.85 (BM8125 and OarFCB128) to 0.95 (HSC) were examined, it was determined that all microsatellites were highly informative for this studied population. Overall mean of observed heterozygosity value was lower than the expected heterozygosity value. The average of FIS value, also known as inbreeding coefficient and described as Wright' F statistics, was 0.061. All microsatellite studied deviated from the Hardy-Weinberg equilibrium ($P < 0.01$). The null allele frequency values obtained from the studied microsatellite loci were below 20%.

It is necessary to understand the processes that cause decreasing genetic diversity such as genetic bottleneck, genetic drift and inbreeding in farm animal populations. For this reason, genetic bottleneck analysis was performed to investigate whether there was a bottleneck in sheep population raised in Algeria and Turkey. Since the mutation pattern of evolution and microsatellites are not clearly known, the data set obtained was tested with 3 different mutation models, Infinite Allele Model (IAM), Stepwise

Mutation Model (SMM), and Two Phase Model of Mutation (TPM) model reported by Cornuet and Luikart (1996), Luikart and Cornuet (1998) and Piry et al. (1999). Sign, Standardized differences and Wilcoxon sign rank tests were used to predict excess of heterozygosity (Table 3).

Table 3. Test for null hypothesis under three microsatellite evolution models for bottleneck analysis

Breeds	Mutation models	Sign test			Standardized differences test		Wilcoxon rank test (one tail for H excess)
		Hee	He	P	T2	P	P
BB	IAM	9.14	13	0.03197	0.368	0.35648	0.04163
	TPM	9.04	7	0.20682	−1.855	0.03183	0.64014
	SMM	8.84	3	0.00251	−5.571	0.00000	0.99246
RB	IAM	9.18	13	0.03337	1.849	0.03223	0.03650
	TPM	9.23	11	0.25394	−0.052	0.47934	0.08441
	SMM	8.77	5	0.04407	−3.713	0.00010	0.91559
BR	IAM	9.09	12	0.09829	1.632	0.05133	0.00754
	TPM	9.05	8	0.38062	−0.622	0.26704	0.57654
	SMM	8.88	3	0.00235	−3.983	0.00003	0.99246
TD	IAM	9.10	11	0.23283	1.797	0.03615	0.00754
	TPM	8.90	10	0.38311	−0.373	0.35474	0.40198
	SMM	8.82	2	0.00039	−3.865	0.00006	0.99934
DM	IAM	9.16	12	0.10504	2.286	0.01113	0.00084
	TPM	8.96	10	0.39470	0.289	0.38616	0.29974
	SMM	8.92	6	0.10245	−2.590	0.00479	0.92429
AW	IAM	8.98	13	0.02658	1.892	0.02925	0.00754
	TPM	9.09	7	0.19955	0.168	0.43338	0.35986
	SMM	8.91	5	0.03746	−2.861	0.00211	0.98233
KK	IAM	8.97	9	0.60387	1.357	0.08737	0.05350
	TPM	8.91	7	0.22826	−0.785	0.21623	0.82043
	SMM	8.89	5	0.03849	−4.231	0.00001	0.98233
KY	IAM	8.97	12	0.08815	2.967	0.00150	0.00021
	TPM	9.00	11	0.21735	1.457	0.07254	0.05350
	SMM	8.88	8	0.41511	−1.038	0.14970	0.66061
TU	IAM	9.04	9	0.58874	1.239	0.10761	0.12619
	TPM	8.94	7	0.22311	−1.045	0.14791	0.70026
	SMM	8.92	5	0.03718	−4.725	0.00000	0.98492

BB: Barbarine, RB: Rembi, BR: Berbere, TD: Taâdmit, DM: D'men, AW: Awassi, KK: Karakas, KY:Karayaka, TU:Tuj, IAM: The infinite allele model, TPM: Two-phase mutation model, SMM: The step- wise mutation model, Hee: Expected number of loci with heterozygosity excess, He: heterozygosity excess

It is indicated that the probability values were lower than 0.05 for the TPM and SMM except IAM. On the other hand, the probability values for one tail for heterozygosity excess obtained using the Wilcoxon rank test were non-significant in two-phase mutation model and the stepwise mutation model while this value was significant the infinite allele model.

As a second method, a mode-shift graph was obtained using allele frequency classes of 16 microsatellite to identify potential bottlenecks in the studied population (Figs. 1 and 2).

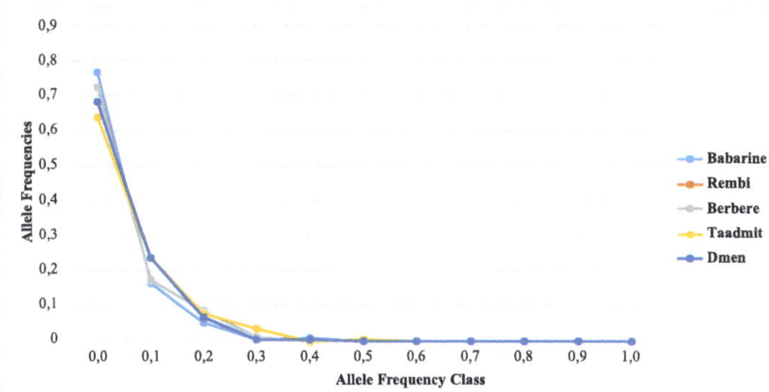

Fig. 1. Mode-shift graph for bottleneck in the Algerian sheep breed

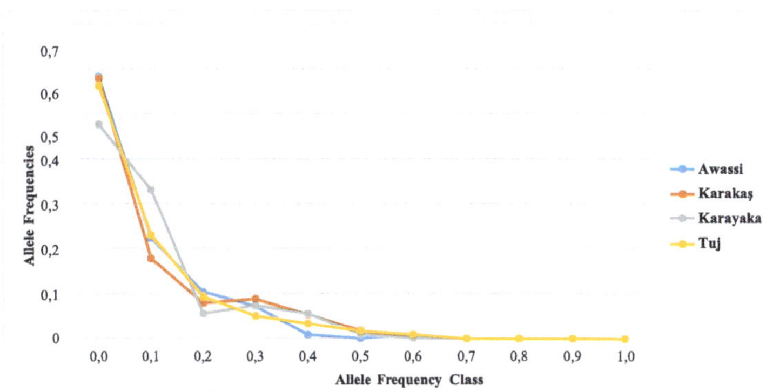

Fig. 2. Mode-shift graph for bottleneck in the Turkish sheep breed

As it can be seen from mode-shift graph, an L-shaped chart consistent with the distribution ranges of the normal frequency class was obtained in the bottleneck test performed.

4 Discussion

Obtained genetic diversity parameters such as mean number of allele and effective allele were considerably higher than those of earlier studies conducted in sheep breeds raised in different countries (Arora and Bhatia 2006; Ben Sassi-Zaidy et al. 2016; Cemal et al. 2013; Guang-Xin et al. 2016; Kdidi et al. 2015; Ocampo et al. 2016; Oner et al. 2014; Yilmaz et al. 2014), while these values and polymorphic information content were similar to those of earlier studies (Abdelkader et al. 2018; Hoda and Marsan 2012; Yılmaz et al. 2015). On the other hand, obtained high heterozygosity and polymorphic information content values supported high genetic variability in sheep breeds studied.

It is expected results that most of the studied loci will deviate from the Hardy-Weinberg equilibrium given that the size of the population is very limited and various breeding activities performed in breeds studied.

F_{IS} values, which is a measure of the deviation of genotypic frequencies from panmixia in populations in terms of heterozygous deficiency or excess, showed that loss of heterozygosity at INRA0132 locus. Similar findings have been expressed in the previous literature conducted in different sheep breeds (Loukovitis et al. 2016; Salamon et al. 2015; Vahidi et al. 2016; Yilmaz et al. 2014; Yılmaz et al. 2015).

Null alleles that is defined as a non-amplifiable allele due to mutations in the PCR binding site, causing only a single allele to peek like a homozygote, thus causing erroneous reading. Observed null allele frequencies for the all microsatellites below the critical value (20%) reported by Dakin and Avise (2004) indicated that these markers studied can be used confidently to identify genetic diversity in these native sheep breeds.

The infinite allele model (IAM) and the stepwise mutation models (SMM) generally give inconsistent results when describing the mutation in microsatellites. Therefore, it is reported that the two-phase mutation model (TPM) is the most useful model to test the heterozygosity excess in the bottleneck tests performed with microsatellites (Dirienzo et al. 1994; Luikart et al. 1998; Piry et al. 1999). On the other hand, it has been reported that the Wilcoxon test, which has high statistical confidence even in bottleneck analysis studies using a limited number of loci (<20), can be used with high confidence in bottleneck studies (Piry et al. 1999).

It can be said that serious demographic bottlenecks have not been experienced in the sheep populations studied given that considering the TPM model of Wilcoxon test results.

The mode shift graph method, which is a qualitative graphical representation of the allele frequency distribution, was first proposed by Luikart et al. (1998). The "L-shaped distribution" of the allele frequency distribution graph is used as a criterion for the bottleneck. If the graph shows the normal distribution (L-shaped), then the mutation-migration balance is concerned. Obtained L-shaped distribution suggests that there is not a genetic bottleneck in the studied populations that is large enough to be considered recently (last 40–80 generations).

5 Conclusion

Consequently, the present study results indicated that genetic diversity in sheep population was significantly high in the gene pool. Our findings revealed that the microsatellite markers used in this study that can be successfully used in genetic diversity and bottleneck studies for these breed. On the other hand, obtained results will help to interpret the genetic structure of indigenous sheep populations and will be of benefit to the efforts for conservation and breeding of these breed. The strong inference that the studied sheep populations have not undergone major bottlenecks is also important for sheep breeders, breeding and conservation programs.

Acknowledgements. We acknowledge Republic of Turkey Ministry of Food, Agriculture and Livestock for supplying animal materials and The Agricultural Biotechnology and Food Safety Application and Research Centre (ADÜ-TARBİYOMER) of Adnan Menderes University for providing laboratory facilities to carry out molecular genetic analysis.

Conflict of interest. We certify that there is no conflict of interest with any financial organization regarding the material discussed in the manuscript.

References

Abdelkader, A.A., Ata, N., Benyoucef, M.T., Djaout, A., Azzi, N., Yilmaz, O., Cemal, I., Gaouar, S.B.S.: New genetic identification and characterisation of 12 Algerian sheep breeds by microsatellite markers. Ital. J. Anim. Sci. **17**(1), 38–48 (2018)

Amills, M., Capote, J., Tosser-Klopp, G.: Goat domestication and breeding: a jigsaw of historical, biological and molecular data with missing pieces. Anim. Genet. **48**, 631–644 (2017)

Arora, R., Bhatia, S.: Genetic diversity of Magra sheep from India using microsatellite analysis. Asian-Australas. J. Anim. Sci. **19**(7), 938–942 (2006)

Ben Sassi-Zaidy, Y., Maretto, F., Charfi-Cheikhrouha, F., Mohamed-Brahmi, A., Cassandro, M.: Contribution of microsatellites markers in the clarification of the origin, genetic risk factors, and implications for conservation of Tunisian native sheep breeds. Genet. Mol. Res. **15**(1), 1–11 (2016)

Cemal, I., Yilmaz, O., Karaca, O., Binbaş, P., Ata, N.: Analysis of genetic diversity in indigenous Çine Çaparı sheep under conservation by microsatellite markers. J. Facu. Vet. Med. Kafkas Univ. **19**, 383–390 (2013)

Cornuet, J.M., Luikart, G.: Description and power analysis of two tests for detecting recent population bottlenecks from allele frequency data. Genetics **144**(4), 2001–2014 (1996)

Dakin, E.E., Avise, J.C.: Microsatellite null alleles in parentage analysis. Heredity **93**, 504–509 (2004)

Dirienzo, A., Peterson, A.C., Garza, J.C., Valdes, A.M., Slatkin, M., Freimer, N.B.: Mutational processes of simple-sequence repeat loci in human-populations. Proc. Natl. Acad. Sci. U.S.A. **91**(8), 3166–3170 (1994)

FAO: Molecular genetic characterization of animal genetic resources. FAO, Rome (2011)

FAO: The State of the World's Animal Genetic Resources for Food and Agriculture. Rome: FAO Commission on Genetic Resources for Food and Agriculture Assessments (2015)

FAOSTAT. http://www.fao.org/faostat/en/#data. Accessed 18 Dec 2017

Guang-Xin, E., Zhong, T., Ma, Y.H., Gao, H.J., He, J.N., Liu, N., Zhao, Y.J., Zhang, J.H., Huang, Y.F.: Conservation genetics in Chinese sheep: diversity of fourteen indigenous sheep (Ovis aries) using microsatellite markers. Ecol. Evol. **6**(3), 810–817 (2016)

Hecker, K.H., Roux, K.H.: High and low annealing temperatures increase both specifity and yield in touchdown and stepdown PCR. Biotechniques **20**, 478–485 (1996)

Hoda, A., Marsan, P.A.: Analysis of Genetic Variation in Animals. InTech Open, UK (2012)

Kalinowski, S.T., Taper, M.L., Marshall, T.C.: Revising how the computer program CERVUS accommodates genotyping error increases success in paternity assignment. Mol. Ecol. **16**, 1099–1106 (2007)

Kdidi, S., Calvo, J.H., Gonzalez-Calvo, L., Ben Sassi, M., Khorchani, T., Yahyaoui, M.H.: Genetic relationship and admixture in four Tunisian sheep breeds revealed by microsatellite markers. Small Rumin. Res. **131**, 64–69 (2015)

Loukovitis, D., Siasiou, A., Mitsopoulos, I., Lymberopoulos, A.G., Laga, V., Chatziplis, D.: Genetic diversity of Greek sheep breeds and transhumant populations utilizing microsatellite markers. Small Rumin. Res. **136**, 238–242 (2016)

Luikart, G., Allendorf, F.W., Cornuet, J.M., Sherwin, W.B.: Distortion of allele frequency distributions provides a test for recent population bottlenecks. J. Hered. **89**(3), 238–247 (1998)

Luikart, G., Cornuet, J.M.: Empirical evaluation of a test for identifying recently bottle- necked populations from allele frequency data. Conserv. Biol. **12**(1), 228–237 (1998)

Luikart, G., Gielly, L., Excoffier, L., Vigne, J.D., Bouvet, J., Taberlet, P.: Multiple maternal origins and weak phylogeographic structure in domestic goats. Proc. Natl. Acad. Sci. U.S.A. **98**, 5927–5932 (2001)

Marshall, T.C., Slate, J., Kruuk, L.E.B., Pemberton, J.M.: Statistical confidence for likelihood-based paternity inference in natural populations. Mol. Ecol. **7**, 639–655 (1998)

Miller, S.A., Dykes, D.D., Polesky, H.F.: A simple salting out procedure for extracting DNA from Human nucleated cells. Nucleic Acids Res. **16**(3), 1215 (1988)

Montgomery, G.W., Sise, J.A.: Extraction of DNA from sheep white blood-cells. NZ J. Agric. Res. **33**(3), 437–441 (1990)

Ocampo, R., Cardona, H., Martinez, R.: Genetic diversity of Colombian sheep by microsat-ellite markers. Chil. J. Agric. Res. **76**(1), 40–47 (2016)

Oner, Y., Ustuner, H., Orman, A., Yilmaz, O., Yilmaz, A.: Genetic diversity of Kivircik sheep breed reared in different regions and its relationship with other sheep breeds in Turkey. Itali. J. Anim. Sci. **13**(3), 588–593 (2014)

Peakall, R., Smouse, P.E.: GenAlEx 6: genetic analysis in Excel. Population genetic software for teaching and research. Mol. Ecol. Notes **6**, 288–295 (2006)

Peakall, R., Smouse, P.E.: GenAlEx 6.5: genetic analysis in Excel. Population genetic software for teaching and research – an update. Bioinformatics **28**, 2537–2539 (2012)

Piry, S., Luikart, G., Cornuet, J.M.: BOTTLENECK: a computer program for detecting recent reductions in the effective population size using allele frequency data. J. Heredity **90**(4), 502–503 (1999)

Rege, J.E.O., Gibson, J.P.: Animal genetic resources and economic development: issues in relation to economic valuation. Ecol. Econ. **45**(3), 319–330 (2003)

Salamon, D., Gutierrez-Gil, B., Simcic, M., Kompan, D., Dzidic, A.: Microsatellite based genetic structure of regional transboundary Istrian sheep breed populations in Croatia and Slovenia. Mljekarstvo/Dairy **65**(1), 39–47 (2015)

Taberlet, P., Coissac, E., Pansu, J., Pompanon, F.: Conservation genetics of cattle, sheep, and goats. C.R. Biol. **334**, 247–254 (2011)

TUIK. http://tuikapp.tuik.gov.tr. Accessed 08 June 2018

Vahidi, S.M.F., Faruque, M.O., Anbaran, M.F., Afraz, F., Mousavi, S.M., Boettcher, P., Joost, S., Han, J.L., Colli, L., Periasamy, K., Negrini, R., Ajmone-Marsan, P.: Multilocus genotypic data reveal high genetic diversity and low population genetic structure of Iranian indigenous sheep. Anim. Genet. **47**(4), 463–470 (2016)

Weir, B.S., Cockerham, C.C.: Estimating F-statistics for the analysis of population-structure. Evolution **38**(6), 1358–1370 (1984)

Wollny, C.B.A.: The need to conserve farm animal genetic resources in Africa: should policy makers be concerned? Ecol. Econ. **45**(3), 341–351 (2003)

Wright, S.: Evolution in mendelian populations (Reprinted from Genetics, Vol 16, Pg 97–159, 1931). Bull. Math. Biol. **52**(1–2), 241–295 (1990)

Yeh, F.C., Yang, R.C., Boyle, T.: POPGENE VERSION 1.31: Microsoft Window-based free ware for population genetic analysis, University of Alberta. Edmonton, Alberta, Canada (2012)

Yilmaz, O., Cemal, İ., Karaca, O.: Genetic diversity in nine native Turkish sheep breeds based on microsatellite analysis. Anim. Genet. **45**(4), 604–608 (2014)

Yılmaz, O., Sezenler, T., Sevim, S., Cemal, İ., Karaca, O., Yaman, Y., Karadağ, O.: Genetic relationships among four Turkish sheep breeds using microsatellites. Kafkas Üniversitesi Veteriner Fakültesi Dergisi **1411**(46), 576–582 (2015)

Zeder, M.A., Hesse, B.: The initial domestication of goats (Capra hircus) in the Zagros mountains 10,000 years ago. Science **287**, 2254–2257 (2000)

Using of Black Soldier Fly (*Hermetia Illucens*) Larvae Meal in Fish Nutrition

Senada Čengić-Džomba(✉), Emir Džomba, Salko Muratović, and Dženan Hadžić

Faculty of Agriculture and Food Sciences, University of Sarajevo, Sarajevo, Bosnia and Herzegovina
s.cengic-dzomba@ppf.unsa.ba

Abstract. Fish consumption is expected to increase in future due to increase in demand for animal products. Fish meal is a mayor ingredient in Salmonidae diets and it considerably increases the cost of fish feed. Replacing fish meal with plant protein was not so successful, primarily due to low biological value of plant protein (composition and ratio of amino acids). Recently, many researchers started to look into the possibility of using insects in fish nutrition. EU Regulation No 813/2017 [1] approved the use of the following seven insects as animal feed: Black soldier fly – BSF (Hermetia illucens), Common housefly (Musca domestica), Banded cricket (Gryllodes sigillatus), Field cricket (Gryllus assimilis), House cricket (Acheta domesticus), Lesser Mealworm (Alphitobius diaperinus) and Yellow Mealworm (Tenebrio molitor). Among them, use of black soldier fly shows the most promising results. BSF contains between 40 and 45% of crude protein and even more than 40% of ether extract - depending on the substrate it was reared on, and stage in the life cycle. Protein from BSF larval meal contains favourable composition of amino acids for fish nutrition. Additionally, fatty acids composition of BSF that depends on the growing substrates can be adjusted to have higher content of ALA, EPA and DHA, and lower content of SFA. The literature data about replacing fish meal with BSF larval meal is controversial. What appears to be the best is to replace up to 50% of fish meal with BSF larval meal in rainbow trout nutrition, as this shows no adverse effect on fish performances, including weight gain rate, specific growth rate, protein efficiency ratio, and protein productive value.

Keywords: Insects · Fish meal replacing · Black soldier fly larval meal

1 Introduction

The accelerated population growth increases the need for animal products, and therefore, for animal feed. According to the FAO estimates [2], the needs for animal feed will increase by 70% before 2050. Since the aquaculture is one of the fastest growing sectors of animal production, within the overall amount of animal feed production, production of fish feed is receiving more and more attention. Quantity of the fish meal produced, which is the most important source of protein for majority of fish species, fails short of meeting the growing market demand. In addition, cost of this feed is not affordable for majority of producers, so combined, these factors impose a real need to

M. Brka et al. (Eds.): AgriConf 2019, IFMBE Proceedings 78, pp. 132–140, 2020.
https://doi.org/10.1007/978-3-030-40049-1_17

find an appropriate replacement for fish meal that would be more affordable. To this end, numerous studies have been conducted, initially focusing on the possibility of replacing the fish meal with protein rich food of plant origin, first of all the soybean meal. The study results showed that the soybean meal cannot be used as adequate replacement for fish meal in diet of carnivorous fish, for two main reasons: presence of antinutritive substances, but also the insufficient quantity of amino acids that contain sulphur. A promising option appears to be the use of insects as an alternative source of nutrients, primarily proteins and fats, in fish nutrition. In their natural habitat, the rainbow trout, being a carnivore, eats a wide range of various aquatic and terrestrial organisms. In addition to smaller fish, fish eggs, crustaceans, snails, worms and plankton, insects also make an important part of their diet – larval and aquatic stages of dragonflies, mayflies, and caddis flies; grasshoppers, ants, and others [3]. Although many different species of insects may be used as feed for rainbow trout, only some of them are suitable for industrial production of insect meal. The studies, which intensified in the latest decades of the 20th century, showed that the black soldier fly, common housefly larvae, termites, silkworms, mealworms and grasshopper are the species that might meet the demands of industrial production.

The EU legislation treats insects and insect products (except live animals) as animal by-products, i.e. products not intended for human use [4, 5]. Restrictions that apply to feeding farm animals (animals raised for human food production and production of feed for animals and other products) also apply to insects, meaning that the processed proteins of insect origin may only be used as feed for pets and wild game, but not in feed for pigs and poultry. The prohibition does not include fats of insect origin [6]. On the other hand, the farmed insects can only be fed by animal-based food and animal origin products that fall into category 3 (fish meal, dicalcium phosphate, and tricalcium phos phate of animal origin, products made of blood of non-ruminants, hydrolysed protein obtained from blood and skin of ruminants, gelatine and collagen obtained from non-ruminants, eggs, milk, colostrum, honey, and rendered fat). Using manure, restaurant waste, meat-bone meal and ruminant-origin protein is not allowed. The regulation EC No 893/2017 [1] approved the using of processed protein originating from farmed insects only in feed of aquatic organisms. The farmed species must not be harmful to health of people, plants or animals, and must not be pathogenic, protected or invasive species. In consideration of the stated requirements, the Regulation precisely identifies seven species the farming of which is allowed: black soldier fly (*Hermetia illucens*), common housefly (*Musca domestica*), banded cricket (*Gryllodes sigillatus*), field cricket (*Gryllus assimilis*), house cricket (*Acheta domesticus*) lesser mealworm (*Alphitobius diaperinus*) and yellow mealworm (*Tenebrio molitor*). In the USA and Canada, use of only one species is allowed – the black soldier fly – in animal feed, and in the USA only for feed for salmonidae fish, while in Canada, the BSF may also be used in aquaculture and poultry feeding [7].

2 Black Soldier Fly

The black soldier fly (*Hermetia illucens*), is a particularly interesting species due to the efficacy with which it converts the organic waste (cow, pig and poultry manure, poultry feed, fruit and vegetable waste, offal and restaurant waste) into nutritiously valuable insect protein that might be used as replacement for the expensive protein rich feed, first of all the fish meal, in diet of certain animal species [8]. In addition to their extraordinary nutritive value, the larvae feeding on manure significantly reduce the content of sodium and phosphorus in the manure, as well as the unpleasant smell at farms, and thus have positive effect on environment [9]. Black soldier fly larvae (BSFL) reared in this way (on the manure) could have a useful role in reducing and recycling nutrients at farms, as the EC Regulation No 893/2017 [1] prohibits their use as animal feed. For feeding fish, the most convenient way is to use dried and ground BSF larvae, full fat and partially defatted.

2.1 Nutrient Composition of Black Soldier Fly Larval (BSFL) Meal

Content of crude protein greatly varies in different insect species, going from 20 to almost 80%: *Analeptes trifasciata*, *Arophalus rusticus* 20.1%; *Melanoplus mexicanus* 77.1% [10]. The average content of crude protein in BSF larvae reared on various substrates ranges between 40 and 45% (Table 1). Newton et al. [11] found the protein content of 42.1%, while Makkar et al. [12] state somewhat higher protein content (56.9%) in the BSF larvae.

Table 1. Some nutrient parameters of BSF larval meal, %

Substrate	Crude protein	Ether extract	Ash	Source
Brewery spent grain	46.7	42.2	5.56	Ruhnke et al. [13]
Chicken feed	41.2	33.6	10	Spranghers et al. [14]
Vegetable waste	39.9	37.1	9.6	Spranghers et al. [14]
Restaurant waste	43.1	38.6	2.7	Spranghers et al. [14]
Chicken feed	40.4	24.2	9.6	Liu et al. [15]
Banana peels	34.9	38		Nyakeri et al. [16]
Faecal sludge	45.4	18.1		Nyakeri et al. [16]
Fruit waste	30.75	40.7		Meneguz et al. [17]

The amino-acid profile of BSFL is characterized by a balanced ratio between essential and non-essential amino-acids. The comparative overview of composition of amino- acids in BSF larvae and fish meal (Table 2) shows almost even presence of essential and non-essential amino acids in the fish meal and in the BSF larvae meal, and this is the main qualitative indicator that places the insect larvae among the high-value feed. In addition, with respect to amino-acid composition, there is an obvious advantage the larvae meal has over the soybean meal; in other words, the content of most of essential amino-acids is significantly higher in BSF larvae meal, except for tryptophan. In contrast to the essential amino-acids, the content of most non-essential amino-acids is higher in the soybean protein.

Table 2. Amino acid profile (g/16 g nitrogen) of the BSFL meal, fish meal and soymeal [12]

Amino acids	Black soldier fly larvae	Fishmeal	Soymeal
Essential			
Methionine	2.1	2.7	1.32
Cystine	0.1	0.8	1.38
Valine	8.2	4.9	4.5
Isoleucine	5.1	4.2	4.16
Leucine	7.9	7.2	7.58
Phenylalanine	5.2	3.9	5.16
Tyrosine	6.9	3.1	3.35
Histidine	3.0	2.4	3.06
Lusine	6.6	7.5	6.18
Threonine	3.7	4.1	3.78
Tryptophan	0.5	1.0	1.36
Non-essential			
Serine	3.1	3.9	5.18
Arginine	5.6	6.2	7.64
Glutamic acid	10.9	12.6	19.92
Aspartic acid	11.0	9.1	14.14
Proline	6.6	4.2	5.99
Glycine	5.7	6.4	4.52
Alanine	7.7	6.3	4.54

In addition to the high protein content, insects are also a rich source of fats, but their content varies significantly (Table 1). The content of fat, as well as the composition of the fatty acids in the BSF larvae, directly depends on the feed of the larvae and the stage in the life cycle of the insect. According to the survey by Liu et al. [15], fat content went up from 4.8% in one-day old larvae, to 28% in larvae that were 14 days old. After the 14th day, the content of fat decreased, and at the stage of early and late pupae it was 8.2% and 7.2% respectively, in order to go up again to 21.6% in post-mortem adults. According to Spranghers et al. [14], fat content in the BSF prepupae fed chicken feed, digestate, vegetable waste and restaurant waste was 336 g kg^{-1}, 218 g kg^{-1}, 371 g kg^{-1} and 386 g kg^{-1} in dry matter respectively. Newton et al. [33] 28% and 34.8% fat per dry weight of BSFL that were fed pig and poultry manure. The same authors state a predominant level of saturated fatty acids that ranged, depending on the substrate, from 648.2 g kg^{-1} to 828.0 g kg^{-1} fatty acids methyl esters. Content of mono and polyunsaturated acids (state MUFA and PUFA) was significantly lower, with monounsaturated fatty acids ranging from 95.4–190.8 g kg^{-1} fatty acids methyl esters, while the content of the n3 PUFA was extremely low ranging from 8.6–23.3 g kg^{-1} of fatty acids methyl esters. The dominant n3 PUFA was the ALA, the content of which ranged 7 do 13.7 g kg^{-1}. Oonincx et al. [18] also states a high content of SFA with dominant C12:0 (27–32%), low level of MUFA with dominant C18:1n9 (23–32%), and significantly higher content of C18:2n6 (2.1–6.6%) compared to the content of C18:3n3 (0.0–0.5%) in larvae fed meals of varying content of protein and fats. Low content of n3 fatty acids, particularly EPA and DHA, may be a restricting factor when it comes to feeding certain species of salt water fish with BSFL due to their inability to convert the ALA into these two acids. As opposed to the rainbow trout, where this conversion is possible, for majority of salt water fish EPA and DHA are essential fatty acids and as such must be taken in with feed. The research done by St-Hilaire et al. [19] suggests that it is possible to modify the amino-acid composition by adjusting the larvae feed. Results of an experiment that tested the content of fatty acids in BSFL fed cow manure and meals with varying shares of cow manure and fish offal show that the increase of fish offal in the meal significantly increases content of the EPA and DHA in the larvae. In larvae fed cow manure alone, the EPA content was 0.03% of total lipid on DM basis, while in the larvae fed meal where the ratio between cow manure and fish offal was 50:50%, the EPA content increased significantly ($P = 0.034$), while the DHA content went up from 0.006 to 0.59% ($P = 0,019$) of total lipid. Similar were the results obtained by Oonincx et al. [20], who tested the effect of adding varying quantities (0%, 1%, 2% and 4%) of linseed oil to the meals on the content of fatty acids in the larvae. The results of this experiment show that the increased consumption of linseed oil results with significant decrease of the SFA content (from 1.1% to 0.8%), and simultaneous significant increase of the ALA content (from 0.5% to 9.7%).

Jia and Hing [21] state the content of ash in live and sun-dried prepupae at levels 2% and 20.35% respectively. Dalle Zotte et al. [22] found – 72.7 g kg^{-1} of ash, 9.7 g kg^{-1} calcium, and 8.3 g kg^{-1} phosphorus in defatted larvae meal. Testing the impact substrate has on the content of different nutrients in prepupae, Spranghers et al. [14] found a significant correlation between content of ash in the substrate and ash in the prepupae ($R^2 = 0.954$; $P = 0.023$). Newton et al. [34] state the following contents of mineral matters in prepupae reared on poultry manure: Ash-14.6%; P-1.51%; Ca-5.00%; Mg-0.39%; K-0.69%. The same authors state mineral content in prepupae reared on pig manure as follows: ash-16.6%; P-0.88%; Ca-5.36%; Mg-0.44%; K-1.16%. Julita et al. [23] tested the nutritive composition of larvae reared on horse manure and sheep manure and found somewhat higher values of mineral component in larvae reared on horse manure (ash-19.03%; calcium-3.03%; and phosphorus-0.82%), while the sheep manure reared larvae had the following values: ash-15.35%; calcium-2.48%; and phosphorus-0.54%.

During the BSF life cycle, the mineral content is not constant, and according to Liu et al. [15], content of most minerals (except sodium, iron and zinc) is higher in prepupal than in larval stage.

According to Schiavone et al. [24], content of chitin in partially defatted BSFL meal was 50 g kg^{-1}, while in highly defatted meal it amounted to 69 g kg^{-1}. Higher content of chitin has adverse effect to digestibility of the BSF meal. Taking into consideration that the chitin content went up with the age of the larvae, in order to achieve better nutritional value, the larvae should preferably be harvested prior to the pre-pupal stage [25].

2.2 Replacing the Fish Meal with the BSF Larvae Meal

The experiment conducted in a certified fish farm in Goettingen [26] tested the possibility of replacing fish meal with black soldier fly (BSF) meal in feed of rainbow trout. The experiment tested three meals, where the percentage of fish meal was replaced with BSF meal (0%, 50% and 75%). In addition to the animal component, the meal included soybean meal, wheat (organically produced), and mineral and vitamin additives. Content of proteins and energy in all three meals was even. Results of this study show that replacing the fish meal with BSF meal in the share of 50% were justified, i.e. had no adverse effect on feed conversion and protein efficiency ratio, while the higher share of BSF (replacing 75% of fish meal) did have adverse effect on all tested production and nutritional parameters (final body weights, feed conversion, weights gain and protein efficiency ratio). Reasons for these lower production parameters when using the higher share of insect meal are linked to the higher content of chitin on one hand, and on the other hand, the increased share of soybean meal.

An experiment conducted in Switzerland [27] tested the impact of replacing fish meal (50%) in the extruded feed for fish using defatted meal obtained from BSF larvae on the production parameters of rainbow trout (*Oncorhynchus mykiss*). Total content of BSF in the meal amounted to 28%. A 7-week long experiment was conducted in an organic certified farm in practical conditions. The results show no significant difference in growth parameters, while some differences were found in some nutritional parameters: protein efficient ratio – lower in BSF meal, protein productive value – significantly lower in BSF treatment, lipid efficiency ratio – significantly higher in BSF treatment, and lipid productive value – significantly higher in the BSF treatment. The meals containing BSF had higher share of saturated fatty acids: 37.4 g/100 g fatty acid methyl esters (FAME), control meal 22.5 g/100 g FAME, and lower content of unsaturated fatty acids (mono and polyunsaturated): BSF- 62.6 g/100 g FAME; control meal 77.5 g/100 g FAME, while the content of amino acids was generally higher in the BSF meals, with the exception of methionine and lysine, the content of which was lower than recommended according NRC [28]. Chemical profile of fish from both treatments at the end of experiment was rather uniform, except for the content of crude protein, which was somewhat lower in case of fish fed the BSF (15.2% vs. 16.2% of fresh matter). Organoleptic evaluation showed no negative impact of adding BSF on the smell and taste of fish fillets.

In an eight-week long experiment, Sealey et al. [29] tested the replacement of 25 and 50% of fish meal with the standard BSF and fish offal enriched BSF prepupae in feed of rainbow trout. The results of this experiment suggest that the feed with standard BSF has significantly reduced the fish growth, but they also suggest that none of the 4 experimental meals had negative impact on sensory parameters.

Replacement of the fish meal with defatted BSFL meal in the diet of rainbow trout has been tested by Gasco et al. [30]. The 78 days long experiment tested replacement of 0, 25, and 50% of fish meal protein with 0, 20, and 40% BSF meal. The initial body mass of the fish was 179 ± 10 g, while the quantity of food was 1.5% of the body mass. Results of this experiment show that BSF had no adverse impact on any of the tested parameters (weight gain (WG), feed conversion rate (FCR), feeding rate (FR), protein efficiency ratio (PER), specific growth rate (SGR), slaughtering performances, post mortem traits and colour parameters).

Xiao et al. [31] tested the replacement of fish meal with the BSF larvae meal in the diet of yellow catfish. The basic meal included fish meal, soybean meal, fish oil and gluten meal. The fish meal protein was replaced with BSF larvae protein in the quantities of 13, 25, 37, 48, 68, 85 and 100%. The greatest weight gain rate (%), specific growth rate (%/day), protein efficiency ratio and protein productive value were obtained in fish fed the meal where 25% of protein was replaced by BSFL meal protein; in other words, the maximal replacement of fish meal protein should not exceed 48%.

Belghit et al. [32] tested the effects of replacing fish meal with partially defatted BSFL meal in the feed of Atlantic salmon (*Salmo salar*) to growth parameters, digestibility and sensory quality of salmon. Fish meal and plant protein feed were used as sources of protein in the ratio 20:80, while in the experimental meals, 33, 66 and 100% of fish meal were replaced with appropriate quantity of BSFL. Fish oil has been added to the feed with lower content of fish meal in order to ensure adequate quantity of polyunsaturated fatty acids. Results of this experiment showed no negative effect of BSFL on the growth parameters and feed intake, but also that complete replacement of fish meal was not acceptable due to negative impact on sensory indicators of fish.

3 Conclusion

Using insects as animals feed is challenging, but also a promising issue in animal nutrition. Nutritive value of insects' meal primarily depends on the substrates. Black soldier fly larval meal could be very interesting source of proteins and fats in fish nutrition. Possibility of replacing fish meal with the BSF larval meal in Salmonidae nutrition depends on the BSF meal composition and its effect on sensory characteristics of fish. Generally, published data suggest great potential of insects generally in animal nutrition. Amino acids profile of insect protein is very appreciated, while fatty acids composition depends on the rearing substrate, so it is possible to manipulate the SFA/PUFA ratio, which is important in Salmonidae nutrition.

Declaration of Conflicting Interests. The authors declare that there is no conflict of interest.

References

1. Commission Regulation (EU) 2017/893: Amending Annexes I and IV to Regulation (EC) No 999/2001 of the European Parliament and of the Council and Annexes X, XIV and XV to Commission Regulation (EU) No 142/2011 as regards the provisions on processed animal protein
2. van Huis, A., van Itterbeeck, J., Klunder, H., Mertens, E., Halloran, A., Muir, G., Vantomme, P.: Edible insects - Future prospects for food and feed security. FAO Forestry Paper 171, Wageningen (2013)
3. United States Department of Agriculture – Natural Resources Conservation Service: Rainbow Trout (*Oncorhynchus Myciss*), Fish and Wildlife Habitat management Leaflet, no. 13, p. 2 (2000)

4. Regulation (EC) No 1069/2009 of the European parliament and of the council: Laying down health rules as regards animal by-products and derived products not intended for human consumption and repealing Regulation (EC) No 1774/2002 (Animal by-products Regulation)
5. Commission Regulation (EU) No 14/2011: Approving non-minor amendments to the specification for a name entered in the register of protected designations of origin and protected geographical indications (Limone di Sorrento (PGI))
6. Regulation (EC) No 999/2001 of the European Parliament and of the Council: Laying down rules for the prevention, control and eradication of certain transmissible spongiform encephalopathies
7. Sogari, G., Amato, M., Biasato, I., Chiesa, S., Gasco, L.: The potential role of insects as feed: a multi-perspective review. Animals **9**, 119 (2019)
8. Sheppard, D.C., Newton, G.L., Thompson, S.A., Savage, S.E.: A value added manure management system using the black soldier fly. Bioresour. Technol. **50**, 275–279 (1994)
9. Newton, G.L., Sheppard, D.C., Burtle, G.: Black Soldier Fly Prepupae: A Compelling Alternative to Fish Meal and Fish Oil. Public Comment on Alternative Feeds for Aquaculture (2008). https://lpelc.org/research-summary-black-soldier-fly-prepupae-a-compelling-alter-native-to-fish-meal-and-fish-oil/last. Accessed 6 June 2019
10. Sanchez-Muros, M.J., Barroso, F.G., Manzano-Afuglario, F.: Insect meal as renewable source of food for animal feeding: a review. J. Clean. Prod. **65**, 16–27 (2014)
11. Newton, G.L., Booram, C.V., Barker, R.W., Hale, O.M.: Dried *Hermetia illucens* larvae meal as a supplement for swine. J. Anim. Sci. **44**, 395–400 (1977)
12. Makkar, H.P.S., Tran, G., Heuzé, V., Ankers, P.: State-of-the-art on use of insects as animal feed. Anim. Feed Sci. Technol. **197**, 133 (2014)
13. Ruhnke, I., Normanta, C., Campbell, D.L.M., Iqbal, Z., Leea, C., Hinch, G.N., Roberts, J.: Impact of on-range choice feeding with black soldier fly larvae (*Hermetia illucens*) on flock performance, egg quality, and range use of free-range laying hens. Anim. Nutr. **4**, 452–460 (2018)
14. Spranghers, T., Ottoboni, M., Klootwijk, C., Ovyn, A., Deboosere, S., De Meulenaer, B., Michiels, J., Eeckhout, M., De Clercqb, P., De Smeta, S.: Nutritional composition of black soldier fly (*Hermetia illucens*) prepupae reared on different organic waste substrates. J. Sci. Food Agric. **97**, 2594–2600 (2017)
15. Liu, X., Chen, X., Wang, H., Yang, Q., Rehman, K., Li, W., Cai, M., Li, Q., Mazza, L., Zhang, J., Yu, Z., Zheng, L.: Dynamic changes of nutrient composition throughout the entire life cycle of black soldier fly. PLoS ONE **12**(8), e0182601 (2017). https://doi.org/10.1371/journal.pone.0182601
16. Nyakery, M.E., Ogola, H., Ayieco, M., Amimo, F.: Valorisation of organic waste material: growth performance of wild black soldier fly larvae (*Hermetia illucens)* reared on different organic wastes. J. Insects Food Feed **3**(3), 193–202 (2017)
17. Meneguz, M., Schiavone, A., Gai, F., Dama, A., Lussiana, C., Renna, M., et al.: Effect of rearing substrate on growth performance, waste reduction efficiency and chemical composition of black solider fly (*Hermetia illucens*) larvae. J. Sci. Food Agric. https://doi.org/10.1002/jsfa.9127 (2018)
18. Oonincx, D.G.A.B., van Broekhoven, S., van Huis, A., van Loon, J.J.A.: Feed conversion, survival and development, and composition of four insect species on diets composed of food by-products. PLoS ONE **10**(12), e0144601 (2015). https://doi.org/10.1371/journal.pone.0144601
19. St-Hilaire, S., Cranfill, K.: Fish offal recycling by the black soldier fly produces a food- stuff high in omega-3 fatty acids. J. World Aquac. Soc. **38**(2), 309–313 (2007)

20. Oonincx, G.A.B., Laurent, S., Veenenbos, M.E., van Loon, J.J.A.: Dietary enrichment of edible insects with omega 3 fatty acids. Insect Sci. **00**, 1–10 (2019)
21. Jia, F.T., Hing, W.N.: Fish feed formulation using black soldier fly pre-pupae from fruit waste. J. Eng. Appl. Sci. **12**(3), 6393–6397 (2017)
22. Dalle Zotte, A., Singh, Y., Michiels, J., Cullere, M.: Black soldier fly (*Hermetia Illucens*) as dietary source for laying quails: live performance, and egg physico-chemical quality. Sens. Profile Storage Stab. Anim. **9**(3), 115 (2019)
23. Julita, U., Suryani, Y., Kinasih, I., Yuliawati, A., Cahyanto, T., Maryeti, Y., Permana, A.D., Fitri, L.L.: Growth performance and nutritional composition of black soldier fly, *Her- metia illucens (L)*, (*Diptera: Stratiomyidae*) reared on horse and sheep manure. In: The 4th International Seminar on Sciences, IOP Conference Series: Earth and Environmental Science, vol. 187, p. 012071. IOP Publishing (2018)
24. Schiavone, A., De Marco, M., Martínez, S., Dabbou, S., Renna, M., Madrid, J., Hernandez, F., Rotolo, L., Costa, P., Gai, F., Gasco, L.: Nutritional value of a partially defatted and a highly defatted black soldier fly larvae (*Hermetia illucens L.*) meal for broiler chickens: apparent nutrient digestibility, apparent metabolizable energy and apparent ileal amino acid digestibility. J. Anim. Sci. Biotechnol. **8**, 51 (2017)
25. Ekman, J.: Production of Fish Feed from Vegetable Waste. Applied Horticultural Research Pty Ltd. Project Number: VG13050. Horticulture Australia Ltd (2014)
26. Stamer, A., Wessels, S., Neidigk, R., Hoerstgen-Schwark, G.: Black soldier fly (*Hermetia illucens*) larvae-meal as an example for a new feed ingredients' class in aquaculture diets. In: Rahmann, G., Aksoy., U. (eds.) Proceedings of the 4th ISOFAR Scientific Conference 'Building Organic Bridges', at the Organic World Congress 2014, Istanbul, Turkey (eprint ID 24223) (2014)
27. Stadtlander, T., Stamer, A., Buser, A., Wohlfahrt, J., Leiber, F., Sandrock, C.: *Hermetia illucens* meal as fish meal replacement for rainbow trout on farm. J. Insects Food Feed **3**(3), 165–175 (2017)
28. National Research Council (NRC): Nutrient Requirements of Fish and Shrimp. National Academies Press, Washington D.C. (2011)
29. Sealey, W.M., Gaylord, T.G., Barrows, F.T., Tomberlin, J.K., McGuire, M.A., Ross, C., St-Hilaire, S.: Sensory analysis of rainbow trout *Oncorhynchus mykiss*, Fed Enriched black soldier fly prepupae *Hermetia Illucens*. J. World Aquac. Soc. **42**(1), 34–45 (2011)
30. Gasco, L., Stas, M., Schiavone, A., Rotolo, L., De Marco, M., Dabbou, S., Renna, M., Malfatto, V., Lussiana, C., Katz, H., Zoccarato, I., Gai, F.: Use of black soldier fly (*Hermetia illucens*) meal in rainbow trout (*Oncorhynchus mykiss*) feeds. In: European Aquaculture Society, Aquaculture Europe 2015. Rotterdam, Netherlands (2015)
31. Xiao, X., Jin, P., Zheng, L., Cai, M., Yu, Z., Yu, J., Zhang, J.: Effects of black soldier fly (*Hermetia illucens*) larvae meal protein as a fishmeal replacement on the growth and immune index of yellow catfish (*Pelteobagrus fulvidraco*). Aquac. Res. **49**(4), 1569–1577 (2018)
32. Belghit, I., Liland, N.S., Gjesdal, P., Biancarosa, I., Menchetti, E., Lic, Y., Waagbøa, R., Krogdahlc, Å., Loc, E.J.: Black soldier fly larvae meal can replace fish meal in diets of sea-water phase Atlantic salmon (*Salmo salar*). Aquaculture **503**, 609–619 (2019)
33. Newton, L., Sheppard, C., Watson, D.W., Burtle, G., Dove, R.: Using the black soldier fly, *Hermetia Illucens*, as a value-added tool for the management of swine manure, pp. 1–17. North Carolina State University, Raleigh (2005)
34. Newton, L., Sheppard, C., Watson, D.W., Dove, R.: Using the black soldier fly *Hermetia illucens*, as a value-added tool for the management of swine manure. In: Report for Mike Williams Director of the Animal and Poultry Waste Management Center. North Carolina State University, Raleigh (2005)

Effect of Different Energy: Protein Ratio on Commercial Laying Hens' Performances in the First Half of Laying Period

Emir Džomba(✉), Senada Čengić-Džomba, and Salko Muratović

Faculty of Agriculture and Food Sciences, University of Sarajevo, Sarajevo, Bosnia and Herzegovina
e.dzomba@ppf.unsa.ba

Abstract. An experiment was conducted to evaluate effect of different dietary levels of energy and protein on commercial laying hen's performance. A hundred twenty Lohmann Brown hens in the first half of production period (30 weeks old) were distributed in a completely randomized experimental design with 2 × 3 factorial arrangement. The first factor was dietary energy at levels of 11.7 and 12.5 MJ ME/kg feed and second one was dietary protein at levels of 16%, 17% and 18% of CP achieving 6 treatments with 5 replications of 4 birds each. The experiment lasted for 6 weeks. Hens performances (egg production, weight of egg, egg mass, feed intake, feed conversion ratio and mortality of hens) and egg structure (albumen, yolk and shell proportion, total egg solids) were evaluated. Increased level of protein has positive effect ($P < 0.05$) on egg weight and albumen proportion in egg. Feed intake decreased and yolk proportion increased at higher energy level in dets. Significant dietary ME X CP interaction on feed intake, feed conversion as well as albumen and yolk proportion were found.

Keywords: Energy and protein · Laying hens · Performances · Egg components

1 Introduction

Manipulation with energy concentration in feed is effective way for regulation of feed intake. Leghorn type hens consumed feed in amount to satisfy energy requirements [1]. So, NRC [2] nutrient requirements of laying hens calculated on energy level of 2900 kcal (approximately 12.1 MJ) metabolizable energy (ME) per kg of feed. Thus, amount of feed consumed by commercial laying hens depend od energy concentration. Therefore NRC [2] recommendation for protein content in feed ranging from 12,5% to 18% depending on amount of consumed feed (from 80 to 120 g/hen/day). Many studies [3–5] were performed to investigate influence of dietary energy: protein ratio on performances of laying hens and egg quality. Increasing energy and/or protein cost leading to manipulation with energy and protein content in diets to achieve optimal hens' performances. Egg production is not affected by dietary energy level if it is above requirements of laying hens [6]. On other hand, data of influence of protein level in diets on hens' performances are controversial. Some studies showed positive effect

M. Brka et al. (Eds.): AgriConf 2019, IFMBE Proceedings 78, pp. 141–147, 2020.
https://doi.org/10.1007/978-3-030-40049-1_18

increasing protein level in diets on egg production and egg weight [7], egg quality including dry and wet albumen solids and protein content in albumen and yolk [8] and feed conversion [9]. Novak et al. [8] found lowered feed conversion with increasing dietary CP in laying hens from 15% to 19,9% while Alagawany [10] found non significant effect of increased dietary CP on dry yolk, yolk protein and organic matter in eggs. Moreover, excess of protein in diet causes increased emission N components (uric acid, ammonia, urea) in biosphere [11]. Apart that, some nutritional strategies to overcome heat stress of poultry are based on decreasing CP in diets and supplementation amino acids because of protein has higher heat increment among nutrients. Generally, energy protein ratio in hens' diets depend on many factors including environmental circumstances [12], production cycle, economic feasibility and environmental issues [13]. Aim of this study was to determine effect of different energy and protein ratio on performances of commercial laying hens kept under common production practice.

2 Materials and Methods

2.1 Birds and Housing

A hundred twenty Lohmann Brown hens in the phase of production (30 weeks old) were distributed in 2 × 3 factorial arrangement according to in completely randomized experimental design. The hens housed in caged system (four hens in each cage allowing 510 cm^2 space/hen). Each cage represented an experimental unit with five replications per a treatment. Feed and water supply were ad libitum. Lightening period was 18 h, ambiental temperature varied from 21 to 26 °C and relative humidity was about 70%. The experiment lasted for 6 weeks.

2.2 Experimental Diets

The hens fed ad libitum by pelleted feed. Totally, six different experimental diets designated to contain two different levels energy (11.7 and 12.5 MJ ME/kg) with three different levels of protein (18.05, 17.05 and 16.0% of crude protein) were use in the experiment. Other parameters of experimental diets were similar (Table 1).

2.3 Data Collection

On daily basis egg production (EP) and mortality were recorded. Feed intake (FI) were estimated weekly as difference between initial amount of feed and weight of feed at the end of week. Egg weight (EW) was determinate throughout two consecutive middle days of each week. Total egg mass (EM) produced was calculated as EM = EP x EW. In the middle of the experiment five eggs from each replication were used to determinate the jolk:albumen:shell ratio. Also, egg components were used to anlayse dry matter content. At the end of the experiments body weight were measured in ten hens from each replicate. Feed conversion ratio (FCR) was calculated as ratio between feed intake and egg mass production (FCR = FI/EM).

Table 1. Ingredients and nutritive values of the experimental diets

	11.7 MJ ME			12.5 MJ ME		
	18% CP	17% CP	16% CP	18% CP	17% CP	16% CP
Ingredients, %						
Corn	51.53	53.50	56.12	50.54	52.99	56.05
Soybean meal	31.16	28.00	25.10	33.69	30.47	25.67
Sunflower meal	2.00	3.00	3.00			3.00
Alfalfa, dehy.	1.01	2.00	2.00		1.00	
Sunflower oil	3.55	3.25	2.91	4.99	4.85	4.53
Limestone	8.08	8.08	7.89	8.12	8.10	8.14
Dicalcium phosphate	1.78	1.74	1.94	1.66	1.66	1.67
Salt Lysine	0.30	0.30	0.30	0.30	0.30	0.30
Methionine	0.01	0.02	0.02		0.01	0.02
Vit.min. premix*	0.08	0.10	0.12	0.10	0.12	0.12
	0.50	0.50	0.50	0.50	0.50	0.50
Chemical composition						
ME, MJ/kg	11.73	11.70	11.68	12.55	12.51	12.50
Crude protein, %	18.12	17.10	15.98	18.17	17.00	16.1
Ether extract, %	6.11	5.88	5.56	7.55	7.44	7.14
Crude fiber, %	3.29	3.46	3.42	2.52	2.72	2.85
Ca, %	3.64	3.64	3.60	3.60	3.60	3.60
P, %	0.43	0.42	0.44	0.40	0.40	0.40
Lysine, %	1.01	0.98	0.99	1.02	0.99	0.97
Methionine + Cystine, %	0.69	0.69	0.69	0.69	0.69	0.69

*Contain per kg: Vit. A (E 672) 2,400,000 i.j., Vit. D3 (E 671) 300,000 i.j., Vit. E (E 3a700, DL-α tocoferol acetat) 2,000 mg, Vit. K3 450 mg, Vit B1 400 mg, Vit. B2 800 mg, Vit. B6 (3a831) 400 mg, Vit. B12 2,000 mcg, Vit. C (E 300) 3,000 mg, Niacin 6,000 mg, Pantothenicacid 1,400 mg, Folic acid 100 mg, Biotin 10 mg, Cholin chlorid 100,000 mg, Fe (E 1, $FeSO_4xH_2O$) 12.25 mg, J (E 2, KJ) 145 mg, Cu (E 4, $CuSO_4 \times 5H_2O$) 800 mg, Mn (E 5, MnO) 15,000 mg, Zn (E 6, ZnO) 10,000 mg, Se (E 8, Na_2SeO_3) 30 mg, Antioxidants (BHA, E 320), (EQ, E 324) 10,000 mg.

2.4 Chemical Analysis and Data Processing

Dry matter content in eggs was calculated after drying homogenized albumen and yolk mass at 105 °C. All data were processed by ANOVA following by testing variance homogeneity using the SPSS software for Windows, version 17.0 and differences among treatments were tested by Tukey multiple rang test.

3 Results and Discussion

Neither energy nor protein levels in diets had influence ($P > 0.05$) on egg production (Table 2). Amount of energy and protein consumed by hens varied from 1.28 to 1.37 MJ and 17.58 to 20.36 g, respectively depending on experimental treatments.

Table 2. Effect of different levels of energy and proteins on laying hens' performances from 28 to 35 weeks

Item		EP, %	EW, g	EM,g	FI, g/day	FCR, g/g	HW, kg
ME, MJ/kg	CP, %						
11.70	16	80.78 ± 0.93	64.19 ± 1.10	53.26 ± 1.02	110.06 ± 0.74	2.07 ± 0.05	1.71 ± 0.01
11.70	17	83.39 ± 0.94	63.59 ± 0.99	52.85 ± 0.98	113.02 ± 0.54	2.16 ± 0.04	1.74 ± 0.01
11.70	18	83.57 ± 0.99	66.07 ± 1.05	54.29 ± 1.08	112.36 ± 0.42	2.08 ± 0.03	1.82 ± 0.01
12.50	16	82.97 ± 0.93	61.02 ± 1.01	51.12 ± 1.08	109.75 ± 0.94	2.16 ± 0.03	1.78 ± 0.01
12.50	17	83.71 ± 0.42	64.88 ± 1.02	54.47 ± 0.98	107.92 ± 0.91	2.00 ± 0.05	1.75 ± 0.01
12.50	18	81.67 ± 1.01	65.12 ± 0.99	53.78 ± 0.99	108.99 ± 0.88	2.05 ± 0.02	1.83 ± 0.01
ME, MJ/kg							
11.7		82.58 ± 0.55	64.16 ± 0.68	53.47 ± 0.68	111.81 ± 0.48^{a}	2.11 ± 0.02	1.77 ± 0.01
12.5		82.78 ± 0.55	63.67 ± 0.68	53.12 ± 0.69	108.91 ± 0.61^{b}	2.07 ± 0.02	1.78 ± 0.01
CP. %							
16		81.87 ± 0.65	62.61 ± 0.73^{a}	52.19 ± 0.73	109.91 ± 0.61	2.12 ± 0.03	1.77 ± 0.01
17		83.55 ± 0.67	64.23 ± 0.84^{ab}	53.66 ± 0.87	110.51 ± 0.74	2.08 ± 0.04	1.75 ± 0.01
18		82.62 ± 0.71	65.94 ± 0.94^{b}	54.03 ± 0.91	110.67 ± 0.77	2.07 ± 0.03	1.81 ± 0.01
ANOVA (P value)							
ME		0.796	0.331	0.724	0.000	0.281	0.604
CP		0.203	0.044	0.263	0.692	0.533	0.102
ME x CP		0.108	0.174	0.286	0.042	0.027	0.916

EP = egg production; EW = egg weight; EM = egg mass; FI = feed intake; FCR = feed conversion ratio; HW = hen weight

a,b values in the same column with different superscripts differ significantly on P < 0.05 level

According to GfE [14] commercial laying hen (1.8 kg body weight) need about 1.33 MJ ME and 19.8 g of crude protein for producing 60 g egg at peak of production. Energy level in experimental diet was some lower but hens was 30-week-old having lower productivity. Similarly, hens consumed enough, even slightly more, protein comparing to requirement. However, it had no influence of egg production and egg mass (Table 2). Similar investigations [5, 15] found positive effect of increased concentrations of protein in diets on egg production. On other hand, Jonquiere et al. [16] did not find significant effect of increased protein on laying hens' performances, even at crude protein concentration of 20% in diets, indicating that dietary protein in surplus have lower digestibility resulting excess nitrogen in manure. It seems, as Khajali et al. [17] reported, that hens can achieve well performance on lower (14–15%) well balanced (in term of amino acids) crude protein concentrations in diets. Contrary to energy, protein level in diets had influence on egg weight ($P < 0.05$). Some researchers [5, 8] also found positive effect of increased dietary protein, especially at lower crude protein concentration. This could be explained by lower intake of amino acids necessary for egg components building. Even in case of methionine, lysine, tryptophan and threonine supplementation egg weight was reduced if totally consumed protein was low [8].

Feed intake decreased for 2.6% by increasing dietary energy concentration ($P < 0.05$) but experimental diets had no effect ($P > 0.05$) on amount of consumed feed

Table 3. Effect of different levels of energy and proteins on egg composition

Item		Egg components, %			Total egg solid
		Albumen	Yolk	Shell	
ME, MJ/kg	CP, %				
11.70	16	62.77 ± 0.67	27.17 ± 0.64	10.06 ± 0.12	25.45 ± 2.80
11.70	17	65.03 ± 0.87	24.79 ± 0.74	10.18 ± 0.36	25.37 ± 0.37
11.70	18	67.87 ± 1.34	22.88 ± 1.11	9.25 ± 0.53	25.29 ± 1.11
12.50	16	63.36 ± 0.55	26.54 ± 0.48	10.10 ± 0.31	24.01 ± 0.43
12.50	17	65.19 ± 0.86	24.41 ± 0.70	10.40 ± 0.36	24.76 ± 1.83
12.50	18	63.56 ± 0.64	26.32 ± 0.59	10.120.15	24.39 ± 0.49
ME, MJ/kg					
11.7		65.22 ± 0.68	24.94 ± 0.56	9.83 ± 0.22	25.37 ± 0.55
12.5		64.04 ± 0.42	25.76 ± 0.38	10.21 ± 0.16	24.39 ± 0.34
CP. %					
16		63.06 ± 0.43[b]	26.86 ± 0.39[a]	10.08 ± 0.16	24.73 ± 0.81
17		65.13 ± 0.59[ab]	24.60 ± 0.49[b]	10.29 ± 0.25	25.07 ± 0.46
18		65.71 ± 0.87[a]	24.60 ± 0.69[b]	9.69 ± 0.28	24.84 ± 0.63
ANOVA (P value)					
ME		0.098	0.167	0.178	0.228
CP		0.009	0.002	0.199	0.939
ME x CP		0.011	0.009	0.435	0.910

[a,b] values in the same column with different superscripts differ significantly on $P < 0.05$ level

for egg mass productions (FCR). Also, increasing level of crude protein in diets had no effect on feed intake and FCR. Interaction of energy and protein in diets caused difference in amount of feed intake and feed conversion ratio ($P < 0.05$). In this study, feed intake has tendency to increase with increasing protein level in diets but this pattern is not so clear and enough pronounced to completely reverse the energy effect. Although protein have quadratic effect on feed intake at constant level of energy [16] according to Pesti [18] feed intake increased with protein level increasing (from 12% to 21%) in diets but decreases by increasing energy content (from 2.6 to 3.2 kcal/kg).

Egg components (albumen, yolk and shell) proportion were not affected ($P > 0.05$) by energy level in diets) but higher level of dietary protein significantly increased albumen percentage in egg. Consequently, yolk proportion decreased because of shell percentage was unchanged. Neither energy nor protein level in diets had no effect on total egg solids (Table 3).

It is worth to highlight significant effect of dietary interaction between energy and protein on albumen proportion in egg ($P = 0.011$). Increasing dietary energy level decreased albumen proportion but the value did not reached significant level ($P = 0.098$). Literature data of dietary energy effect on albumen weight are inconsistent: Whitehead et al. [19] found significant positive effect of supplemental fat on albumen weight but according to Grobas et al. [20] this effect is due to nature of fat (fat composition) but no due to dietary energy level.

On other hand, 18% of crude protein in diets increased albumen proportion comparing to other protein levels and this increasing correspond to increasing of egg weight (Table 2). Positive Effect of dietary protein on increasing of egg albumen is well documented [8, 21]. Gunawardana [5] found higher albumen weight in egg of hens fed by higher dietary protein levels but albumen proportion was unchanged. Higher intake of in diets with 17% and 18% of crude protein led to increased lysine and sulfur amino acids intake what is favorable for albumen weight [22].

4 Conclusion

The results of this study showed that increasing dietary level of energy from 11.5 to 12.5 MJ ME/kg has minor effect on hens' performances, except on feed intake. Increasing protein concentration, in level which is close to recommendation for particular productivity, significantly influence on albumen proportion in egg what is related to higher weight of laid egg. Other investigated parameters were similar. In conclusion, protein intake in higher amount potentially could ensure higher quantity of specific egg compounds such lysozyme to use in industry and medicine.

Declaration of Conflicting Interests. The authors declare that there is no conflict of interest.

References

1. Harms, R.H., Russell, G.B., Harlow, H., Ivey, F.J.: The influence of methionine on commercial laying hens. J. Appl. Poult. Res. **7**, 45–52 (1998)
2. National Research Council (NRC): Nutrient Requirements of Poultry, 9th rev. edn. National Academies Press, Washington D.C. (1994)
3. Ding, Y., Bu, X., Zhang, N., Li, L., Zou, X.: Effects of metabolizable energy and crude protein levels on laying performance, egg quality and serum biochemical indices of Fengda-1 layers. Anim. Nutr. **2**, 93–98 (2016)
4. Hassan, Md.R., Choe, H.S., Jeong, Y.D., Hwangbo, J., Ryu, K.S.: Effect of dietary energy and protein on the performance, egg quality, bone mineral density, blood properties and yolk fatty acid composition of organic laying hens. Ital. J. Anim. Sci. **12**(1), e10 (2013). https://doi.org/10.4081/ijas.2013.e10
5. Gunawardana, P., Roland, Sr., D.A., Bryant, M.M.: Effect of energy and protein on performance, egg components, egg solids, egg quality, and profits in molted Hy-Line W-36 hens. J. Appl. Poult. Res. **17**, 432–439 (2008). https://doi.org/10.3382/japr.2007-00085
6. Leeson, S., Summers, J.D., Caston, L.J.: Response of layers to low nutrient density diets. J. Appl. Poult. Res. **10**, 46–52 (2001)
7. Calderon, V.M., Jensen, L.S.: The requirement for sulfur amino acid by laying hens as influenced by the protein concentration. Poult. Sci. **69**, 934–944 (1990)
8. Novak, C., Yakout, H.M., Scheideler, S.E.: The effect of dietary protein level and total sulfur amino acid: lysine ratio on egg production parameters and egg yield in Hy-Line W-98 hens. Poult. Sci. **85**, 2195–2206 (2006)

9. Bunchasak, C., Poosuwan, K., Nukraew, R., Markvichitr, K., Choothesa, A.: Effect of dietary protein on egg production and immunity responses of laying hens during peak production period. Int. J. Poult. Sci. **4**, 701–708 (2005)
10. Alagawany, M.: Protein and total sulphur amino acids relationship in feeding laying hens. Ph.D. thesis, Agriculture Faculty, Zagazig University, Zagazig, Egypt (2012)
11. Ferket, P.R., van Heugten, E., van Kempen, T.A.T.G., Angel, R.: Nutritional strategies to reduce environmental emissions from nonruminants. J. Anim. Sci. **80**, E168–E182 (2002)
12. Dias, A.N., Bueno, C.F.D., Couto, F.A.P., Rodrigues, P.A., Nogueira, W.C.L., Faria Filho, D.E.: Crude protein and metabolizable energy levels for layers reared in hot climates. Braz. J. Poult. Sci. **14**(3), 159–232 (2012)
13. Koreleski, J., Swiatkiewicz, S.: Laying performance and nitrogen balance in hens fed organic diets with different energy and methionine levels. J. Anim. Feed Sci. **18**, 305–312 (2009)
14. Ausschuß für Bedarfsnormen der Gesellschaft für Ernährungsphysiologie: Empfehlungen zur Energie- und Nährstoffversorgung der Legehennen und Broiler. DLG - Verlag, Frankfurt am Main (1999). 145 p.
15. Keshavarz, K.: Further investigations on the effect of dietary manipulations of nutrients on early egg weight. Poult. Sci. **74**, 62–74 (1995)
16. Junqueira, O.M., de Laurentiz, A.C., da Silva Filardi, R., Rodrigues, E.A., Casartelli, E.M.: Effects of energy and protein levels on egg quality and performance of laying hens at early second production cycle. J. Appl. Poult. Res. **15**, 110–115 (2006)
17. Khajali, F., Khoshouie, E.A., Dehkordi, S.K., Hematian, M.: Production performance and egg quality of Hy-line W36 laying hens fed reduced-protein diets at a constant total sulphur amino acid: lysine ratio. J. Appl. Poult. Res. **17**, 390–397 (2008). https://doi.org/10.3382/japr.2008-0000215
18. Pesti, G.M.: Response surface approach to studying the protein and energy requirements of laying hens. Poult. Sci. **70**, 103–114 (1991)
19. Whitehead, C.C., Bowman, A.S., Griffin, H.D.: The effects of dietary fat and bird age on the weights of eggs and egg components in the laying hen. Br. Poult. Sci. **32**, 565–574 (1991)
20. Grobas, S., Mendez, J., de Blas, C., Mateos, G.G.: Influence of dietary energy, supplemental fat and linoleic acid concentration on performance of laying hens at two ages. Br. Poult. Sci. **40**, 681–687 (1999)
21. Rabello, C.B.V., Pinto, A.L., Silva, E.P., Lima, S.B.P.: Níveis de óleo de soja na dieta de poedeiras comerciais criadas em região de alta temperatura. Revista Brasileira de Ciências Agrárias **2**(2), 174–182 (2007)
22. Novak, C., Yakout, H., Scheideler, S.: The combined effect of dietary lysine and total sulfur amino acid level on egg production parameters and egg components in Dekalb Delta laying hens. Poult. Sci. **83**, 977–984 (2004)

The Effect of Dietary Supplementation of Rumen Protected Glucose on Metabolic Parameters and Milk Quality in Dairy Cows

Stipo Benak[1], Mislav Đidara[2], Vesna Gantner[2](✉), and Marcela Šperanda[2]

[1] Belje Plus- a Feed Factory, Darda, Croatia
[2] Faculty of Agrobiotechnical Sciences Osijek, Josip Juraj Strossmayer University of Osijek, Osijek, Croatia
vgantner@fazos.hr

Abstract. Blood glucose concentrations in high productive dairy cows is a master regulator of hormones and metabolites which are in relation with milk production, reproductive performances and health. Low blood glucose postpartum is caused by inadequate glucose entry rate relative to whole-body demand as opposed to the alternative possibility that postpartum cows have a lower regulatory set point for blood glucose. The aim of this research was to evaluate the addition of rumen protected glucose on metabolic and milk parameters in dairy cows at the peak of lactation. Twenty-four dairy cows average 52.83 (SD = 28.94) days (control) and 54.58 (SD = 27.78) days in milk (experimental) were used for the dietary trial with rumen protected glucose addition in concentration of 200 g/day/cow. Biochemical parameters glucose, total protein, albumin, urea, insulin-like growth factor-1 (IGF-1), non-esterified fatty acids (NEFA), β-hydroxybutyrate, (BHB) and alkaline phosphatase (ALP) were measured at the beginning of the trial and 21st day after dietary treatment. Parameters for milking control were daily milk yield (DMY), milk fat content (MFC), milk protein content (MPC), urea in milk, lactose in milk, fat/protein index and somatic cell count (SCC). No significant differences of biochemical parameters and milk performances between control group and the experimental group were found. Feeding bypass glucose showed 57% reduction in the number of milk somatic cells in the milk of multiparty cows at the end of the experiment. Dietary addition of rumen protected glucose in dosage of 200 g/day/cow in the period of lactation peak showed no effect on milk production and metabolic profile.

Keywords: Rumen protected glucose · Biochemical parameters · Milk performances · Dairy cow

1 Introduction

Milk production in dairy cows' demands nutrients in high quantities. One of the limiting nutrient for milk production is a glucose as a component for lactose synthesis. Due to the rumen microbial degradation of glucose sources from feed, high producing dairy cows are in the constant hypoglycaemic state (Drackley et al. 2001). Low blood glucose postpartum is caused by inadequate glucose entry rate relative to whole-body

M. Brka et al. (Eds.): AgriConf 2019, IFMBE Proceedings 78, pp. 148–154, 2020.
https://doi.org/10.1007/978-3-030-40049-1_19

demand as opposed to the alternative possibility that postpartum cows have a lower regulatory set point for blood glucose (Lucy et al. 2013). Although gluconeogenesis provides glucose for the milk synthesis, additional direct source of glucose would give relief to the metabolic pathways. Encapsulation of glucose sources by fatty acids protects glucose from rumen degradation since microbial population negligible digest fat. In this way glucose is absorbed in the small intestines and becomes directly available for milk synthesis and other metabolic processes. This is especially important for weeks 3 to 6 postpartum during which milk production is extensive and the demand for glucose exceeds the capacity for gluconeogenesis (Herdt 2000). During this period due to the high fat mobilisation, the lack of glucose and the pore adaptive responses can lead to a hyperketonaemia which has detrimental effect on milk production. All of these marks the importance for an alternative source of glucose, and the bypass products as a feed additives have potential to do this (Sauls et al. 2018). The aim of this research was to evaluate the addition of rumen protected glucose on metabolic and milk parameters in dairy cows at the peak of lactation.

2 Materials and Methods

The trial was conducted on the dairy farm Zeleno polje, Croatia (lat.: 45°46′12″N, long.: 18°30′16″E), from 27th of February until 19th March 2019 on two groups of high productive Holstein-Friesian dairy cows (N = 24). Animals were chosen according to parity (12 were from the first parity, 12 multiparous). Cows were in average 52.83 (SD = 28.94) days in control and 54.58 (SD = 27.78) days in the experimental group. Cows were fed with the total mixed ratio (Table 2). In the experimental group was added 200 g/day/cow by-pass glucose (100 g glucose protected by 100 g fatty acids) and therefore by-pass fat was reduced for the energy balance in the control groups. Feeding was twice a day, milking control was carried out before the trial and after 21 days of the trial. Parameters for milking control were daily milk yield (DMY), milk fat content (MFC), milk protein content (MPC), urea in milk, lactose in milk, fat/protein index and somatic cell count (SCC). The blood samples were taken before the trial and 21st day of the trial in the Vacutainer tube with sodium fluorid as anticoagulants for glucose and with lithium heparin for other parameters: total protein, albumin, urea, insulin-like growth factor-1 (IGF-1), non-esterified fatty acids (NEFA), β-hydroxybutyrate, (BHB) and alkaline phosphatase (ALP). Analysis were done by automatic biochemical analyzer AU 400 (Beckman Coulter, Germany). In accordance to daily milk yield cows were divided into two classes: high productive (DMY > 40 kg), or low productive (DMY $\leq$ 40 kg). The variation in analysed traits (biochemical parameters in cows' blood and daily milk traits) due to treatment was tested by least square analyses of variance for using the PROC GLM procedure in SAS (SAS Institute Inc., 2000). Following statistical model was used:

$$THI = 1.8 \times Ta - (1 - RH) \times (Ta - 14.3) + 32 \quad (1)$$

Where Ta is average temperature in degrees Celsius and RH is relative humidity as a fraction of the unit.

$$y_{ijklmn} = \mu + b_1(d_i/305) + b_2(d_i/305)^2 + b_3 \ln(305/d_i) + b_4 \ln^2(305/d_i)^2 + S_j + A_k + T_l + e_{ijklm} \quad (2)$$

Where

yijklm = estimated daily milk trait (yield or contents);
μ = intercept;
b1, b2, b3, b4 = regression coefficients;
di = days in milk (i = 6 to 305 day);
S_j = fixed effect of calving season class j (j = 1/2005 to 12/2012);
Ak = fixed effect of age at first calving class k (k = 21 to 36 month),
Tl = fixed effect of THI class (l = 0 (normal condition – values under the given threshold) or l = 1 (heat stress condition – values equal and above the given threshold)),
eijklm = residual.

Table 1. Composition and chemical composition of diets

Ingredients, %	Control (% DM)	Experiment (% DM)
Alfalfa hay	7.6%	7.5%
Wheat straw	0.6%	0.6%
Alfalfa silage	11.3%	11.3%
Corn silage	31.2%	31.0%
Beat pulp silage	4.6%	4.6%
Corn meal	19.0%	18.9%
Energizer RP 10	1.9%	1.5%
Molasses	4.6%	4.5%
Mineral mix mmp2	0.4%	0.4%
Protein concentrate mix GJKM 40%	18.9%	18.8%
Bypass glucose		0.8%
Dry matter, %)	52.2%	52.3%
Crude protein (CP), % DM	17.10	17.10
Soluble protein, % CP	34.6	34.60
ADF, % DM	17.75	17.68
aNDFom, % DM	28.55	28.40
peNDF, % DM	20.63	20.55
EE, % DM	5.36	5.33
ME, MJ/kg	10.96	10.86

3 Results and Discussion

The Statistical model showed no significant differences of biochemical parameters between control group and the one fed with bypass glucose (Table 2). Established blood glucose levels were almost identical in both groups. Lower blood protein level was recorded in group fed with bypass glucose and the albumin was a bit higher in that group. Concentration of triglycerides did not differ between groups. An almost 7% increase in cholesterol concentration was established in experimental group fed with bypass glucose. Concentration of blood urea was also 5% higher in experimental group. In contrast to that NEFA levels were 33% lower in group fed with bypass glucose. Levels of BHB showed almost no difference between control and experimental group. Concentration of IGF-1 was merely 3% higher in experimental group, unlike the ALP activity with 22% higher values in the experimental group (Table 1).

Table 2. Biochemical parameters in blood of dairy cows fed bypass glucose

Parameter	Treatment (Lsmeans)	
	Control	Experiment
Glucose, mmol/L	3.537	3.534
Protein, g/L	84.267	82.592
Albumin, g/L	33.150	34.175
Triglycerides, mmol/L	0.131	0.130
Cholesterol, mmol/L	6.171	6.589
Urea, mmol/L	4.293	4.516
NEFA, mmol/L	0.327	0.219
BHB, mmol/L	0.530	0.524
IGF-1, ng/mL	17.506	17.998
ALP, U/L	58.250	70.892

Similar results were reported by Yan et al. (2016) who found significantly decrease of NEFA and BHB concentration in cows fed with 300 and 400 g glucose addition. Insulin-like growing factor has a critical endocrine role controlling nutrient metabolism in dairy cattle. Furthermore, Piechotta et al. (2014) reported that antepartal IgF-1 concentrations indicate differences in the metabolic adaptive capacity.

No significant differences of milk quality parameters between groups was established by applied statistical model (Table 3). Content of milk fat was nominally higher in control group and the milk protein in the experimental group. Lactose content was almost identical in both groups. Milk fat to milk protein ratio was higher in control group and so was the milk urea concentration with 10% higher value in control group.

Table 3. Milk quality parameters of dairy cows fed bypass glucose

Parameter	Treatment (Lsmeans)	
	Control	Experiment
Daily milk yield, L	42.6	39.7
Milk fat, %	4.27	4.19
Milk protein, %	2.91	3.09
Lactose, %	4.65	4.68
Milk fat/protein ratio	1.48	1.36
Milk urea, mg/dL	23.7	21.5

The results are in relation to Yan et al. (2016) who reported non-significant increasing of milk yield in the experimental group with 200 g glucose addition. In our research there is the tendency of increasing milk protein content.

Statistical model showed no significant differences of biochemical parameters between the beginning and the end of the experiment when taken into account fixed effects of treatment and parity od dairy cows (Table 4). Most profound difference was recorded for the ALP activity in blood of experimental group multiparous cows with more than 50% increase in activity after 21 days of treatment. Cholesterol levels in that group also increased by 26%. Contrary to that concentration of urea in that group decreased by 19% and NEFA by more than 70%. At the same time multiparous cows of the control group showed only 36% decrease of the NEFA level. IGF-1 level recorded in that group of cows was almost 10% lower at the end of the experiment, while in the experimental group of multiparous cows was more than 5% higher.

Table 4. Difference of biochemical parameters in blood of dairy cows fed bypass glucose between day 0 and day 21 of experiment

	GLUC, mmol/L	ALP, U/L	PROT, g/L	ALB, g/L	CHOL, mmol/L	UREA, mmol/L	NEFA, mmol/L	BHB, mmol/L	IGF-1, ng/mL
	mean	mean	mean	mean	mean	mean	mean	mean	mean
C, heifers, day 0	3.11	109.43	86.78	35.75	5.52	5.73	0.26	0.76	18.91
C, heifers, day 21	3.48	54.97	84.68	33.33	6.21	4.12	0.21	0.54	16.67
diff. 0–21	11,90%	−49,78%	−2,42%	−6,77%	12,50%	−27,92%	−19,23%	−28,95%	−11,85%
E, heifers, day 0	3.48	82.58	84.65	33.85	5.38	5.25	0.32	0.52	18.65
E, heifers, day 21	3.53	71.38	82.35	33.42	6.47	4.26	0.22	0.45	16.72
diff. 0–21	1,44%	−13,56%	−2,72%	−1,27%	20,26%	−18,86%	−31,25%	−13,46%	−10,35%
C, cows, day 0	3.27	55.48	86.03	35.35	5.58	5.65	0.71	0.77	20.25
C, cows, day 21	3.60	61.53	83.85	32.97	6.13	4.46	0.44	0.52	18.34
diff. 0–21	9,79%	10,90%	−2,53%	−6,73	10,04	−21,06	−36,62	−32,47	−9,43
E, cows, day 0	3.37	46.57	86.90	34.67	5.32	5.89	0.72	0.59	18.32
E, cows, day 21	3.54	70.40	82.83	34.93	6.71	4.77	0.22	0.60	19.26
diff. 0–21	5,04%	51,17%	−4,68%	0,78%	26,13%	−19,02%	−70,83%	1,69%	5,13%

C = control group; E = experimental bypass glucose group; GLUC = glucose; ALP = alkaline phos phatase; PROT = total proteins; ALB = albumin; CHOL = cholesterol; NEFA = non-esterified fatty acids; BHB = beta-hydroxy butyric acid; IGF-1 = insulin-like growth factor-1

In this investigation glucose concentrations did not differ between primiparous and multiparous cows, independent of the treatment which in opposite to Ruoff et al. (2017) who found grate risk of ketosis with severe hypoglycaemia in multiparous cows within

6 wk of lactation. They reported hypoglycaemia in multiparous cows in wk 1 to 2, which was not situation in our experimental model. IGF-1 concentration is earlier biomarkers of inadequate metabolic adaptation to the high energy demands (Piechotta et al. 2015). Lower IGF-1 concentration antepartum lead to higher risk of ketosis. In this investigation the level of IGF-1 did not differ according to treatment and there was a mild rise in multiparous cows (Table 4), which may be because of greater risk of negative energy balance reported my Rouff et al. (2017).

Statistical model showed no significant differences of milk quality parameters between the beginning and the end of the experiment when taken into account fixed effects of treatment and parity od dairy cows (Table 5).

Table 5. Difference of milk quality parameters of dairy cows fed bypass glucose between day 0 and day 21 of experiment

	DMY kg/d	Fat, %	Prot. %	Lact. %	SCC × 1000	F/P	Urea mg/dL
	mean	mean	mean	mean	mean	mean	mean
C, heifers, day 0	31.07	3.67	2.91	4.72	237.17	1.26	21.33
C, heifers, day 21	33.66	4.23	2.88	4.64	257.80	1.47	22.62
diff. 0–21	8,34%	15,26%	1,03%	1,69%	8,70%	16,67%	6,05%
E, heifers, day 0	32.90	3.70	3.02	4.73	333.67	1.23	24.33
E, heifers, day 21	33.50	3.83	3.09	4.68	612.67	1.25	20.13
diff. 0–21	1,82%	3,51%	1,99%	−1,06%	83,62%	1,63%	−17,26%
C, cows, day 0	46.63	3.64	3.20	4.61	252.50	1.11	21.83
C, cows, day 21	50.90	4.32	2.93	4.67	207.00	1.49	24.74
diff. 0–21	9,16%	18,68%	−8,13%	1,30%	−18,02%	33,33%	13,33%
E, cows, day 0	46.40	4.59	3.19	4.62	754.67	1.44	24.33
E, cows, day 21	47.14	4.61	3.10	4.67	321.60	1.49	23.18
diff. 0–21	1,59%	0,44%	−2,82%	1,08%	−57,39%	3,47%	−4,73%

C = control group; E = experimental bypass glucose group; DIM = days in milk; DMY = daily milk yield; Prot. = milk protein; Lact. = lactose; SCC = somatic cell count; F/P = milk fat protein ratio

Feeding bypass glucose showed 57% reduction in the number of milk somatic cells in the milk of multiparty cows at the end of the experiment. At the same time somatic cell count increased in the milk of the heifers by 83%. Contrary to experimental group cows, both heifers (16%) and multiparty cows (33%) of control group showed increase of milk fat to protein ratio at the end of the experiment compared to the beginning. This was mostly because of the fat content increase in both control heifers and multiparty cows group by 15% and 18% respectively.

4 Conclusions

Dietary addition of rumen protected glucose in dosage of 200 g/day/cow in the period of lactation peak showed no effect on milk production and metabolic profile. Better effect could be seen if rumen protected glucose was given from the beginning of the lactation and/or in the higher dosages.

References

Drackley, J.K., Overton, T.R., Douglas, G.N.: Adaptations of glucose and long-chain fatty acid metabolism in liver of dairy cows during the periparturient period. J. Dairy Sci. **84**(Suppl.), E100–E112 (2001)

Herdt, T.H.: Ruminant adaptation to negative energy balance: influences on the etiology of ketosis and fatty liver. Vet. Clin. North Am. Food Anim. Pract. **16**, 215–230 (2000)

Lucy, M.C., Escalante, R.C., Keisler, D.H., Lamberson, W.R., Mathew, D.J.: Glucose infusion into early postpartum cows defines an upper physiological set point for blood glucose and causes rapid and reversible changes in blood hormones and metabolites. J. Dairy Sci. **96**, 5762–5768 (2013)

Piechotta, M., Holzhausen, L., Araujo, M.G., Heppelmann, M., Sipka, A., Pfarrer, C., Schuberth, H.J., Bollwein, H.: Antepartal insulin-like growth factor concentrations indicating differences in the metabolic adaptive capacity of dairy cows. J. Vet. Sci. **15**, 343–352 (2014)

Piechotta, M., Mysegades, W., Ligges, U., Lilienthal, J., Hoeflich, A., Miyamoto, A., Bollwein, H.: Antepartal insulin-like growth factor 1 and insulin-like growth factor binding protein 2 concentrations are indicative of ketosis in dairy cows. J. Dairy Sci. **98**(5), 3100–3109 (2015)

Ruoff, J., Borchardt, S., Heuwieser, W.: Short communication: associations between blood glucose concentration, onset of hyperketonemia, and milk production in early lactation dairy cows. J. Dairy Sci. **100**(7), 5462–5467 (2017)

SAS Institute Inc.: SAS User's Guide, version 8.2 edition. SAS Institute Inc., Cary (2000)

Sauls, J.A., Banuelos, S., Atanasov, B., Bradford, B., Stevenson, J.S.: Physiologic responses to feeding rumen-protected glucose to lactating dairy cows. Kansas Agric. Exp. Stat. Res. Rep. **4** (10), 3 (2018)

Blood Acid-Base Balance of Istrian Goats

Zvonko Antunović[1(✉)], Željka Klir[1], Vatroslav Šerić[2], Krunoslav Zmaić[1], and Josip Novoselec[1]

[1] Faculty of Agrobiotechnical Sciences Osijek, Josip Juraj Strossmayer University of Osijek, V. Preloga 1, 31000 Osijek, Croatia
zantunovic@fazos.hr
[2] Clinical Hospital Osijek, Josipa Huttlera 4, 31000 Osijek, Croatia

Abstract. The aim of the research was to determine of acid-base balance in the blood of Istrian goats during summer season and compare with the other breeds of goats in Mediterranean environment. The research included 21 Istrian goats average age 4 years. Feeding of goats was with pasture and when returned to stable were fed a grain mixture (0.3 kg/goat/day) and meadow hay (*ad libitum*). In the blood of goats concentrations acid-base indicators (blood pH, BE(efc), blood gas analysis: partial pressure of carbon dioxide (pCO_2), partial pressure of oxygen (pO_2), total pressure of carbon dioxide (tCO_2), as well as electrolytes: $Ca2^+$, Na^+, K^+, Cl^- and $HCO3^-$ - bicarbonate) as well as total protein and albumin concentration were determined. Afterwards, anion gap (AG), strong ion differences (SID), z-values, weak anions and cations ($ATOT_{tp}$ and $ATOT_{alb}$), strong ion gap (SIGtp, SIGalb) and unmeasured anions (XA) were estimated. Indicators of acid-base balance in blood of Istrian goats did not differ compared to reference values, except blood pH value (7.32) which was at lower limit of reference values, and higher values of pCO_2 and tCO_2 (11.49 kPa and 28.8 mmol/L) possibly related to pulmonary ventilation due to higher air temperature during summer season.

Keywords: Acid-base status · Blood · Istrian goat

1 Introduction

Istrian goat is critically endangered Croatian breed. In the past, this breed of goats has grown far more than today. During the 20th century, unfavorable social, economic and sociological events and unfavorable legal regulations for the breeding industry led to almost complete extinction of the goat population in Istria [1]. Thanks to the attachment, the livestock traditions and the enthusiasm of individuals' breeders, a smaller number of Istrian goats have been preserved until today [2]. Today, around 100 breeding goats of this breed are cultivated throughout the growing area of Istria. According to Croatian Farming Agency data, a breeding population of Istrian goats in the Republic of Croatia is very small (36 head: 33 goats and 3 bucks), which is only 0.53% of the total population of all registered breeding goats in Croatia [3]. For its conservation, a lot of work is needed with a better organization of breeders and the work of professional services, as well as the possibility of better animal monitoring,

M. Brka et al. (Eds.): AgriConf 2019, IFMBE Proceedings 78, pp. 155–162, 2020.
https://doi.org/10.1007/978-3-030-40049-1_20

which will contribute to the determination of their blood indicators where we also include acid-base equilibrium indicators [4]. Acid-base balance and electrolytes values are important for monitoring of nutritional status and health [5]. Acid-base balance is vital to maintaining an animal's blood pH and its health state [6]. In recent years, the determination of acid-base balance of blood includes more parameters, like calculation of the anion gap (AG), strong ion difference (SID), z values, base excess and determination of organic acids (lactate, keto acids) and inorganic anions (sulphates, phosphates, etc.) [7]. Values of acid-base balance have not yet been determined for goats and the precise negative charge of albumin and globulins and their dissociation constant is unknown [8]. Gärtner et al. [9] suggested using more information about acid-base balance by calculating strong ion differences (SID), weak anions and cations ($ATOT_{tp}$, $ATOT_{alb}$), strong ion gap (SIG_{tp}, SIG_{alb}) or unmeasured anions (XA).

The aim of the present research is to determine the acid-base balance of Istrian goats and compare with the other breeds of goats in Mediterranean environment.

2 Materials and Methods

2.1 Experimental Goats and Their Management

The research included 21 Istrian goats. Goats were 4 years old, healthy and in good condition. The research was conducted during the summer feeding season (in July, average daily air temperatures was 25 °C; in comparison with average air temperature from 1961 to 2018 which was 21.1 °C) on the family farm in Istria, Croatia. Goats were grazing extensive Mediterranean pastures and they returned to a stall to be fed with hay (*ad libitum*) and with approximately 0.3 kg/day of grain mixture (50% corn, 30% barley, 20% wheat), as well as water, salt and meadow hay *ad libitum*.

Committee for Animal Welfare of Faculty of Agrobiotechnical Sciences Osijek established that the present research is carried out under the legal provisions according to Animal Protection Act (Official gazette No. 133 since 2006, No. 37 since 2013 and No. 125 since 2013). Animal care and conditions of the research followed the recommendations of European Union directive 86/609/EEC [10].

2.2 Blood Collection and Analyses

Blood was collected for analysis in the morning, before goats were grazing. Blood was taken from the jugular vein from sterile vacuum tubes containing Li-heparin and in and analysed by automatic analyser Rapid Lab 348, that works on the base of ion -selective electrodes. Following parameters were determined in plasma: pH, partial pressure of carbon dioxide (pCO_2), partial pressure of oxygen (pO_2), total pressure of carbon dioxide (tCO^2), acutal base excess- Cbase(B), standard base excess - Cbase(Ecf) and electrolytes ($Ca2^+$, Na^+, K^+, Cl^- and $HCO3^-$ - bicarbonate). Serum was taken from the blood sampled in sterile vacuum tubes without anticoagulant. In serum, total proteins and albumin were determined, while globulin content was calculated as the difference between total protein and albumin. Strong ion difference (SID) was calculated following Eq. (1) according to Stewart [11], z value was calculated following Eq. (2) according to Whitehair et al. [12] and anion gap (AG) following Eq. (3) according to Kaneko et al. [13]:

$$\mathrm{SID} = [(\mathrm{Na}^+ + \mathrm{K}^+) - \mathrm{Cl}^-] \tag{1}$$

$$\mathrm{z\,value} = \mathrm{SID}/\mathrm{Na} \tag{2}$$

$$\mathrm{AG} = [(\mathrm{Na}^+ + \mathrm{K}^+) - (\mathrm{Cl}^- + \mathrm{HCO3}^-)] \tag{3}$$

Weak anions and cations (ATOTtp and ATOTalb) were calculated by multiplying the total protein (g/dL) or albumin by 2.9 [14, 15]. Strong ion gap (SIG) calculations were based on the serum concentrations of albumin ($\mathrm{ATOT_{alb}}$, $\mathrm{SIG_{alb}}$) and total proteins ($\mathrm{ATOT_{tp}}$, $\mathrm{SIG_{tp}}$) according to Gärtner et al. [9] following Eqs. (4) and (5). In these calculations pKa = 7.06 was used according to Constable [16]. Following Eq. (6) by Gärtner et al. [9] unmeasured anions (XA) were calculated:

$$\mathrm{SIG_{alb}} = \mathrm{ATOT_{alb}}/\left(1 + 10(\mathrm{pKa} - \mathrm{pH})\right) - \mathrm{AG}, \tag{4}$$

$$\mathrm{SIG_{tb}} = \mathrm{ATOT_{tb}}/\left(1 + 10(\mathrm{pKa} - \mathrm{pH})\right) - \mathrm{AG}, \tag{5}$$

$$\begin{aligned}\mathrm{XA} = {} & \mathrm{cNa}^+ + \mathrm{cK}^+ + \mathrm{cCa}^{2+} + \mathrm{cMg}^{2+} - \mathrm{cCl}^- - \mathrm{cHCO_3} - [0.141 \times \mathrm{albumin} \times (\mathrm{pH} - 5.42)] \\ & - [0.04 \times \mathrm{globulin} \times (\mathrm{pH} - 5.58)] - [\mathrm{phosphate} \times 0.309 \times (\mathrm{pH} - 0.469)].\end{aligned} \tag{6}$$

2.3 Statistical Analysis

The results of indicators of acid-base balance in plasma of goats were performed by MEANS procedure of statistical software SAS 9.4® [17]. Results were expressed as mean, minimum (Min.) and maximum (Max.) values, standard deviation (SD) and standard error of mean (SEM).

3 Results and Discussion

Table 1 presents indicators of acid-base balance in plasma of Istrian goats and Table 2 calculated values of acid-base balance in plasma of Istrian goats

Table 1. Indicators of acid-base balance in plasma of Istrian goats.

Indicators	Mean	SD	Min.	Max.	SEM
Total proteins (g/L)	67.48	6.52	56.7	77.2	1.54
Albumin (g/L)	28.08	2.80	22.8	32.7	0.66
Globulin (g/L)	39.40	5.00	31.90	46.30	1.18
Na^+, mmol/L	147.79	4.43	137.3	154.4	1.04
K^+, mmol/L	4.99	0.49	3.98	5.95	0.12
Ca^{2+}, mmol/L	1.17	0.07	1.01	1.27	0.02
Cl^-, mmol/L	109.44	2.85	105	117	0.67
pH	7.32	0.05	7.16	7.40	0.01
pCO_2, kPa	11.49	19.37	5.40	89.0	4.57
tCO_2, kPa	28.28	2.72	24.10	32.8	0.64
pO_2, kPa	8.04	2.45	4.5	13.8	0.58
HCO_3^-, mmol/L	26.53	2.66	22.8	31.1	0.63
Cbase(B), mmol/L	0.67	2.59	-4.1	5.6	0.61
Cbase(Ecf), mmol/L	1.13	2.63	-3.2	5.9	0.62
sO_2, %	80.4	13.22	57.8	101.5	3.11

SD-standard deviation; SEM-standard error of mean; pCO_2-partial pressure of carbon dioxide, pO_2 -partial pressure of oxygen, HCO_3-bicarbonate, tCO_2 -total pressure of carbon dioxide, Cbase(B)-acutal base excess, Cbase(Ecf)-standard base excess

Table 2. Calculated values of indicators of acid-base balance in plasma of Istrian goats

Indicators	Mean	SD	Min.	Max.	SEM
SID, mmol/L	43.34	6.97	24.65	55.35	1.64
AG, mmol/L	17.69	3.72	1.23	25.55	0.01
z values	0.29	0.04	0.18	0.36	0.01
$ATOT_{tp}$, mmol/L	23.15	2.24	19.45	26.48	0.53
$ATOT_{alb}$, mmol/L	17.47	1.74	14.18	20.34	0.41
SIG_{tp}, mmol/L	−1.54	0.57	−2.87	−0.89	0.13
SIG_{alb}, mmol/L	−1.88	0.75	−3.79	−1.19	0.18
XA	5.44	4.95	−5.18	14.38	1.17

SD-standard deviation; SEM- standard error of mean; SID-strong ion difference, AG-anion gap, $ATOT_{tp}$-weak anions and cations multiplying the total protein by 2.9, $ATOT_{alb}$-weak anions and cations multiplying the albumin by 2.9, SIG_{alb}-strong ion gap based on the serum concentrations of albumin, SIG_{tp}-strong ion gap based on the serum concentrations of total protein, XA-unmeasured anions.

There were no significant differences for most indicators of blood acid-base balance in comparison with the reference values according to Castillo et al. [18], except for pH values and K^+ concentration (Table 3). The established pH value was at the lower limit of the reference values, and more values were determined for pCO_2 and tCO_2. Changes in pH values have direct influence on the rates of reaction and many basic biological processes in the organism. Values below 7.0 and above 7.7 are life threatening [19]. Values of pH in blood depend on the relative concentration of carbonic acid and base bicarbonates in blood [20]. Similarly changes for pH values and pCO_2 observed Redlberger et al. [8] in study with goats aged 6–56 weeks. A small decrease of pH and increase in blood CO2 stimulates pulmonary ventilation, so CO_2 is washed out by an increase in expiration. A decreased extracellular CO2 causes a decrease blood pH [21]. The same author states that at pH decreases the first reaction is the HCO^{3-} buffer resulting in the formation of additional CO_2 and depletion HCO^{3-}. The present study did not determine the occurrence of depletion of HCO^{3-} in the blood of Istrian goats. Therefore, according to the recommendations of Castillo et al. [22], in assessing the indicators of acid-base balance of the blood, moving anion gap should be taken with caution because its values are very dependent on the movement of content bicarbonate, albumin and pCO_2. Higher pCO_2 and tCO_2 resulted from production of this gas during cell metabolism, its transportation to the lungs with blood and excretion with expired air [23]. The same author stated that according to the Henderson-Hasselbalch equation the pCO_2 is a measure of the respiratory component, affecting the changes in the pH of the plasma. In the blood of Murciana-Granadina goats in Spain, Gonzalo et al. [24] found higher pH value (7.36), lower HCO3 (23.0 mmol/L), Na (142.5 mmol/L), K (4.18 mmol/L) and value of AG (10.85 mmol/L) as well as similar Cl concentration (107.50 mmol/L) in comparison with present research. Similar values of Na, Cl and $ATOT_{tp}$, as well as lower content of pCO_2, tCO_2, pO_2, sO_2, AG, K^+, SID z-values as well as higher values of Cbase(Ecf) in high pregnant Alpine goats in Croatia found Antunović et al. [25]. Dobranić et al. [26] in the blood of crossbreed and Boer goats, but in lactation, in Croatia found a slightly lower concentration of K^+ (4.76 and 4.69 mmol/L), similar concentration of Na^+ (147.45 and 147.61 mmol/L) and slightly higher concentration of Cl^- (111.75 and 113.19 mmol/L). In the blood of Alpine goats in Croatia during lactation Antunović et al. [7] determined higher pH values (7.37–7.39) and lower pCO_2 (6.1–6.75 kPa) and pO_2 (3.85–5.39 kPa). Mentioned minor changes of acid-base indicators in the blood of goats showed a significant resistance on the occurrence of metabolic disease, which is consistent with the conclusions by Drackley et al. [27].

In general sense, metabolic activity modifies acid-base status, although it is difficult to estimate the degree of this contribution to the metabolic component of acid-base balance [18].

Table 3. Indicators of acid-base balance in blood of goats in other investigations

Data sources						
Indicator	Kaneko et al. [13] [1]	Castillo et al. [18] [2]	Antunović et al. [28] [3]	Antunović et al. [25] [4]	Antunović et al. [7] [5]	Redlberger et al. [8] [6]
pH	7.32–7.54	7.353	7.36	7.425	7.36–7.38	7.38
Na^+ (mmol/L)	142–155	148.0	145.48	147.60	135.43–152.38	145
K^+ (mmol/L)	3.5–6.7	4.10	3.62	4.30	3.53–4.70	
Cl^-(mmol/L)	99–110.3	108.0	109.39	109.30	101.94–108.41	104.5
AG (mmol/L)	10–20	17.88	14.87	13.63	14.68–23.81	8.9–21.9
SID (mmol/L)	–	40.0	–	42.60	29.94–49.39	44
pCO_2 (mmHg)	–	48.76	5.9 kPa	6.33 kPa	6.1–6.85	6.6
tCO_2 (mmol/L)	25.6–29.6	–	–	30.09	–	
pO_2 (kPa)	–	–	7.76	4.85	3.85–5.39	
HCO_3^- (mmol/L)	–	26.14	24.62	30.32	25.26–26.98	28.5

[1]goats, [2]non-pregnant goats aged 4–5 years, [3]lactating goats, [4]pregnant goats aged 4 years, [5]lactating goat aged 4 years, [6]goat kids aged 6–56 week of life

4 Conclusion

Indicators of acid-base balance in the blood of Istrian goats did not differ compared to reference values, except blood pH value which was at lower limit of reference values, and higher values of pCO_2 and tCO_2. These changes indicated possible relation to pulmonary ventilation due to higher air temperature during summer season.

References

1. Ivanković, A., Mioč, B., Šubara, G., Dovč, P., Širić, I., Ramljak, J.: Genetic characterization of Istrian goat: the key-point for a long-term conservation. Mljekarstvo **68**(2), 85–97 (2018)
2. Mioč, B., Ivanković, A., Širić, I., Držaić, V.: Exterior characteristics of Istrian goat. In: Marić, S., Lončarić, Z. (eds.) Proceedings of 48th Croatian and 8th International Symposium on Agriculture, Dubrovnik, Croatia, pp. 765–769 (2013)
3. CAA-Croatian Agricultural Agency: Annual report 2018. Breeding of sheep, goats and small animals, Križevci, p. 154 (2018)
4. Antunović, Z., Novaković, K., Klir, Ž., Novoselec, J.: Phenotypic characteristics of different age categories of Istrian goat. In: Mioč, B., Širić, I. (eds.) Proceedings of 54th Croatian and 14th International Symposium on Agriculture, Vodice, Croatia, pp. 429–433 (2019)
5. Bouda, J., Núñez-Ochoa, L., Ávula-González, E., Doubek, J., Fuente-Martinez, B., Aguilar-Bobadilla, J.: Blood acid-base and plasma electrolyte values in healthy ostriches: the effect of age and sex. Res. Vet. Sci. **87**, 26–28 (2009)
6. Fetmann, M.J.: Fluid and electrolyte metabolism. In: Thrall, M.A. (ed.) Haematology and Veterinary Clinical Biochemistry, Roca, Sao Paulo, pp. 311–334 (2007)
7. Antunović, Z., Šperanda, M., Novoselec, J., Đidara, M., Mioč, B., Klir, Ž., Samac, D.: Blood metabolic profile and acid-base balance of dairy goats and their kids during lactation. Veterinarski arhiv **87**(1), 43–55 (2017)

8. Redlberger, S., Fisher, S., Köhler, H., Diller, R., Reinhold, P.: Age-dependent physiological dynamics in acid-base balance, electrolytes, and blood metabolites in growing goats. Vet. J. **229**, 45–52 (2017)
9. Gärtner, T., Zoche-Golob, V., Redlberger, S., Reinhold, P., Donat, K.: Acid-base assessment of post-parturient German Holstein dairy cows from jugular venous blood and urine: a comparison of the strong ion approach and traditional blood gas analysis. PLoS ONE **14**(1), e0210948 (2019)
10. Council of the European Union: Council Directive 86/609/EEC of 24 November 1986 on the approximation of laws, regulations and administrative provisions of the Member States regarding the protection of animals used for experimental and other scientific purposes. Off. J. Eur. Communities **358**, 1–28 (1986)
11. Stewart, P.A.: Modern quantitive acid-base chemistry. Can. J. Phys. Pharmac. **61**, 1441–1461 (1983)
12. Whitehair, K.A., Haskins, S.C., Whitehair, J.G., Pascoe, P.J.: Clinical applications of quantitative acid-base chemistry. J. Vet. Intern. Med. **9**, 1–11 (1995)
13. Kaneko, J.J., Harvey, J.W., Bruss, M.L.: Clinical Biochemistry of Domestic Animals, 6th edn., p. 931. Elsevier/Academic Press, Amsterdam (2008)
14. Constable, P.D.: Clinical assessment of acid-base status: strong ion difference theory. Vet. Clin. North A Food Anim. Pract. **15**, 297–311 (1999)
15. Waller, A., Lindinger, M.I.: Physiochemical analysis of acid-base status during recovery from high-intensity exercise in Standardbred racehorses. Equine Comp. Exerc. Physiol. **2**, 119–127 (2005)
16. Constable, P.D.: Calculation of variables describing plasma non-volatile weak acids for use in the strong ion approach to acid-base balance in cattle. Am. J. Vet. Res. **63**(4), 482–490 (2002)
17. SAS 9.3®, SAS Institute Inc., Cary, NC, USA
18. Castillo, C., Hernandez, J., Benedito, J.L., Lopez-Alonso, M., Miranda, M., Gutierrez-Panizo, C., Sotillo, J.: Quantitative evaluation of acid-base balance on milk producing goats: effect of sex and milk yield. Vet. Med. **45**, 241–246 (2000)
19. Carlson, G.P.: Luid, electrolyte and acid-base balance. In: Kaneko, J., Harvey, J., Bruss, M. (eds.) Clinical Biochemistry of Domestic Animals, 5th edn, pp. 485–516. Elsevier/Academic Press, Amsterdam (1997)
20. Coppock, C.E., Grant, P.A., Portzer, S.J., Charles, D.A., Escobosa, A.: Lactating dairy cow responses to dietary sodium, chloride and bicarbonate during hot weather. J. Dairy Sci. **65**, 566 (1982)
21. Kist, A.E.: The effect of a 48 hour fast on acid-base balance and blood gas values on dairy cows. Research Project Veterinary Medicine University Utrecht, p. 23 (2011)
22. Castillo, C., Garcia-Partida, P., Hernandez, J., Benedito, J.L.: Application of Fencl's equations in the acid-base balance changes related to pregnancy on ewe. Vet. Med. **43**, 27–31 (1998)
23. Sobiech, P., Stopyra, A., Kuleta, Z., Zbanysek, M., Milewski, S.: Acid-base balance parameters of arterial, venous and capillary blood in sheep. Bull. Vet. Inst. Pulawy **49**, 125–127 (2005)
24. Gonzalez, F.H.D., Hernandez, F., Madrid, J., Martinez-Subiela, S., Ceron, J.J., Tecles, F.: Acid–base and electrolyte status during early induced pregnancy toxaemia in goats. Vet. J. **193**, 598–599 (2012)
25. Antunović, Z., Novoselec, J., Šperanda, M., Đidara, M., Klir, Ž., Kopačin, T.: Blood acid-base balance and hematological parameters in pregnant goats from organic farming. Krmiva **53**(4), 151–156 (2011)

26. Dobranić, T., Samardžija, M., Đuričić, D., Harapin, I., Vince, S., Grančer, D., Prvanović, N., Grizelj, J., Karađole, M., Bedrica, Lj., Cvitković, D.: The metabolic profile of Boer goats during puerperium. In: XVI Congress of the Mediterranean Federation for Health and Production of Ruminants, Zadar, Croatia, pp. 403–408 (2008)
27. Drackley, J.K., Kim, Y.K., Strang, B.D., Young, J.W.: Metabolic response of lactating goats to feed restriction and dietary 1,3 butanediol. J. Dairy Sci. **72**, 204 (1989)
28. Antunović, Z., Šperanda, M., Senčić, Đ., Šerić, V., Šperanda, T., Gutzmirtl, D.: Blood metabolic profile of dairy goats in organic production. Krmiva **48**(5), 231–236 (2006)

Persistency of the Effect of Heat Stress in Simmental Cows in Eastern Croatia

Vesna Gantner[1(✉)], Denis Kučević[2], Mirna Gavran[1], Ranko Gantner[1], Maja Gregić[1], Tina Bobić[1], Marcela Šperanda[1], and Muhamed Brka[3]

[1] Faculty of Agrobiotechnical Sciences Osijek, Josip Juraj Strossmayer University of Osijek, Osijek, Croatia
vgantner@fazos.hr

[2] Faculty of Agriculture, Department of Animal Science, University of Novi Sad, Novi Sad, Serbia

[3] Faculty of Agriculture and Food Sciences, Department for Animal Breeding, University of Sarajevo, Sarajevo, Bosnia and Herzegovina

Abstract. Aiming to determine the persistency of effect of heat stress in Simmental cows reared in Eastern Croatia regarding the parity and susceptibility to heat stress, 539,993 test-day records from 59,821 cows collected in the period from January 2005 to December 2012 were analysed. The cows with determined statistically significant decrease in daily milk production at set temperature-humidity index (THI) threshold value (68 and 72) were included in the further analyses. The persistency of heat stress effect was defined as the drop in daily milk traits in the subsequent milk recordings (1st and 2nd) after the heat stress. The obtained results indicate that cows heat stressed at lower THI threshold value had higher drop in daily milk yield in subsequent milk recordings. Furthermore, cows in first parity experience higher drop comparing to older cows. The negative effect of heat stress on daily milk contents that existed in all cows was more pronounced in the 2nd milk recording. Finally, the lower was THI threshold value (cows more susceptible to heat stress), the stronger and more persistent was heat stress effect. Given the very likely increase in the frequency of days with heat stress in the future, it is necessary to develop a proper mitigation strategy for dairy cattle production.

Keywords: Daily milk production · Simmental cows · Heat stress · Persistency

1 Introduction

Current animal production is characterized by high animal productivity and, on the other hand, by low stress resistance. Moreover, this high production must be realized in an environment that is changing to animals less comfort zone due to changes of climate. Accordingly, to IPCC (2007) the climate change will significantly influence animal production worldwide. Gauly et al. (2013) point out the heat stress of highly productive cows as one of the crucial problems of European milk production in the future while Segnalini et al. (2013) emphasize the need for a mitigation strategy to reduce the negative effects of warming on farm animals in the Mediterranean region.

M. Brka et al. (Eds.): AgriConf 2019, IFMBE Proceedings 78, pp. 163–168, 2020.
https://doi.org/10.1007/978-3-030-40049-1_21

Hansen (2013) stated that heat stress of dairy cows will become an acute problem regardless the climate changes since the increasing production results in higher animal sensitivity to heat stress. Dairy cows characterized by high productivity lose the thermoregulation ability at temperatures 25–29 °C. Kadzere et al. (2002) emphasised that the intensive genetic selection for milk production has resulted in larger frames and larger gastrointestinal tracts in high productive cows that enable them to digest more feed and consequently creates more metabolic heat and reduces the animals' ability to regulate normal temperature at heat stress. The same authors concluded that the thermoneutrality shifts to lower temperatures due to the increased milk yield, feed intake and metabolic heat. Heat stress environment in dairy cows results in reduced dry matter intake, milk production (West et al. 1999; Casa and Ravelo 2003) and reproductive performances (Ravagnolo et al. 2000; Bohmanova et al. 2007). Also, heat stress environment affects milk composition, somatic cell counts (SCC) and mastitis frequencies (Ravagnolo et al. 2000; Bouraoui et al. 2002; St-Pierre et al. 2003; West 2003; Correa-Calderon et al. 2004; Gantner et al. 2011; Collier et al. 2012; Hammami et al. 2013; Smith et al. 2013). All of this changes lead to considerable loss of profit dairy farms (St-Pierre et al. 2003). Heat stress could be expressed on many ways while the most common measure in dairy cattle is the temperature-humidity index (THI) that includes ambient temperature and relative humidity (Kibler 1964). Since the effect of heat stress in high productive dairy cows persisted even after termination of heat stress condition (Johnson et al. 1962), the aim of this research was to determine the persistency of heat stress effect in Simmental cows in respect to the parity (I, II, III) and susceptibility to heat stress (THI = 68/72).

2 Materials and Methods

For statistical analysis, 539,993 individual test-day records (TDR) collected in period from January 2005 to December 2012 on 59,821 Simmental cows reared on 6,242 farms were used. TDR were collected during the regular milk recording performed by an alternative milk recording method (AT4/BT4) on farms in Eastern Croatia. Additionally, at each milk recording, ambient temperature and relative humidity were recorded as the basis for calculation of daily temperature-humidity index (THI) using equation by Kibler (1964):

$$\mathrm{THI} = 1.8 \times \mathrm{Ta} - (1 - \mathrm{RH}) \times (\mathrm{Ta} - 14.3) + 32 \tag{1}$$

Where Ta is average temperature in degrees Celsius and RH is relative humidity as a fraction of the unit. Test day records with lactation stage in (<5 days and >305 days), age at first calving in (<21 and >36 months), missing or parity >3, and missing or nonsense Ta and RH value were deleted from the dataset. Only animals with minimum 3 test day records per parity were taken into analysis. Regarding the parity, cows were divided into 3 classes: I, II and III. Data were provided by the Croatian Agricultural Agency. Basic statistical parameters of daily milk traits (daily milk yield, DMY; daily fat content, DFC; daily protein content, DPC) regarding the parity class (I, II, III) is presented in Table 1.

Table 1. Basic statistical parameters of daily milk traits (milk yield, kg, DMY; fat content, %, DFC; protein content, %, DPC) per parity (n = 539,993)

Parity/Trait	DMY (kg)	DFC (%)	DPC (%)
I	15.52 ± 5.1	4.14 ± 0.83	3.38 ± 0.41
II	16.54 ± 6.02	4.16 ± 0.87	3.45 ± 0.43
III	16.58 ± 6.01	4.11 ± 0.88	3.37 ± 0.44

The variability of daily milk traits due to heat stress environment was tested by least square analyses of variance for set THI threshold value (68 and 72) separately for each parity class (I, II, III) using the PROC MIXED procedure in SAS (SAS Institute Inc. 2000). The following mixed model was used:

$$y_{ijklmn} = \mu + b_1(d_i/305) + b_2(d_i/305)^2 + b_3 \ln(305/d_i) + b_4 \ln^2(305/d_i) + S_j + A_k + T_l + e_{ijklm} \quad (2)$$

Where

y_{ijklmn} = estimated daily milk trait (yield or contents);
μ = intercept;
b1, b2, b3, b4 = regression coefficients;
d_i = days in milk (i = 6 to 305 day);
S_j = fixed effect of calving season class j (j = 1/2005 to 12/2012);
A_k = fixed effect of age at first calving class k (k = 21 to 36 month),
Tl = fixed effect of THI class (l = 0 (normal condition – values under the given threshold) or l = 1 (heat stress condition – values equal and above the given threshold)),
e_{ijklm} = residual.

The significance of the differences between the THI classes was tested by t-test. Only cows with determined statistically significant decrease in daily milk yield were included in the further analyses. The persistency of heat stress effect was defined as the drop in daily milk traits in the subsequent milk recordings (1st (within 35 days) and 2nd (between 35 and 70 days)) after the heat stress. The daily milk trait measured on the recording day when heat stress occurred was used as the reference level. The persistency of the heat stress effect was analyzed separately for each parity.

3 Results and Discussion

The amount of drop in daily milk traits in subsequent milk recordings after detection of heat stress at different THI values (68 and 72) is presented in Tables 2 and 3.

In the first parity Simmental cows, that experienced statistically significant ($P < 0.05$) decrease of daily milk yield at THI = 68, milk yield dropped for 1.074 kg/day in the 1st subsequent milk recording the daily, while in the 2nd subsequent milk recording the drop in milk production was even higher (1.384 kg/day). The higher drop in the second milk recording was also observed for daily fat and protein content (0.302% per day and 0.287% per day, respectively). The lowest drop in daily milk yield recorded in the 1st subsequent milk recording, followed by the drop decrease in the 2nd subsequent milk recording was determined in third parity cows.

Table 2. Drop in daily milk traits in the subsequent milk recordings after detection of heat stress in accordance to the parity (I, II, and III) when THI = 68

Parity	1st milk recording after heat stress			2nd milk recording after heat stress		
	DMY (kg)	DFC (%)	DPC (%)	DMY (kg)	DFC (%)	DPC (%)
I	1.074	0.192	0.081	1.384	0.302	0.111
II	0.473	0.193	0.077	0.414	0.287	0.101
III	0.431	0.2	0.066	0.218	0.307	0.111

*DMY – daily milk yield (kg); DFC – daily fat content (%); DPC – daily protein content (%)

The effect of heat stress in cows that experience statistically significant ($P < 0.05$) decrease of daily milk yield at THI = 72 (Table 3) persisted in the 1st and 2nd subsequent milk recordings. Primiparous cows had higher and more persistent drop in daily milk traits compared to multiparous. The analysis of daily milk contents, fat and protein, showed a higher drop in daily fat than protein content. Also, the amount of drop increased in the 2nd compared to the 1st subsequent milk recording in all parities and at both THI threshold values.

Table 3. Drop in daily milk traits in the subsequent milk recordings after detection of heat stress in accordance to the parity (I, II, and III) when THI = 72

Parity	1st milk recording after heat stress			2nd milk recording after heat stress		
	DMY (kg)	DFC (%)	DPC (%)	DMY (kg)	DFC (%)	DPC (%)
I	0.793	0.192	0.097	1.153	0.341	0.148
II	0.277	0.186	0.087	0.234	0.34	0.142
III	0.345	0.23	0.089	0.204	0.333	0.146

*DMY – daily milk yield (kg); DFC – daily fat content (%); DPC – daily protein content (%)

According to many studies (e.g., Kadzere et al. 2002; Bohmanova et al. 2007; Collier et al. 2006; Hansen 2013; Gantner et al. 2017), the THI threshold values depend on a variety of factors, for example: production level, parity, breed, region. Du Preez et al. (1990a, b) determined that the heat stress affects milk production and feed intake when THI values exceed 72. Bouraoui et al. (2002) defined 69 as the THI threshold value, while Bernabucci et al. (2010) and Collier et al. (2012) set it at 68. Vitali et al.

(2009) emphases that cows are in increased risk of death when THI reaches 80. The results of this research implicate that the lower is THI threshold value, the higher is the drop in daily milk yield in subsequent milk recordings. Furthermore, the highest drop of daily milk yield was determined in primiparous with a tendency of drop increase in the 2nd subsequent milk recording. On the other hand, multiparous had a lower drop of daily milk yield with a tendency to decrease the drop in the 2nd subsequent milk recording. These results indicate that the negative effect of heat stress was more pronounced and more persistent in younger cows (in I parity) and cows that were more susceptible to heat stress (heat stressed at the lower THI threshold values). Finally, as cows are less resistant to the heat stress (more susceptible, I parity), the effect of heat stress was more severe (the drop in daily production was higher and more persistent).

4 Conclusions

The research results indicate that cows heat stressed at a lower THI threshold value had a higher drop of daily milk yield in the subsequent milk recordings. Furthermore, younger cows (I parity) experience a higher drop compared to the older ones (II and III parity). The negative effect of heat stress on daily milk contents that existed in all cows was more pronounced in the 2nd subsequent milk recording. Finally, as cows are less resistant to the heat stress (more susceptible, first parity), the heat stress effect is more severe (the drop was higher and more persistent).

References

Bernabucci, U., Lacetera, N., Baumgard, L.H., Rhoads, R.P., Ronchi, B., Nardone, A.: Metabolic and hormonal acclimation to heat stress in domestic ruminants. Animal **4**, 1167–1183 (2010)

Bohmanova, J., Misztal, I., Cole, J.B.: Temperature-humidity indices as indicators of milk production losses due to heat stress. J. Dairy Sci. **90**, 1947–1956 (2007)

Bouraoui, R., Lahmar, M., Majdoub, A., Djemali, M., Belyea, R.: The relationship of temperature humidity-index with milk production of dairy cows in a Mediterranean climate. Animal Res. **51**, 479–491 (2002)

Casa, A.C., Ravelo, A.C.: Assessing temperature and humidity conditions for dairy cattle in Cordoba Argentina. Int. J. Biometeorol. **48**, 6–9 (2003)

Collier, R.J., Dahl, G.E., van Baale, M.J.: Major advances associated with environmental effects on dairy cattle. J. Dairy Sci. **89**, 1244–1253 (2006)

Collier, R.J., Hall, L.W.: Quantifying heat stress and its impact on metabolism and performance. Department of Animal Sciences, University of Arizona, Tucson, Arizona (2012)

Correa-Calderon, A., Armstrong, D., Ray, D., de Nise, S., Enns, M., Howison, C.: Thermoregulatory responses of Holstein and Brown Swiss heat-stressed dairy cows to two different cooling systems. Int. J. Biometeorol. **48**, 142–148 (2004)

Du Preez, J.H., Giesecke, W.H., Hattingh, P.J.: Heat stress in dairy cattle and other livestock under Southern African conditions. I. Temperature-humidity index mean values during the four main seasons. Onderstepoort J. Vet. Res. **57**, 77–86 (1990a)

Du Preez, J.H., Hatting, P.J., Giesecke, W.H., Eisenberg, B.E.: Heat stress in dairy cattle and other livestock under Southern African conditions. III. Monthly temperature-humidity index mean values and their significance in the performance of dairy cattle. Onderstepoort J. Vet. Res. **57**, 243–248 (1990b)

Gantner, V., Bobić, T., Gantner, R., Gregić, M., Kuterovac, K., Novaković, J., Potočnik, K.: Differences in response to heat stress due to production level and breed of dairy cows. Int. J. Biometeorol. **61**(9), 1675–1685 (2017)

Gantner, V., Mijić, P., Kuterovac, K., Solić, D., Gantner, R.: Temperature-humidity index values and their significance on the daily production of dairy cattle. Mljekarstvo **61**(1), 56–63 (2011)

Gauly, M., Bollwein, H., Breves, G., Brügemann, K., Dänicke, S., Das, G., Demeler, J., Hansen, H., Isselstein, J., König, S., Lohölter, M., Martinsohn, M., Meyer, U., Potthoff, M., Sanker, C., Schröder, B., Wrage, N., Meibaum, B., von Samson-Himmelstjerna, G., Stinshoff, H., Wrenzycki, C.: Future consequences and challenges for dairy cow production systems arising from climate change in Central Europe—a review. Animal **7**, 843–859 (2013)

Hammami, H., Bormann, J., M'hamdi, N., Montaldo, H.H., Gengler, N.: Evaluation of heat stress effects on production traits and somatic cell score of Holsteins in a temperate environment. J. Dairy Sci. **96**, 1844–1855 (2013)

Hansen, P.J.: Genetic control of heat stress in dairy cattle. In: Proceedings 49th Florida Dairy Production Conference, Gainesville, 10 April 2013 (2013)

Intergovernmental Panel on Climate Change – IPCC: Climate change 2007: the physical science basis. Contribution of Working Group I to the Fourth Assessment Report of the Intergovernmental Panel on Climate Change. Cambridge University Press, Cambridge (2007)

Johnson, H.D., Ragsdale, A.C., Berry, I.L., Shanklin, M.D.: Effect of various temperature humidity combinations on milk production of Holstein cattle. Res. Bull. Missouri Agric. Exp. Station **791** (1962)

Kadzere, C.T., Murphy, M.R., Silanikove, N., Maltz, E.: Heat stress in lactating dairy cows: a review. Livestock Prod. Sci. **77**, 59–91 (2002)

Kibler, H.H.: Environmental physiology and shelter engineering. LXVII. Thermal effects of various temperature-humidity combinations on Holstein cattle as measured by eight physiological responses. Research Bulletin, University of Missouri, College of Agriculture, Agricultural Experiment Station, 862 (1964)

Ravagnolo, O., Misztal, I., Hoogenboom, G.: Genetic component of heat stress in dairy cattle, development of heat indeks function. J. Dairy Sci. **83**, 2120–2125 (2000)

SAS Institute Inc.: SAS User's Guide, version 8.2 edition. SAS Institute Inc., Cary (2000)

Segnalini, M., Bernabucci, U., Vitali, A., Nardone, A., Lacetera, N.: Temperature humidity index scenarios in the Mediterranean basin. Int. J. Biometeorol. **57**, 451–458 (2013)

Smith, D.L., Smith, T., Rude, B.J., Ward, S.H.: Short communication: comparison of the effects of heat stress on milk and component yields and somatic cell score in Holstein and Jersey cows. J. Dairy Sci. **96**, 3028–3033 (2013)

St-Pierre, N.R., Cobanov, B., Schnitkey, G.: Economic loses from heat stress by US livestock industries. J. Dairy Sci. **86**, 52–77 (2003)

Vitali, A., Sagnalini, M., Bertocchi, L., Bernabucci, U., Nardone, A., Lacetera, N.: Seasonal pattern of mortality and relationships between mortality and temperature humidity index in dairy cows. J. Dairy Sci. **92**, 3781–3790 (2009)

West, J.W., Hill, G.M., Fernandez, J.M., Mandebvu, P., Mullinix, B.G.: Effect of dietary fiber on intake, milk yield, and digestion by lactating dairy cows during cool or hot, humid weather. J. Dairy Sci. **82**, 2455–2465 (1999)

West, J.W.: Effects of heat-stress on production in dairy cattle. J. Dairy Sci. **86**, 2131–2144 (2003)

The Variation of Air Temperature, Humidity and Wild Boar Population in Hunting Area in Eastern Croatia in Period 2008–2018

Mirna Gavran(✉), Maja Gregić, and Vesna Gantner

Faculty of Agrobiotechnical Sciences Osijek,
Josip Juraj Strossmayer University of Osijek, Osijek, Croatia
mgavran@fazos.hr

Abstract. Climate change drives population growth of wild boar directly by relaxing the negative effect of cold winters on survival and reproduction, and indirectly by increasing food availability. Air temperature is one of the main climatic elements when observing the climate of an area. Furthermore, humidity is an important factor for normal life of wild animals. Considering the great importance of the wild boar population in Croatia, the aim of this study was to determine the relationship between weather conditions and population size (regarding the categories: offspring, young, middle-aged, and mature) of wild boar population in hunting ground in Eastern Croatia during the analysed period from year 2008 to year 2018. Based on the conducted research following could be pointed out: during the analysed period from year 2008 till year 2018 the mean yearly air temperature varied in interval from 11–13 °C, during the analysed period from year 2008 till year 2018 the mean yearly humidity varied in interval from 76–84%, the lowest total number of offspring was determined in year 2017. Since the offspring are extremely susceptible to inadequate environmental conditions especially low temperature the decrease of boar population could be expected during the period characterised by extreme environmental conditions.

Keywords: Wild boar · Population · Air temperature · Humidity

1 Introduction

In accordance to Blüchel (2000), populations of wild boar (*Sus scrofa*) are found in most of Europe, north to the Baltic and North Seas, but are extinct in the British Isles, although reintroduction is being considered. They also live in central and south Asia, in North Africa, and as introductions in parts of North and South America and Australia. Males (boars) can weigh more than 250 kg (550 lbs) in Europe and be as tall as 1 m (39 in.) at the shoulder (Blüchel 2000). It could be found on the entire territory of the Republic of Croatia and on the islands (which is undesirable). After the Second World War, it was estimated that there were around 300 wild boars in Croatian hunting grounds. Today there are over 10,000 wild boars in Croatia (Darabuš and Jakelić 2002). The natural limiting factor for the spread of wild boar is the climate - low winter temperatures and a thick snow cover. Recent years we have witnessed a gradual warming with milder winters. Due to severe climatic conditions, offspring can die (Janicki et al. 2007).

M. Brka et al. (Eds.): AgriConf 2019, IFMBE Proceedings 78, pp. 169–173, 2020.
https://doi.org/10.1007/978-3-030-40049-1_22

Accordingly, to Vetter et al. (2014), the air temperature has a great influence on the natural milieu of hunting ground, the water level of the water surface and the health of the game in it. In most organisms the temperature of the body depends on the ambient temperature, which limits their spread to the biosphere. In accordance to Olczak et al. (2015), wild pigs build stronger nests in severe conditions. Also, resting behaviour and reproduction may be disturbed by high temperatures. From climatic disasters, to wild boars is the most dangerous frost without snow. When the land freezes, wild boar cannot root, and rooting is an important way to find food. Considering the great importance of the wild boar population in Croatia, the aim of this study was to determine the relationship between weather conditions and population size (regarding the categories: offspring, young, middle-aged, and mature) of wild boar population in hunting ground in Eastern Croatia during the analysed period from year 2008 to year 2018.

The first written data on the analyzed hunting ground originated in year 1721 when Karlo VI donated Baron Prandau land and woods south of the Drava River during the serving in the war against the Turks. During the First World War and immediately afterwards, the poaching destroyed most of the wildlife funds. Also in 1938, wildlife funds were affected by very long and sharp winter. By the Second World War, game funds have increased. In the years of the Second World War no information on the state of hunting was recorded, which is understandable in terms of conditions. Accordingly, to the inventory of 01.01.1955 it was determined 40 throats of wild boar population (Gavran Ltd. 2019b). Today the hunting ground, situated in eastern Croatia, makes a natural whole of 9,364 hectares and the same populations of wildlife game resides in the whole area. The cultivated land surface in the hunting grounds is used for sowing the thickets for wild game, primarily grasslands, in order to raise the hunting area's nutritional capacity and the number of wild game (Gavran Ltd. 2019a).

2 Materials and Methods

Accordingly, to Katušin (2015), determining the age of wild boar before being shot is essential to proper management of the aforementioned species. Wild boar from the moment of farrowing to April 1 next year is a *squeaker*. The squeaker is the most obvious to evaluate for physical growth and size; short tail without tussock. Squeakers are brown colors including two dark stripes on each side of the body. In the winter hair dominate dark or brown tones, which is an indisputable sign for identification of squeakers. Squeaker is always with the sow, playful, curious, often incautious. Wild boar from April 1 to the end of the second year of life is called *Pig of the sounder*. The strongest pig of the sounder reaches a weight of about 50 kg in December and January. Wild boars in the second year of life (annuals) can be simply estimated. Pig of the sounder in such age group enters in late March, early April. They normally live in independent droves, still, sometimes join sows with juveniles (mixed herd). Pig of the sounder living alone is almost always males. At that age, tusks are already noticeable. Pig of the sounder has a long tail. At the end of the year, the annuals are physically strengthened, and sometimes they are larger than older sows. Estimating age in nature by body size is therefore not completely reliable. Also, it is possible to estimate the age of wild boar according to the incisors (tusks). According to Sertić (2008), in order to determine the climate of the hunting ground, it is necessary to know and analyze a few elements directly affecting it: precipitation, air temperature, wind, humidity and insolation. The organization of measurements and monitoring of the climatic phenomena is carried out in the network of meteorological stations. Meteorological stations are organizational units of meteorological services which have the task of performing meteorological observations (measurement and monitoring) according to the established unique criteria. A visual rating of a certain size is considered as observation, while the measurement refers either to the reading of the instrument value or to the registration of the appropriate type of instrument. The climate condition could affect the wild game indirectly and directly by the average and extreme temperatures, winds, insolation and the amount and shape of rainfall and humidity. Air temperature is one of the main climatic elements when observing the climate of an area. Given the temperature conditions, it is necessary to determine the annual strokes of this climatic element. Standard meteorological data on air temperature show an air temperature of 2 m above ground level. Thus, a range of 12 mid-monthly air temperature values obtained from climatological measurements at 7, 14 and 21 h is a long-term flow of temperature. The following values are followed: maximum daily, minimum daily, average monthly and average annual. Each air temperature corresponds to the maximum amount of water vapor in which the air is saturated. Accordingly, the relative humidity of the air is a meteorological phenomenon that shows the percentage of air saturated with an aqueous vapor at a certain temperature of air. If, for example, the relative humidity of the air is 50%, in the air is only half the amount of water vapor that the air could contain at that temperature. Knowing this meteorological element is generally important because the formation of fog, cloud, rain, roses, frosts, and other hydrometeors depends on the amount of moisture in the air. In this paper, monthly precipitation sums, mean temperatures and average humidity were acquired for the respective region and periods. Data were provided by Croatian Meteorological and Hydrological Service.

3 Results and Discussion

The relationship between weather conditions (air temperature and humidity) and population size of wild boar population in hunting ground in Eastern Croatia during the analyzed period from year 2008 to year 2018 is shown in Figs. 1 and 2.

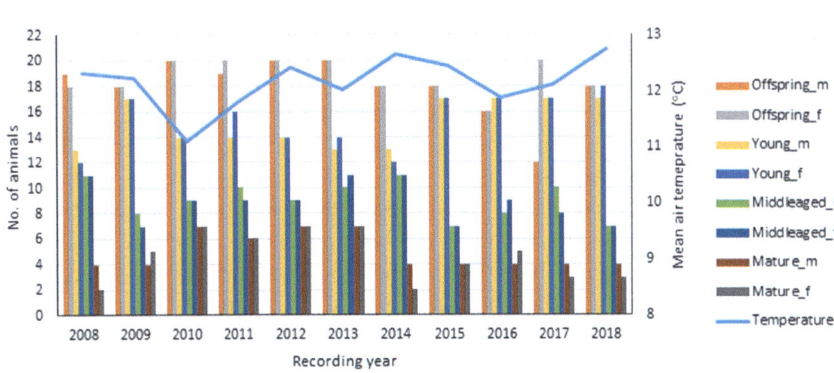

Fig. 1. The influence of temperature on number of wild boar population in period from 2008–2018.

During the analyzed period from year 2008 till year 2018 the mean yearly air temperature varied in interval from 11–13 °C. The highest number of offspring, of both gender, males and females, was recorded in years 2010, 2012 and 2013, when mean air temperature ranged between 11–12.5 °C. During the year 2017, when the mean air temperature amounted 12 °C the smallest number of male offspring was recorded. Furthermore, in year 2018, the highest number of young animal, males and females, was recorded, when the mean year air temperature was higher comparing to other years and amounted 13 °C. The highest number of middle-aged animals, females and males, was determined in year 2008 and 2014, while in the same years the smallest number of mature animals, males and females was determined. During the years 2008 and 2014 the mean yearly air temperature was slightly above 12 °C.

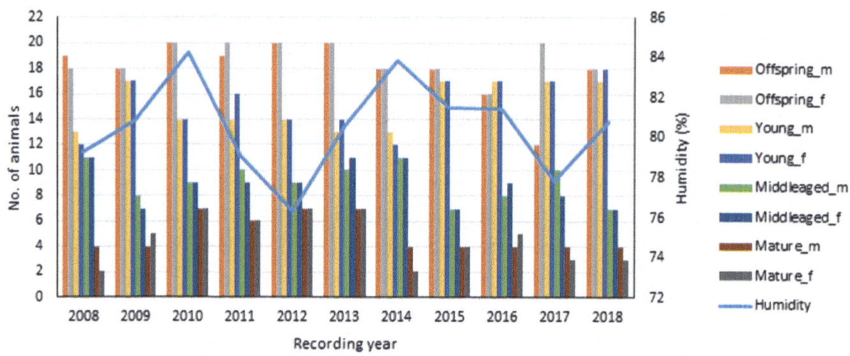

Fig. 2. The influence of humidity on number of wild boar population in period from 2008–2018.

In the researched hunting ground during the analysed period from year 2008 till year 2018 the mean yearly humidity varied in interval from 76–84%. The lowest mean humidity was recorded during the year 2012 when the number of offspring, of both gender, males and females, was highest and amounted 20 animals. At the same year, equal number of female and male animals was determined, by category as follows: young 14, middle aged 9, and mature 7 animals. The highest mean yearly humidity was determined in years 2010 and 2014. In year 2014, the lowest number of mature animals, especially boars was recorded.

Accordingly, to Geisser and Reyer (2005) the climate conditions are known to be significant factors for the dynamics of population of many ungulate species. Furthermore, conditions of the environment affect survival of offspring as well as reproduction of mature animals. In the population of wild boar, the variation of air temperature is essential from the aspect of survival of the newborn piglets since they are usually born in the period between April and June. The newborn piglets are extremely susceptible to low temperature and frosts during the spring period. Furthermore, during the first 2 years of life, the mortality of offspring can reach up to 90%. In case of increasing temperatures during the winter and spring mortality are likely to be reduced.

Population of wild boar tends to increase if the average temperature during the winter reaches a certain threshold (Conservation Magazine 2015). Defined thresholds values are higher in the warmer south region comparing to the region at the cooler north. Stated could be explained by the adaptation of population to the characteristics of certain region. Accordingly, to Conservation Magazine (2015) population of wild boars are increasing throughout the continent, while the rate of increase differs regarding the region.

4 Conclusion

Based on the conducted research following could be pointed out:

- during the analysed period from year 2008 till year 2018 the mean yearly air temperature varied in interval from 11–13 °C,
- during the analysed period from year 2008 till year 2018 the mean yearly humidity varied in interval from 76–84%,
- the lowest total number of offspring was determined in year 2017.

Since the offspring are extremely susceptible to inadequate environmental conditions especially low temperature the decrease of boar population could be expected during the period characterised by extreme environmental conditions.

References

Blüchel, G.K.: Game and Hunting (2000)

Darabuš, S., Jakelić, Z.I.: Osnove lovstva, Hrvatski lovački savez, Zagreb (2002)

Gavran d.o.o: O nama (2019a). http://www.gavran.hr/O-nama.aspx

Gavran d.o.o: Povijest (2019b). http://www.gavran.hr/Povijest-(1).aspx

Janicki, Z., Slavica, A., Konjević, D., Severin, K.: Zoologija divljači, Veterinarski fakultet, Zagreb (2007)

Katušin, D.: Metode određivanja dobi divlje svinje (Sus scrofa) prije i poslije odstrijela. Završni rad. Veleučilište u Karlovcu (2015)

Olczak, K., Nowicki, J., Klocek, C.: Pig behaviour in relation to weather conditions – a review. Department of Swine and Small Animal Breeding, Institute of Animal Science, University of Agriculture, Al. Mickiewicza 24/28, 30-059 Kraków, Poland (2015)

Sertić, D.: Uzgoj krupne divljači i uređivanje lovišta. Veleučilište u Karlovcu (2008)

Geisser, H., Reyer, H.U.: The influence of food and temperature on population density of wild boar Sus scrofa in the Thurgau (Switzerland) (2005)

Conservation Magazine: Climate change brings Europe more boars. University of Washington (2015)

Vetter, S.G., Ruf, T., Bieber, C., Arnold, W.: The burst of wild boar populations in Europe: how local adaptation mediates the effects of climate change in widespread ungulate (2014). Conference Paper

Determination of Meat Quality in Extensively Reared Akkaraman Sheep Breed

Ahmet Engin Tüzün[1] and Onur Yilmaz[2(✉)]

[1] Koçarli Vocational High School, Aydin Adnan Menderes University, Aydin 090100, Turkey

[2] Department of Animal Science, Faculty of Agriculture, Aydin Adnan Menderes University, Aydin 090100, Turkey

oyilmaz@adu.edu.tr

Abstract. The aim of the current study was to examine the effect of breed on meat quality of Akkaraman sheep breed reared in extensive production systems in Turkey. Animal material for the study consisted of 32 male lambs belonging to Akkaraman sheep breed in three and six month age groups raised solely on pasture after weaning. Post weaning 16 of each animal were slaughtered at 90 days and 16 of each animal was slaughtered at 180 days of age. The meat quality factors such as cooking loss and shear force were determined instrumentally on the *M. longissimus dorsi* (MLD) and *M. semitendinosus* (MST). Some muscle characteristics belong to *M. longissimus dorsi* (MLD) such as backfat thickness, muscle depth, muscle width and muscle area were determined in two different age group. Fatty acid composition were determined only in *M. longissimus dorsi*. The effect of age and muscle type on cooking loss and shear force were significant ($P < 0.01$). Muscle area in 3 and 6 month old lambs were 9.72 and 15.95 cm^2, respectively, while muscle depth for the same ages were 3.24 and 3.88 cm, respectively. Lambs with three months of age group showed slight superiority in terms of muscle area, muscle depth and muscle width in Musculus longissimus dorsi (MLD) samples. Monounsaturated fatty acids (MUFAs) composition was significantly influenced by age differences while conjugated linoleic acid (CLA), saturated fatty acids (SFAs) and polyunsaturated fatty acids (PUFAs) was not statistically significant on the age differences. The study results clearly showed that age significantly affects some of muscle fatty acid composition and muscle characteristics.

Keywords: Lamb · Fatty acids · MLD

1 Introduction

Population growth and nutritional problems remain up to date in the world and as well as in Turkey. Meat production is the most important source of income in the sheep industry throughout Turkey (De Rancourt et al. 2006; Montossi et al. 2013; Gürsoy, 2006). There are various factors affecting quality and quantity of meat production in sheep breeding that has an important role on supplying the animal protein needs (Karaca et al. 1999; Sanudo et al. 1998; Priola et al. 2001; Zygoyiannis 2006). Mainly, these may be classified as genetic and environmental factors such as breed, sex,

M. Brka et al. (Eds.): AgriConf 2019, IFMBE Proceedings 78, pp. 174–184, 2020.
https://doi.org/10.1007/978-3-030-40049-1_23

climate, slaughter hygiene and procedure (Priola et al. 2001; Sanudo et al. 1998). Meat structure, biochemical changes in muscle occurred before and after slaughtering, technological and organoleptic properties of meat are influenced by these factors (Gardener et al. 1999; Hopkins and Fogarty 1998, Beriain et al. 2000).

In recent years, consumer demand has focused on lean meat. Negative management and feeding conditions cause excessive fat in lamb, and this reduces consumer demand. Therefore, the implementation of efforts to improve the quality of meat, as well as to determine the current status of the local breeds are also very important. Lamb meat quality is determined using some parameters such as carcass and meat quality characteristics. It is difficult to develop meat quality characteristics due to the technological, financial, and biological limitations for breeders. Lamb carcass is desirable to include a high proportion of polyunsaturated fatty acid, low fat, high parts of valuable carcass ratio, high-quality tenderness, color, juiciness and sensory characteristics (Sanudo et al. 1998; Priola et al. 2001).

Akkaraman sheep breed are widely raised in Central Anatolia. The aim of this study was to determine some meat quality properties of Akkaraman lambs to evaluate the effects of different age muscle types on this parameter.

2 Material and Method

Animal material for the study consisted of 32 male lambs in three and six month age groups raised in extensive condition belonging to Akkaraman sheep breeds.

Animals were all weaned at 30 days after being gradually introduced to pasture. Post weaning 16 of each animal was slaughtered at 90 days and 16 of each animal was slaughtered at 180 days of age. Lambs were transported to the abbattoir one day prior to slaughter and rested in a paddock to avoid transport stress affecting carcass quality. Prior to slaughter lambs were placed on a 12 h fast with free access to water after which they were weighed, recorded and humanely slaughtered under veterinary supervision. M. longissimus dorsi (MLD) and M. semitendinosus (MST) samples were taken from the left side of each carcass (Fig. 1).

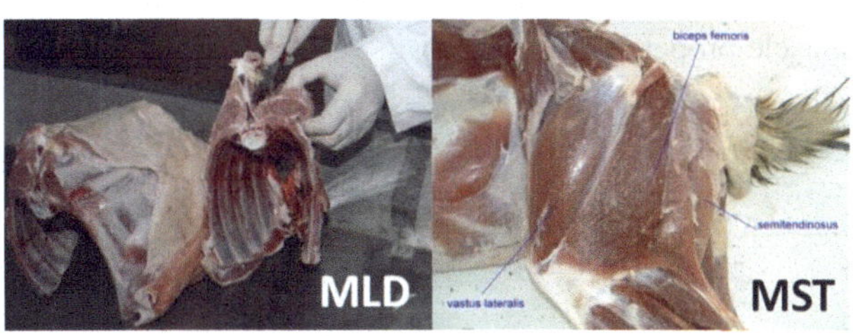

Fig. 1. Sampling location of Musculus longissimus dorsi (MLD) and Musculus semitendinosus (MST) muscle

Section taken from the 12th and 13th rib area of the M. longissimus dorsi (MLD) was drawn on parchment paper and back fat thickness (BF), muscle depth

(MD), muscle width (MW), muscle area (MA) were measured with planimeter (USHIKATA X-PLAN 380 DIII, Japan) (Fig. 2).

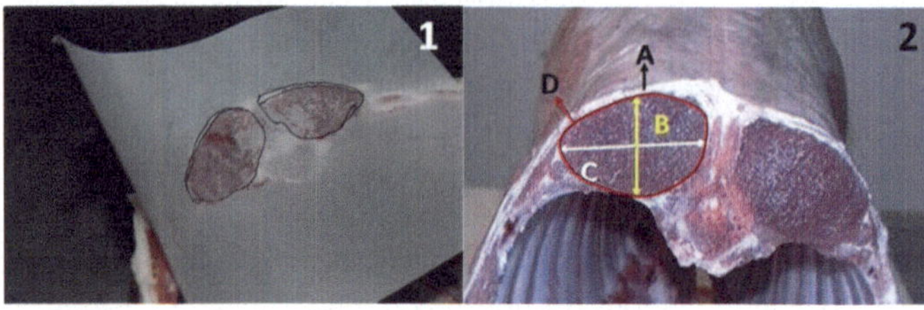

Fig. 2. Drawings of MLD muscle (1) and measurement points (2) (**A:** Backfat thickness, **B:** muscle depth, **C:** muscle width, **D:** muscle area)

Cooking loss (%) was performed as described by Hoffman et al. (2003). Shear force (kg/cm2) values were obtained using a Zwik/Roell texture analysis test device equipped with a V-shaped knife from a Warner Bratzler (WB) shear force apparatus. Other parameters measured were; hot carcass weight and slaughter weight.

Fat in meat samples was extracted based on method described by Folch et al. (1957), Boselli et al. (2001) and AOAC (2012). Fatty acid composition of muscle sampled from M. Longissimus dorsi was performed with gas chromatography based on the method reported by Christie (1992). Fatty acids in the range of C10:0, C24:0, and conjugated linoleic acid (CLA) were identified in fatty acid composition. Additionally, saturated fatty acids (SFAs), monounsaturated fatty acids (MUFAs), polyunsaturated fatty acids (PUFAs), and polyunsaturated fatty acids/saturated fatty acids ratio (P/S) values were calculated from the obtained values.

The UNIVARIATE procedure of SAS (1999) statistical package program was used to check normality of the data. The result of this analysis showed that the data for all the measured characteristics were normally distributed. Afterwards, all characteristics investigated were analysed using the Generalized Linear Models (GLM) procedure of SAS (1999) software. The mathematical model used for the least-squares analysis was as follows,

Model for muscle properties of animals,

$$\mathcal{Y}_{\mathrm{ijk}} = \mu + \alpha_i + b_j + \beta_1 \left[X_{ij} - \bar{X}\right] + e_{ijk}$$

Model for hot carcass weight of animals,

$$\mathcal{Y}_{\mathrm{ijk}} = \mu + \alpha_i + b_j + \beta_2 \left(A_{\mathrm{ij}} - \bar{A}\right) + e_{ijk}$$

Model for fatty acids composition,

$$\gamma_{ij} = \mu + a_i + e_{ij}$$

Y_{ijk} = Observations for muscle traits

μ = Overall mean of the trait
X_{ij} = Hot carcass weight of animals
$\bar{X}$ = Mean hot carcass weight of animals
A_{ij} = Slaughter weight of animals
e_{ijk} = Random errors with the assumption of N $(0, \sigma^2)$
a_i = Fixed effect of age (i = 3 and 6 months)
b_j = Fixed effect of birth type (j = Singleton, twin)
β_1 = Regression coefficient of hot carcass weight
β_2 = Regression coefficient of salughter weight
$\bar{A}$ = Mean slaughter weight of animals
The phenotypic correlations between variables were also obtained using the PROC CORR procedures in SAS (1999).

3 Results

The descriptive statistical means and the standard deviation of birth weight, live weight, average daily gain, slaughter weight, hot carcass weight and cold carcass weight are summarized in Table 1.

Table 1. Basic statistics of the birth weight (BW), live weight (LW), average daily gain (ADG), slaughter weight (SW), hot carcass weight (HCW), cold carcass (CCW) according to age of lambs

		3-Months			6-Months	
Variable	N	Mean ± SD	CV (%)	N	Mean ± SD	CV (%)
BW	16	3.64 ± 0.377	10.35	16	3.58 ± 0.414	11.58
LW	16	19.89 ± 1.858	9.34	16	39.19 ± 1.529	3.9
ADG	16	186.73 ± 18.78	10.06	16	198.25 ± 7.72	3.9
SW	16	19.89 ± 1.858	9.34	16	39.19 ± 1.529	3.9
HCW	16	9.12 ± 1.026	11.25	16	19.16 ± 1.015	5.3
CCW	16	7.40 ± 0.982	13.27	16	17.40 ± 0.659	3.79
Age	16	87.00 ± 1.826	2.1	16	179.63 ± 2.42	1.35

BW: Birth weight, LW:Live weight, ADG: Average daily gain, SW: Slaughter weight, HCW: Hot carcass weight, CCW: Cold carcass weight

The coefficients of variation (CV) obtained for birth weight indicate a high degree of care in the selection of the study material. The lambs were slaughtered at 90 and 180 days of age. Whilst the initial Birth Weights (BW) of all lambs were similar (X = 3.65 kg.), Akkaraman mean SW's for the two age groups were 19.89 kg. and 39.19 kg. Average Daily Gain (ADG) for three and six month age Akkaraman lambs were 186.73 g. and 198.25 g, respectively.

The least squares mean and standart errors for shear force and cooking loss according to age groups are given in Table 2.

Table 2. Least squares mean and standard errors for cooking loss and shear force, according to different age groups

Factors	N	CL (%)	SF (kg/cm^2)
Age		**P = 0.021**	**P = 0.000**
3 Months	16	41.91 ± 0.264	4.40 ± 0.291
6 Months	16	43.32 ± 0.315	5.08 ± 0.381
Muscle Type		**P = 0.000**	**P = 0.000**
MLD	32	41.80 ± 0.003	3.53 ± 0.330
MST	32	43.50 ± 0.003	5.95 ± 0.336
Overall		**42.60 ± 0.002**	**4.74 ± 0.236**

CL: Cooking loss, SF: Shear force

The effect of age and muscle type on cooking loss and shear force were significant (P < 0.01). The cooking loss values obtained were within the expected limits. The highest shear force value in the studied muscle regions was obtained in Musculus Semitendinosus.

The least squares mean and standard errors for MLD properties in two different age groups are given in Table 3.

Table 3. Least square means and standard errors belong to MLD properties in two different age groups

Factors	N	MA (cm^2)	MD (cm)	MW (cm)	BFT (mm)
Age		**P = 0.003**	**P = 0.313**	**P = 0.043**	**P = 0.036**
3-Months	16	9.72 ± 0.943	3.24 ± 0.307	3.93 ± 0.404	1.14 ± 0.372
6-Months	16	15.95 ± 1.006	3.88 ± 0.327	5.67 ± 0.431	2.81 ± 0.397
Birth type		**P = 0.166**	**P = 0.019**	**P = 0.837**	**P = 0.174**
Singleton	22	13.12 ± 0.209	3.73 ± 0.068	4.78 ± 0.090	2.08 ± 0.083
Twin	10	12.55 ± 0.322	3.40 ± 0.105	4.82 ± 0.138	1.86 ± 0.127
Reg Linear		**P = 0.000**	**P = 0.157**	**P = 0.006**	**P = 0.021**
HCW		1.320 ± 0.153	0.208 ± 0.072	0.186 ± 0.055	0.168 ± 0.048
Overall	**32**	**12.84 ± 0.184**	**3.56 ± 0.060**	**4.80 ± 0.079**	**1.97 ± 0.073**

MA: Muscle area, MD: Muscle depth, MW: Muscle width, BFT: Back fat thickness, HCW: Hot carcass weight, SW: Slaughter weight

Muscle area in 3 and 6 months of ages were 9.72 and 15.95 cm2 respectively, while muscle depth for the same ages were 3.24 and 3.88 cm, respectively. Results of the analysis carried out on MLD did not show a significant difference between birth type for the 3 and 6 month age group, except MD. There was significant differences (P < 0.05) in BFT between the breeds in both age groups. The coefficients of regression of hot carcass weight on all the muscle characteristics were also seen to be significant in both age groups, except musle depth. Birth type was not statistically significant on the muscle characteristics studied in both age groups (P > 0.05).

The phenotypic correlation coefficients between slaughter weights, hot carcass weight and muscle characteristics were given in Table 4.

Table 4. Phenotypic correlation coefficients between slaughter weight, hot carcass weight and *Musculus longissimus dorsi* properties

Age (Months)		SW	HCW	MA	MD	MW
3	HCW	0.937***				
	MA	0.923***	0.890***			
	MD	0.609*	0.641**	0.554*		
	MW	0.621**	0.648**	0.729**		0.517*
	BFT	0.714**	0.670**	0.713**	0.330 ns	0.545*
6	HCW	0.839***				
	MA	0.787***	0.551*			
	MD	0.251 ns	0.097 ns	0.367 ns		
	MW	0.549*	0.441 ns	0.571*	-0.191 ns	
	BFT	0.681**	0.434 ns	0.677**	0.171 ns	0.583*

MA: Muscle area, MD: Muscle depth, MW: Muscle width, BFT: Back fat thickness, HCW:
Hot carcass weight, SW: Slaughter weight, *:P < 0.05, **: P < 0.01, ***:P < 0.001, ns: non-significant

Correlation between SW, HCW and MLD properties showed expected significance for the 3 month age groups. The results of the analysis performed for Conjugated linoleic acid (CLA) and its isomers are represented in Table 5.

Table 5. Least square means and standard errors for Conjugated Linoleic Acids (CLAs) and its isomers in two different ages

Factor	N	t9-t11*	*t10-t12**	*t9-c11**	*t10-c12**	*c10-t12**	*c9-c11**	*c10-c12**	CLA
Ages		**P = 0.040**	**P = 0.393**	**P = 0.000**	**P = 0.258**	**P = 0.000**	**P = 0.055**	**P = 0.542**	**P = 0.169**
3-months	16	0.18 ± 0.014	0.13 ± 0.012	0.01 ± 0.001	0.07 ± 0.008	0.01 ± 0.001	0.07 ± 0.009	0.11 ± 0.009	0.59 ± 0.034
6-months	16	0.22 ± 0.014	0.12 ± 0.012	0.02 ± 0.001	0.06 ± 0.008	0.02 ± 0.001	0.10 ± 0.009	0.12 ± 0.009	0.65 ± 0.034
General	**32**	**0.20 ± 0.010**	**0.13 ± 0.008**	**0.02 ± 0.001**	**0.06 ± 0.006**	**0.01 ± 0.001**	**0.09 ± 0.006**	**0.12 ± 0.007**	**0.62 ± 0.024**

*Conjugated linoleic acid isomers, t9-t11: trans9-trans11, t10-t12: trans10-trans12, t9-c11: trans9-cis11, t10-c12: trans10-cis12, c10-t12: cis10-trans12, c9-c11: cis9-cis11, c10-c12: cis10-cis12

Although some of conjugated linoleic acid (CLA) isomers such as trans9-trans11, trans9-cis11 and cis10-trans12 were statistically significant in three and six month age groups. Total conjugated linoleic acid (CLA) values were higher in six month age groups.

The least squares mean and standart errors for saturated fatty acids (SFAs) according to age groups are given in Table 6. The effect of the ages on C10:0, C12:0,

Table 6. Least square means and standard errors for Saturated Fatty Acids (SFAs) in two different sheep breeds

Factor	N	C10:0	C12:0	C14:0	C15:0	C16:0	C17:0	C18:0	C20:0	C22:0	C24:0	SFA
Breed		P = 0.012	P = 0.005	P = 0.017	P = 0.564	P = 0.012	P = 0.049	P = 0.185	P = 0.003	P = 0.194	P = 0.156	P = 0.057
3-months	16	0.25 ± 0.013	0.42 ± 0.058	4.90 ± 0.461	0.70 ± 0.054	25.87 ± 0.491	2.09 ± 0.132	24.17 ± 1.019	0.19 ± 0.014	0.03 ± 0.01	0.02 ± 0.003	58.63 ± 0.849
6-months	16	0.20 ± 0.013	0.17 ± 0.058	3.24 ± 0.461	0.74 ± 0.054	27.72 ± 0.491	2.47 ± 0.132	26.13 ± 1.019	0.25 ± 0.014	0.05 ± 0.01	0.03 ± 0.003	61.00 ± 0.849
General	**32**	**0.22 ± 0.009**	**0.29 ± 0.041**	**4.07 ± 0.326**	**0.72 ± 0.038**	**26.80 ± 0.347**	**2.28 ± 0.093**	**25.15 ± 0.720**	**0.22 ± 0.010**	**0.04 ± 0.007**	**0.02 ± 0.002**	**59.81 ± 0.600**

SFA: Saturated fatty acids

C14:0, C16:0, C17:0 and C20:0 fatty acids belonging to saturated fatty acids was significant. On the other hand, Palmitic acid (C16:0, 26.80%) and stearic acid (C18:0, 25.15%) were obtained at very high level in two different age group.

The effect of the ages on C10:0, C12:0, C14:0, C16:0, C17:0 and C20:0 fatty acids belonging to saturated fatty acids was significant. On the other hand, Palmitic acid (C16:0, 26.80%) and stearic acid (C18:0, 25.15%) were obtained at very high level in two different age group.

Oleic acid (C18:1ω9) in the monounsaturated fatty acid (MUFA) in M. longisimus dorsi muscle was observed with a high percentage of 32.95%. Age group showed a significant effect on the only C15:1ω5 (Table 7).

Table 7. Least square means and standard errors for Monounsaturated Fatty Acids (MUFAs) in two different age groups

Factor	N	C14:1*ω5*	C15:1*ω5*	C16:1*ω7*	C17:1*ω8*	C18:1*ω9*	C20:1*ω9*	MUFA
Breed		P = 0.267	P = 0.021	P = 0.424	P = 0.927	P = 0.050	P = 0.576	P = 0.044
3-months	16	0.20 ± 0.012	0.27 ± 0.017	2.22 ± 0.110	1.06 ± 0.048	34.48 ± 1.054	0.12 ± 0.017	38.35 ± 1.051
6-months	16	0.22 ± 0.012	0.33 ± 0.017	2.10 ± 0.110	1.06 ± 0.048	31.42 ± 1.054	0.11 ± 0.017	35.23 ± 1.051
General	**32**	**0.21 ± 0.008**	**0.30 ± 0.012**	**2.16 ± 0.078**	**1.06 ± 0.034**	**32.95 ± 0.746**	**0.11 ± 0.012**	**36.79 ± 0.743**

MUFA: Monounsaturated fatty acids

Total polyunsaturated fatty acids (PUFAs) values of fatty acid composition determined in MLD in three and six months age were 3.58% and 4.42%, respectively. Results of all polyunsaturated fatty acids considered except C18:2ω6 and C22:5ω6 in all age group did not show a significant difference (Table 8).

Table 8. Least square means and standard errors for Polyunsaturated Fatty Acids (PUFAs) in two different age groups

Factor	N	C18:2*ω6*	C18:3*ω6*	C22:4*ω6*	C22:5*ω6*	PUFA
Breed		P = 0.009	P = 0.418	P = 0.120	P = 0.043	P = 0.082
3-months	16	0.70 ± 0.078	0.09 ± 0.017	1.33 ± 0.226	0.90 ± 0.187	3.58 ± 0.332
6-months	16	0.40 ± 0.078	0.07 ± 0.017	1.84 ± 0.226	1.46 ± 0.187	4.42 ± 0.332
General	**32**	**0.55 ± 0.055**	**0.08 ± 0.012**	**1.58 ± 0.16**	**1.18 ± 0.132**	**4.00 ± 0.235**

PUFA: Polyunsaturated fatty acids

PUFA/SFA ratio and fat (%) determined in Longissimus dorsi were 0.06 and 7.57 in three months age and 0.07 and 7.91 in six months age, respectively (Table 9).

Table 9. Least square means and standard errors for PUFA/SFA and fat (%) in two different age groups

Factor	N	PUFA/SFA	Fat (%)
Breed		P = 0.181	P = 0.767
3-months	16	0.06 ± 0.005	7.57 ± 0.807
6-months	16	0.07 ± 0.005	7.91 ± 0.807
General	**32**	**0.07 ± 0.004**	**7.74 ± 0.571**

4 Discussion

Meat content constitutes a significant proportion of the water (70–80%) as with all other foods. It is desirable to keep the water in the meat structure due to economic and technological. In addition, removing of water from the tissue has adverse effects on sensory properties of meat such as tenderness and juiciness (Hamm 1986; Honikel 1988). Water and cooking loss are affected by breed, sex, chemical composition of meat, muscle type, surface area of meat, cooking temperature and duration. The mean of cooking loss values obtained in the present study was 42.60%. This value were higher than the reports of Lanza et al. (2003) and Ekiz et al. (2009) used same muscle and cooking temperature with the current study in order to measure cooking loss. Shear force values, which are essential for quality meat and consumer demand, was lower than previous studies (Çelik and Yilmaz 2010; Abdullah and Rasha 2000). Considering the results, it can be said to be considerably tender meat sampled from all studied ages in terms of shear force.

Fatty acid composition associated with meat flavour and the nutritional value of meat is an important role in the definition of meat quality. Although fatty acid composition obtained from studied breeds was similar to literature, it is reasonable to mention some differences from the literature (Demirel et al. 2006; Vatansever and Demirel 2009; Marmer et al. 1984; Enser et al. 2000). The differences between the previous literature and the present study were mainly due to the many factors such as breed, feeding, age that affect fatty acid composition. SFA, PUFA and P/S ratio showed no significance in terms of age groups used in this study.

High oleic acid (C18:1), known monounsaturated fatty acid, value led to an increase MUFA value. In addition, the low PUFA caused to low P/S ratio. This situation is explained by the intensive feeding of the animal notwithstanding the pasture during the trial. Although lamb fat tissue has more than 100 varieties of fatty acids, the palmitic acid (C16:0), stearic acid (C18:0) and oleic acid (C18:1) are prominent fatty acids in all (Beriain et al. 2000). The percentage of the fatty acids, obtained from the present study, was approximately 81% in the ratio of total fatty acids.

C16:0, C18:0, and C18:1 fatty acids and P/S ratio results in our study were in agreement when compared to Kivircik and Sakiz lambs (Demirel et al. 2006; Romedi and Yilmaz, 2010, Vacca et al., 2008). However, C18:0 and C18:1 fatty acids were significantly higher than reported values by Vacca et al. (2008).

Although the MUFA values in our study were higher, PUFA and P/S ratio were, in fact, lower than values reported by Vatansever and Demirel (2009). Mean CLA, SFA, MUFA and PUFA values were lower than reported by Diaz et al. (2005).

5 Conclusion

The results showed that lamb meat quality was variable, and influenced by many factors such as ages and muscle types. The leanness of lamb meat should be promoted because consumers are more conscious about their diet and tend to change their eating habits to consume leaner meat. Lamb meat is the preferred red meat option in most parts of Turkey, and is also the main source of income rather than wool or milk for the sheep industry in the country. Therefore, there is still a need to obtain data to identify and improve the meat production ability of sheep breeds in qualitative and quantitative terms.

References

Abdullah, A.Y., Rasha, Q.: Effect of slaughter weight and aging time on the quality of meat from Awassi ram lambs. Meat Sci. **82**(3), 309–316 (2009)

AOAC.: Association of Official Analytical Chemist. Official methods of analysis of the association of official analytical chemists, 19. Arlington (2012)

Beriain, M.J., Purroy, A., Treacher, T., Bas, P.: Effect of animal and nutritional factors and nutrition on lamb meat quality. In: Ledin, I., Morand-Fehr, P. (eds.) Sheep and Goat Nutrition: Intake, Digestion, Quality of Products and Rangelands, pp. 75–86. Zaragoza, Spain (2000)

Boselli, E., Velazco, V., Caboni, M.F., Lercker, G.: Pressurized liquid extraction of lipids for the determination of oxysterols in egg-containing food. J. Chromatogr. A **917**, 239–244 (2001)

Çelik, R., Yilmaz, A.: Certain meat quality characteristics of Awassi and Turkish Merino × Awassi (F1) lambs. Turkish J. Vet. Anim. Sci. **34**(4), 349–357 (2010)

Christie, W.W.: Gas Chromatography and Lipids: A Practical Guide. The Oily Press, Ayrshire (1992)

Demirel, G., Özpinar, H., Nazli, B., Keser, O.: Fatty acids of lamb meat from two breeds fed different forage: concentrate ratio. Meat Sci. **72**, 229–235 (2006)

Enser, M., Richardson, R.I., Wood, J.D., Gill, B.P., Sheard, P.R.: Feeding linseed to increase the n-3 PUFA of pork: fatty acid composition of muscle, adipose tissue, liver and sausages. Meat Sci. **55**, 201–212 (2000)

De Rancourt, M., Fois, N., Lavín, M.P., Tchakérian, E., Vallerand, F.: Mediterranean sheep and goats production: an uncertain future. Small Ruminant Res. **62**(3), 167–179 (2006)

Diaz, M.T., Ivarez, I.A., De la Fuente, J., Sanudo, C., Campo, M.M., Oliver, M.A., Furnols, M. F., Montossi, F., San Julia´n, R., Nute, G.R., Caneque, V.: Fatty acid composition of meat from typical lamb production systems of Spain, United Kingdom, Germany and Uruguay. Meat Sci. **71**, 256–263 (2005)

Ekiz, B., Yilmaz, A., Özcan, M., Kaptan, C., Hanoğlu, H., Erdoğan, İ., Yalçintan, H.: Carcass measurements and meat quality of Turkish Merino, Ramlic, Kivircik, Chios and Imroz lambs raised under an intensive production system. Meat Sci. **82**, 64–70 (2009)

Folch, J., Lees, M., Sloane Stanley, G.H.: A simple method for the isolation and purification of total lipides from animal tissues. J. Biol. Chem. **226**(1), 497–509 (1957)

Gardener, G.E., Kennedy, L., Milton, J.T.B., Pethick, D.W.: Glycogen metabolism and ultimate pH of muscle in merino, first-cross, and second-cross wether lambs as affected by stress before slaughter. Aust. J. Agric. Res. **50**, 175–181 (1999)

Gürsoy, O.: Economics and profitability of sheep and goat production in Turkey under new support regimes and market conditions. Small Ruminant Res. **62**(3), 181–191 (2006)

Hamm, R.: Functional properties of the myofibrillar system and their measurements. In: Bechtel, P.J. (ed.) Muscle as Food, pp. 135–199. New York Academic Press, Inc. (1986)

Hoffman, L.C., Muller, M., Cloete, S.W.P., Schmidt, D.: Comparison of six crossbred lamb types: sensory, physical and nutritional meat quality characteristics. Meat Sci. **65**, 1265–1274 (2003)

Honikel, K.O.: How to measure the water holding capacity of meat? Meat Sci. **49**, 447–457 (1988)

Hopkins, D.L., Fogarty, N.M.: Diverse lamb genotypes – 2. Meat pH, colour and tenderness. Meat Sci. **49**, 477–488 (1998)

Lanza, M., Bella, M., Priolo, A., Fasone, V.: Peas (Pisum sativum L.) as an alternative protein source in lamb diets: growth performances, and carcass and meat quality. Small Ruminant Res. **47**, 63–68 (2003)

Marmer, W.N., Maxwell, R.J., Williams, J.E.: Effects of dietary regimen and tissue site on bovine fatty acid profiles. J. Anim. Sci. **59**, 109–121 (1984)

Montossi, F., Fonti Furnols, M., del Campo, M., San Julián, R., Brito, G., Sañudo, C.: Sustainable sheep production and consumer preference trends: compatibilities, contradictions, and unresolved dilemmas. Meat Sci. **95**, 772–789 (2013)

Karaca, O., Yikilmaz, H., Cemal, İ., Atay, O.: Çine Tipi ve Menemen x Çine Tipi (F1) melezi kuzularin kimi gelişme özellikleri. Uluslararasi Hayvancilik'99 Kongresi, 21-24 Eylül 1999, pp. 776–770. Ege Üniversitesi Ziraat Fakültesi, Bornova, İzmir, Turkey (1999)

Priolo, A., Micol, D., Agabriel, J.: Effects of grass feeding systems on ruminant meat colour and flavour. A Rev. Anim. Res. **50**(3), 185–200 (2001)

Romedi, Ç., Yilmaz, A.: Certain meat characteristics of Awassi and Turkish Merino x Awassi (F1) lambs. Turk. J. Vet. Anim. Sci. **34**(4), 249–257 (2010)

Sanudo, C., Sánchez, A., Alfonso, M.: Small ruminants production systems and factors affecting lamb meat quality. Meat Sci. **49**(1), 29–64 (1998)

SAS.: The SAS system. Version 8. Copyright (c) 1999 by SAS Institute Inc. Cary, NC, USA (1999)

Vacca, G.M., Carcangiu, V., Pazzola, M., Mura, M.C., Luridiana, S., Tilloca, G.: Productive performance and meat quality of MouflonxSarda and SardaxSarda suckling lamb. Meat Sci. **80**, 326–334 (2008)

Vatansever, L., Demirel, G.: Fatty acid content and composition of Turkish beef and lamb at retail. Turk. J. Vet. Anim. Sci. **33**(1), 71–75 (2009)

Zygoyiannis, D.: Sheep production in the world and in Greece. Small Ruminant Res. **62**(1–2), 143–147 (2006)

The Effect of Horse Age and Competitive Season on Stress Indicators in Jumping Horses in Running Track

Maja Gregić[1(✉)], Mirjana Baban[1], Pero Mijić[1], Tina Bobić[1], Dragan Dokić[2], Muhamed Brka[3], and Vesna Gantner[1]

[1] Faculty of Agrobiotechnical Sciences Osijek, Josip Juraj Strossmayer University of Osijek, Osijek, Croatia
mgregic@fazos.hr
[2] Općina Erdut, Bana Josipa Jelačića 4, Dalj, Croatia
[3] Faculty of Agriculture and Food Sciences, Department for Animal Breeding, University of Sarajevo, Sarajevo, Bosnia and Herzegovina

Abstract. Aiming determination of the effect of horse age and competitive season (May, July, September) on stress indicators in jumping horses in running track, research was performed on 14 studs (7 younger, and 7 older) trained and bred in the same conditions. The stress indicators (heartbeats, cortisol, glucose, and lactate concentrations in saliva) were measured or sampled before, during and at the end of the training - running track. The significance of the differences in stress indicators due to horse age regarding the competitive season was tested by t-test (PROC GLM, SAS/STAT). Based on conducted research it could be concluded that all analyzed stress indicators: heartbeats, and cortisol, glucose, and lactate concentrations in saliva, during the entire competitive season, were higher in younger (less experienced) comparing to older (more experienced) horses. Furthermore, all analyzed stress indicators were highest in July that is in the peak of competitive season. Since every physical activity causes a certain amount of stress, monitoring of stress indicators, especially heart-beats could be used for routine evaluation of horse preparedness for a particular activity.

Keywords: Jumping horses · Running track · Stress indicators

1 Introduction

Equestrian sport is an Olympic discipline under the Federation Equeste International – FEI (FEI 2014). In order to attain adequate results in equestrian sports, horse requires adequate and good training. Accordingly, to Mills and McDonnell (2005) training of horses implies achieving concentration and fitness for a certain type of equestrian sport with special emphasis on preservation of health, psychological and physical condition as well as longevity of horse. Training could be defined as a constant repetition of physical and mental actions to encourage the development of adjustments to those actions, which would lead to the efficiency increase in the same actions (Hick and Hick 2000). Since the equestrian sports are physically very demanding for animals, it could result in prevalence of stress condition in horses (Becker-Birck et al. 2013). The horse

M. Brka et al. (Eds.): AgriConf 2019, IFMBE Proceedings 78, pp. 185–190, 2020.
https://doi.org/10.1007/978-3-030-40049-1_24

with a great tolerance to stress and quick adaptation to physical efforts is a desirable one in a show jumping sports. Furthermore, in order to optimize the performances, it is very important to evaluate the level of stress in individual animal. Stress in horses can be evaluated by observing and analysis of the physiology and, also, the concentrations of certain substances in the organism (Snow and Valberg 1994). The cardiovascular parameters (heartbeats/minute), cortisol, eye temperature, and behaviour are good indicators of short-term stress (König et al. 2017). Accordingly, to Harbig (2006) the heart-beats can vary from 50 to 240 heartbeats/minute depending of horses' age and level of physical efforts (step or severe effort). Christensen et al. (2014) stated that the salivary cortisol (widely used measure to evaluate stress response in horses) is affected by the time of day, season, as well as body condition score. Furthermore, the concentration of glucose and lactate in saliva could show the body resistance to the physical stress or training effect (Kroner 2006). The concentration of glucose and lactate in saliva, is affected by a variety of factors such as: fitness, age, sex, breeding conditions, and equestrian sport discipline. The aim of this study was to determine the effect of horse age (younger and older) and competitive season (May, July, September) on stress indicators in jumping horses as a result of training on the running track.

2 Materials and Methods

The research was performed on a Holstein studs and a Croatian sports horse breeding type studs that are used in jumping horse sports and were trained and bred in the same conditions. All horses were healthy and accommodated in the same conditions, in separate stalls. Feeding was the same, slightly different between winter and summer season. One person was responsible for sport and fitness conditions. During the winter period, the horses regularly spent 20 min daily on a running track, lunging and periodically in saddle, while in the preparatory, and competitive period, the training included a running track, lunging, work under the saddle and show jumping, as well as independent motoric movements in outlet. Accordingly, to age, horses were divided in two groups by seven animals: younger (4 to 5 years old) and older (8 to 9 years old). Furthermore, the research was performed during the competitive season in following months: May, July and September. The stress indicators, heartbeats, cortisol, glucose, and lactate, were measured or sampled before, during and at the end of the training - running track (Schmidt et al. 2010; Becker-Birck et al. 2013). The running track is a part of the training that gives a horse correct lateral developments in parallel movements. The equipment of the jumping horses was individual, adapted to each horse. For the measurements of heartbeats polar equipment (RS800CX N G3), equine electrode set for H2 sensor, and equine health check FT1 were used. Saliva samples in horses were taken by a cotton tampon (Salivette Cortisol, code blue) that was fixed by the paean, pushed into the oral cavity, held there about one minute and stored adequately. Tampon with saliva was processed in the laboratory and the saliva sample was stored at −20 C until the analysis. The cortisol concentrations in the saliva was determined by ELISA 96 tests (DES6611). The lactate concentration in saliva was determined accordingly to Lactate Photometry Method on a biochemical analyser (Beckman Coulter AU400, Japan). Furthermore, the glucose concentration in saliva was by

Enzymatic Method at the same biochemical analyser. Basic variability of measured stress indicators in jumping horses during the training in parkour is presented in Table 1.

Table 1. Basic statistic of stress indicators in jumping horses in running track

Stress indicators	Mean	SD	CV	Min	Max
Heartbeats/min	47.06	27.16	57.70	25.00	129.00
Cortisol (ng/ml)	0.91	0.68	74.39	0.02	3.99
Glucose (mg/dl)	40.37	51.35	127.21	0.25	366.65
Lactate (mmol/l)	0.79	0.56	70.45	0.10	3.23

For statistical analysis SAS/STAT (SAS Institute Inc., 2002) was used. The significance of the differences in stress indicators (heartbeats, cortisol, glucose, and lactate) between the horse age (younger, and older) regarding the competitive season (May, July, September) was tested by t-test (PROC GLM).

3 Results and Discussion

LSmeans of analysed stress indicators that is heartbeats, cortisol, glucose, and lactate concentrations in saliva regarding the horse age (younger, and older) separately for each competitive season: May, July, September are shows in Tables 2, 3 and 4.

Table 2. LSmeans of stress indicators in jumping horses in running track regarding the age in competitive season – May

Stress indicators/age	Heartbeats/min	Cortisol (ng/ml)	Glucose (mg/dl)	Lactate (mmol/l)
Younger	47.132^{A}	0.977^{A}	50.167^{A}	0.934^{A}
Older	45.275^{A}	0.736^{B}	26.982^{B}	0.505^{B}

LSmeans marked with different letters (A, B) differ statistically significant ($p < 0.05$)

All measured stress indicators during the competitive season – May were higher in younger horses with statistically significant ($p < 0.05$) difference in concentrations of cortisol, glucose and lactate in saliva, while the difference in heartbeats in amount of 1.857 heartbeats/min) wasn't statistically significant ($p > 0.5$).

Table 3. LSmeans of stress indicators in jumping horses in running track regarding the age in competitive season – July

Stress indicators/age	Heartbeats/min	Cortisol (ng/ml)	Glucose (mg/dl)	Lactate (mmol/l)
Younger	48.945^{A}	1.144^{A}	57.923^{A}	1.073^{A}
Older	47.604^{A}	0.841^{B}	27.615^{B}	0.669^{B}

Lsmeans marked with different letters (A, B) differ statistically significant ($p < 0.05$)

During the competitive season – July, significantly ($p < 0.05$) higher concentrations of cortisol, glucose and lactate was determined in saliva of younger horses, while the difference in heartbeats existed but wasn't statistically significant ($p > 0.5$).

Table 4. LSmeans of stress indicators in jumping horses in running track regarding the age in competitive season – September

Stress indicators/age	Heartbeats/min	Cortisol (ng/ml)	Glucose (mg/dl)	Lactate (mmol/l)
Younger	48.308^{A}	1.033^{A}	53.912^{A}	1.000^{A}
Older	45.088^{A}	0.743^{B}	25.605^{B}	0.556^{B}

Lsmeans marked with different letters (A, B) differ statistically significant ($p < 0.05$)

During the competitive season – September similar trends were determined as in May and July. Higher values of all analyzed stress indicators were determined in younger horses, with statistically significant ($p < 0.05$) difference in concentrations of cortisol, glucose and lactate in saliva. In younger horses, the lowest values of all stress indicators were measured at the beginning of competitive season, in May, while the highest values of all stress indicators in all horses regardless the age were measured in July. Determined values of stress indicators implicate highest stress level in the peak of competitive season comparing to the beginning and the end of season.

The results allow to emphasize that all measured stress indicators (heartbeats, and concentration of cortisol, glucose and lactate in saliva) were higher in younger comparing to older horses during the entire competitive (May, July and September). Higher values of heartbeats/min in younger horses comparing to older ones, whether in a relaxation state or in any kind of physical action, was also observed by Klimke (2012). Which found that. Ille et al. (2013, 2014) in research with experienced and unexperienced jumping horses found out similar results. This could be explained by the fact that in older horses, during the years of training, the heart rate has become more uniform and balanced. The determined higher concentrations of cortisol in saliva of younger horses are in agreement with the results of Covalesky et al. (1992) and Ille et al. (2013, 2014) who determined higher plasma cortisol in inexperienced comparing to experienced horses. Similarly, to results of this research, Covalesky et al. (2002) find out the lowest values of plasma lactate in experienced horses. Furthermore, the lactate concentrations could be used as a parameter of horse fitness, which in equestrian sports means a result (Kroner 2006; Couroucé 2002; Munoz et al. 2002). A horse with a lower lactate level is fitter, meaning that the muscle tissue produces less lactic acid which is converted in blood into the lactate ion. Given the above, König et al. (2017) recommended the concentration of lactate in saliva as indicator for evaluation of horse work efficiency. Horses react individually to stress, but with years of training they achieve routine and become more uniform in performance. Therefore, the older horses have lower values of analyzed stress indicators. Similar trend in uniformity in actions achieved by horses through training was reported by Rose and Evans (1990), Clayton (1991), Brings et al. (1998), Röthing and Prohl (2003) and Voswinkel (2009).

4 Conclusions

Based on conducted research it could be concluded that all analyzed stress indicators: heartbeats, and cortisol, glucose, and lactate concentrations in saliva, during the entire competitive season, were higher in younger (less experienced) comparing to older (more experienced) horses. Furthermore, all analyzed stress indicators were highest in July that is in the peak of competitive season. Since every physical activity causes a certain amount of stress, monitoring of stress indicators, especially heartbeats could be used for routine evaluation of horse preparedness for a particular activity.

References

Becker-Birck, M., Schmidt, A., Lasarzik, J., Aurich, J., Möstl, E., Aurich, C.: Cortisol release and heart rate variability in sport horses participating in equestrian competitions. J. Vet. Behav. Clin. Appl. Res. **8**, 87–90 (2013)

Brings, J., Platen, P., Hoffmann, E.: Testverfahren zur Beurteilung der Ausdauer- und Sprintleistungsfähigkeit im Frauenhandball. Leistungssport **6**, 26–31 (1998)

Christensen, J.W., Beekmans, M., van Dalum, M., van Dierendonck, M.: Effects of hyperflexion on acute stress responses in ridden dressage horses. Physiol. Behav. **128**, 39–45 (2014)

Clayton, H.M.: Conditioning Sport Horses. Sport Horse Publications, Saskatoon (1991)

Courouce, A.: Endurance and sprint training. A. Lindner: Conference on Equine Sports. Medicine and Science, pp. 190–202. Wageningen Pers, Wageningen (1998)

Courouce, A., Chretien, M., Valette, J.P.: Physiological variables measured under field conditions according to age and state of training in French trotters. Equine Vet. J. **34**, 91–97 (2002)

Harbig, S.: Leistungsmonitoring von Hochleistungsvielseitigkeitspferden im Wettkampf und Training: Untersuchungen zur Herzfrequenz Diss., Tierärztl. Hochschule, Hannover (2006)

Hick, C., Hick, A.: Kurzlehrbuch Physiologie, pp. 41–77, 153–160. Urban & Fischer Verlag, München (2000)

Ille, N., Lewinski, M., Erber, R., Wulf, M., Aurich, J., Möstl, E., Aurich, C.: Effects of the level of experience of horses and their riders on cortisol release, heart rate and heart-rate variability during a jumping course. Animal Welfare **22**, 457–465 (2013). Universities Federation for Animal Welfare

Ille, N., Aurich, C., Erber, R., Wulf, M., Palme, R., Aurich, J., Lewinski, M.: Physiological stress responses and horse rider interactions in horses ridden by male and female riders. Comp. Exerc. Physiol. **10**(2), 131–138 (2014)

Klimke, R.: Grundausbildung für Reitpferde. Franckh Kosmos Verlag, Auflage: 7 (2012)

Kroner, K.: Blut- und Speichelparameter beim Kaltblutpferd in Ruhe und bei Zugarbeit. Inaugural-Dissertation, München (2006)

König, V., Borstel, U., Visser, E.K., Hall, C.: Indicators of stress in equitation. Appl. Anim. Behav. Sci. **190**, 43–56 (2017)

Mills, D.S., Mc Donnell, S.M.: The Domestic Horse: The Origins, Development, and Management of its Behavior. Cambridge University Press, Cambridge (2005)

Munoz, A., Riber, C., Santisteban, R., Lucas, R.G., Castejon, F.M.: Effect of training duration and exercise on blood-borne substrates, plasma lactate and enzyme concentrations in andalusian, anglo-arabian and arabian breeds. Equine Vet. J. **34**(Suppl. 34), 245–251 (2002)

Schmidt, A., Biau, S., Möstl, E., Becker-Birck, M., Morillon, B., Aurich, J., Faure, J.-M., Aurich, C.: Changes in cortisol release and heart rate variability in sport horses during long- distance road transport. Domest. Anim. Endocrinol. **38**, 179–189 (2010)

Snow, D.H., Valberg, S.J.: Muscle anatomy, physiology and adaptations to exercise and training. In: Hodgson, D.R., Rose, R.J. (Hrsg.) The Athletic Horse, pp. 145–179. WB Saunders Company, Philadelphia (1994)

Rose, R.J., Evans, D.L.: Cardiovascular and respiratory function in the athletic horse. In: Gillespie, J.R., Robinson, N.E. (eds.) Equine Exercise Physiology 2, pp. 1–19. ICEEP Publications, Davis (1990)

Röthing, P., Prohl, R.: Sportwissenschaftliches Lexikon. 7. Auflage, Hoffmann Verlag, Schorndorf (2003)

Voswinkel, L.: Einfluss der bewegungsaktivität auf wachstums- und ausdauerparameter beim pferd. Aus dem Institut für Tierzucht und Tierhaltung der Agrar- und Ernährungswissenschaftlichen Fakultät der Christian-Albrechts-Universität zu Kiel (2009)

Agricultural Economics

Stakeholder and Social Networks: Basis for Developing Pasture Management Strategies

Emelj Tuna(✉) and Aleksandra Martinovska Stojcheska

Faculty of Agricultural Sciences and Food, University Ss Cyril and Methodius, Skopje, North Macedonia
emelj.tuna@fznh.ukim.edu.mk

Abstract. Pastures in Shara Mountain are managed by the Public enterprise for management of Pastures (PEMP) in the Republic of North Macedonia. The national pasture development strategy is still not ascertained, thus many structural questions remain unsolved. Strengthening the capacities for pasture management is very significant, hence the need to identify all relevant stakeholders on horizontal and vertical level that can contribute in the proper information diffusion and raise public awareness among all affected actors in the pasture management network. In this respect, the aim of this paper is to describe certain aspects of social complexity of pasture management structures in order to provide grounds for developing sustainable management strategy. Combining methods such as social network analysis (SNA) with stakeholder analysis can often contribute to the richness in the analysis and add to the understanding of relations. Therefore, the analysis primarily aims to identify the stakeholders in the context of mountain grazing areas of Shara Mountain, but also present the relations and information sharing network of the actors at regional and local levels on horizontal level, and the vertical and horizontal interplay between the actors at different levels such as their institutional and commercial associates. Mapping each stakeholders' positions and roles in the communication network and information transfer is expected to provide a base for developing strategy tailored to each stakeholders' future involvement in the pasture management system.

Keywords: Information · Relations · Pasture management · Stakeholder · Social networks · Shara mountain

1 Introduction

Majority of the pastures in Shara Mountain are managed by the Public Enterprise for Management of Pastures (PEMP), however a comprehensive strategy on pasture development is still not established. PEMP faces many structural and infrastructural problems such as absence of cadaster plan of pastures, absence of register of pastures, no information about the actual quality (nutritional value) of pastures in use, and accurate records on pasture users. Other problems of the enterprise include the complexity of monitoring and reporting of the right for pasture use, which points out to the need of establishing closer ties and support from the related institutions, in order for the enterprise to be able to execute inspection controls as well as controls of the current on-

M. Brka et al. (Eds.): AgriConf 2019, IFMBE Proceedings 78, pp. 193–200, 2020.
https://doi.org/10.1007/978-3-030-40049-1_25

field situation with the pastures and their actual conditions. The lack of monitoring system and communication with the related institutions is also causing discrepancies in the data on pastures, offered by the State Statistical Office, and the actual situation, leading to problems in tailoring specific and substantial strategies and business plans for pasture management. The official statistics report 745 thousand hectares of pastures in North Macedonia, which comprise 59% of the total country agricultural area [1]. Additionally, the pastures are fragmented, and there is a need for a system of enlargement, regionalization of the pastures in order to achieve development and modernization of the pasture conditions.

The challenges faced by PEMP point to the urgent need for strengthening the capacities for pasture management. The first step is thus to identify all of the relevant stakeholders in the network for pasture use and management. Failing to identify stakeholders' interests is one of the most important reason that many development strategies fail. Additionally, communication networks can play vital role in categorizing and understanding stakeholder relationships [2]. Nevertheless, this approach is focused only on the individual stakeholders, and in order to see the structure of their relations, evaluate their power and influence, we overcome the limitation of stakeholder analysis by combining it with Social network analysis (SNA) [3].

The aim of this analysis is to present the relations and information sharing network of the regional and local actors at horizontal level, and the vertical and horizontal interplay between the actors at different levels such as their institutional and commercial associates (among pasture users and representatives from different state institutions, distribution/trading partners and other identified stakeholders).

Depicting informal and formal relations and networks of information sharing among various actors requires specific data, data gathering and analyzing approaches. In this regard, Stakeholder Analysis and Social Network Analysis were applied for analyzing the structural characteristics that underline the governance in the pasture use and management. Therefore, the analysis primarily aims to identify and select the relevant stakeholders in the context of mountain grazing areas of Shara, and can serve as basis for developing proper strategies. Moreover, it presents the relations and information sharing network of the involved actors at regional and local levels and their vertical and horizontal interaction.

2 Methodology

Combining methods as stakeholder analysis with social network analysis can often contribute to the richness in the analysis and understanding of the relations [3]. This study includes: (1) Stakeholder profiles as a tool for initial characterization of the relevant stakeholders (background information, functions and roles within the scope of this project and their objectives), and (2) Sociogram of stakeholders' network (UCI-NET and NetDraw) to represent stakeholder relationships specific to the problem.

Stakeholder analysis approaches have increasingly been adapted from primarily business management use to policy, development and natural resource management applications [2]. It is an important technique for identification of all key stakeholders (primary and secondary) and their needs. The aim is to develop a strategic view of the human and institutional landscape, and the relationships between the different stakeholders and the issues they care about most. The key identified stakeholders are presented in a stakeholder matrix in order to classify their role and power positions. The objective is to identify the actors, but also build their profile for further development of their relations and their networks. Stakeholder analysis is also recognized as an approach that provides an understanding on the behavior, intentions, interrelations and interest of the most important: individuals and organizations as well as their influence and resources on the decision-making or implementation process of certain projects and activities [4].

Social Network Analysis (SNA) is a specific methodological approach which requires distinctive type of questions in order to construct and map relations among the pre-defined network of actors. The structure of the relations in the different types of networks could contribute to the understanding and predicting of the behavior of the existing actors (stakeholders) [5]. Our SNA was performed using adjacency matrices where the actors' relations in the network were coded (1 in the case of an existing relation and 0 when the relation was absent). The total network of all the participants in the information channel network) and the structure of the information network are presented through sociograms. The SNA data was analyzed using UCINET, specialized software tool for analyzing social structures, and their visualization presented using NetDraw [6].

In addition to desk research and expert opinions, primary data were collected through face-to-face interviews with 50 farmers from the Polog region, most of which Shara Mountain pastures users. In order to achieve this task, it was necessary to include in the survey, as many of the farmers/pasture users of the ascertained region. As full list of users was not readily rarely available, and farmers, sheep and livestock breeders operate in specific conditions and are often difficult to reach, it was important to include PEMP in providing detailed contacts. The survey was conducted through a structured questionnaire, specially designed for the purpose and the selected method of analysis. A "Name generating table" was used in order to construct a map (sociogram) of all the relations and information flow. Each surveyed pasture user was asked to nominate three to five connections (other sheep and livestock breeders, farmers, trading partners, government institutions, associations), with whom they discuss or share information on important issues regarding pastures use and management, but also information on livestock breeding. The number of nominations is usually given as motive for more nominations, since limiting this number could lead to measurement errors [7, 8].

3 Results

3.1 Identifying Stakeholders and Stakeholders Profiles

The first task of the Stakeholder Analysis is to produce a list of key stakeholders, farmers and pasture users in the Shara Mountain area. The stakeholders are classified in three groups: institutional actors, user collectives (associations) and individual stakeholders. The analysis identified six different groups of stakeholders (Fig. 1): (1) Direct pasture users (green); (2) Institutions and institutional representatives (orange);

(3) Trading partners (buyers, veterinary services) (dark blue), (4) Processors (light blue), (5) Associations (yellow) and Local government (grey). The size of the circle corresponds to the number of nominations that each stakeholder was given by the farmers, and their position is assigned in accordance to their importance in the pasture management strategies.

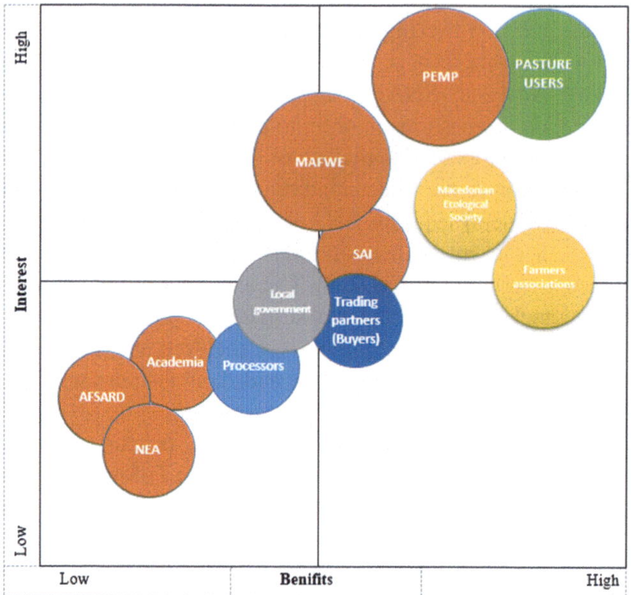

Fig. 1. Stakeholders in accordance to their interests and benefits from pasture management strategies

In terms of information, PEMP holds both the highest power and interest of gathering and sharing information on the needs of their main users. As indicated also by the SNA, the PEMP representatives, are the nucleuses that can influence and are expected to be the main creator, driver and pursuer of the strategies for capacity building and improved pasture management of the mountain grazing areas of Shara Mountain.

In order to identify the most relevant actors in the information network of the mountain grazing areas of Shara, SNA provides mapping or graphical representation of all the existing actors and their relations (Fig. 2). After the nominations, beside the interviewed farmers, the network included 60 additional livestock farmers from the region, 5 institutions/institutions' representatives and 14 trading partners and other stakeholders, or in total a network of around 120 actors. The generated network is resembling two-mode network which gives insight on the horizontal relations with peers, and the vertical relations with other relevant actors in the information network sharing.

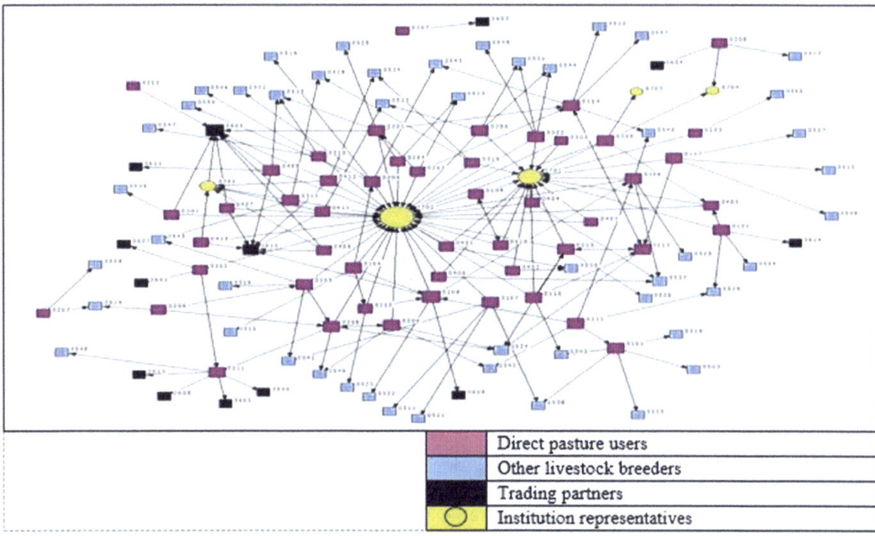

Fig. 2. Major groups of stakeholders (node size representing the number of ties: larger node - larger number of connections)

The representatives of PEMP are constructing the cohesive sub-group of the network, so called the k-core of the information network. The core nodes are connected to each other and others, and the periphery nodes are only connected to the core nodes (actors). These are the nodes where the highest level of social capital is concentrated, which can be seen as an additional confirmation of their importance and power in this network and the reason why this group of people should be managed closely with the aim to be fully engaged in the information transfer and effectively involved in the development of pasture management strategies.

While in Stakeholder Analysis, stakeholders were identified through a combination of desk research, interviews with institution representatives and experts. SNA is based on questionnaires, and the stakeholders are identified, based solely on the views and perceptions of the interviewed pasture users, included in this analysis. They identified primarily the important stakeholders on horizontal level (other farmers, important for the information sharing network), but also the stakeholders they communicate with pasture management related issues on vertical level (institutions, government, associations etc.).

SNA includes two important characteristics of social action: first, every actor (stakeholder) is part of a certain system, which influences its actions and decisions; and second, every actor's position in the network represents its influence, power, or in this case, ability to transfer information [4]. The information network of farmer relations (horizontal level) in the sample of Shara mountain pasture users is constituted of large number of 113 components which suggests that the network is very fragmented, although it was a relatively bounded network of farmers. This fragmentation is confirmed also by the fragmentation measure which is very close to 1 (0.975), a fact that additionally influences the low level of density of this network, and is one of the primary indicators of social cohesion of the network (Fig. 3).

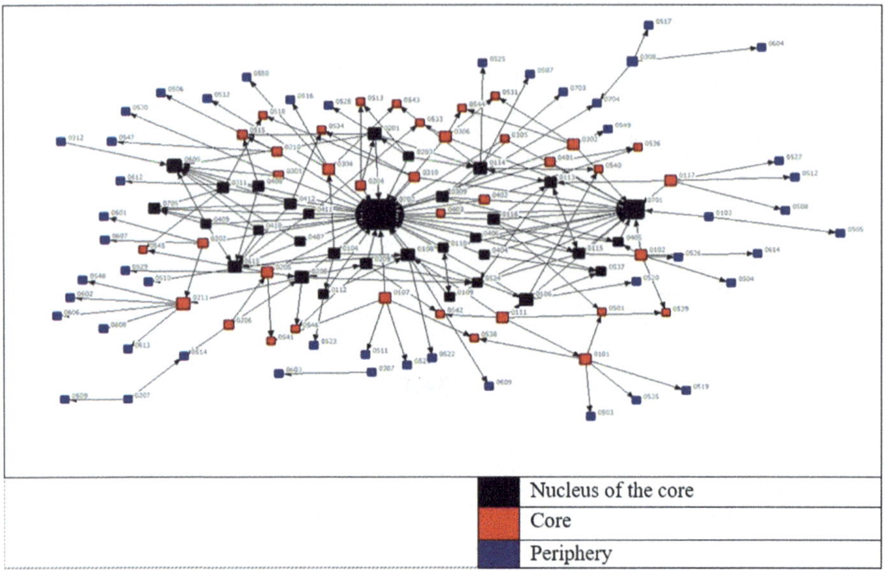

Fig. 3. Core/periphery structure of the information network

Social ties are rarely distributed equally among the actors in the networks, and in terms of identification of the influential actors in the information network, most of the interviewed farmers reported relations of highest strength (average 5.8 on a scale of 1 to 10, 1 – no relation/10 frequent relation), with the representatives of PEMP. These representatives participate in most of the identified subgroups and are the nodes with the highest degree and eigenvector values measuring their direct and indirect connections. Node 0702 (Head of PEMP unit) with the largest number of nominations (37 in-degree relations), and node 0701 (Director of PEMP), with 19 in-degree relations (nominations), hold central position in the information sharing network, which indicates that those are the persons mostly consulted on questions related to pastures. These are the individuals that act as information brokers in this network, and as such have the ability and power to regulate the flow of information. They also occur in the cohesive subgroups, which are mostly based on informal relations with relatives or other, geographical close individuals, which confirms the presumption that being in the same place influences the development of similarities especially when it comes to decisions and interests.

Few other institutional actors were also identified in this network. Again, most of the relations and information sharing are based on informal basis, and none of these actors has prominent role in this information sharing network. Similar situation was present in the case of the other nominated farmers with the highest value of in-degree (number of received relations) not higher than 7.

4 Conclusions

Due to the current deficiencies in management structure and pasture management, and in order to create a formal network for information sharing of all stakeholders in the Shara Mountain pastures', the main objective was to map the stakeholders in the use and management of the Shara Mountain pastures'. More specifically the aim was to describe and analyze certain aspect of social complexity of the pasture management structures in order to provide base for developing sustainable management strategies.

In terms of institutional setup, a structure of institutions from the government should be acting and supporting PEMP as the main and responsible body for pasture management and the broker in the information transfer, and vice versa. However, PEMP and the overall system of information and cooperation face problems in term of missing basic resources such as full database of primary information, lack of human and technical capacity for monitoring and assessment of the factual on-field conditions of the pastures, spatial disposition and use etc. Therefore, establishing tighter communication and cooperation links among the related institutions is necessary, since these links are in most instances either missing or are not functional. Additionally there is a need for collaboration of all the concerned actors in achieving correct interpretation of related legislative related to pastures and pasture management.

The Stakeholder analysis identified many different stakeholders that could contribute to development strategies for the Shara Mountain pastures. The key stakeholders are the beneficiaries – the direct pasture users, and PEMP as the major institutional body in the pasture management in the country. The line Ministry of Agriculture Forestry and Water Economy and its bodies are also important for establishing firm policy base, enforcement and monitoring of the rules stipulated in the Law, and assist in setting solid base for setting a long-term strategy for pasture management in the country. Besides governmental institutions, it is important to emphasize the role of farmers' associations, which could be an important medium for information transfer and communication of common interests. However, such associations are still on a rudimentary level of organization and cooperation; hence supporting farmers' associations and agricultural cooperatives is necessary for acquiring and conducting projects for modernization and advancement of the pastures and their use. Along with the advisory segment and the academia, associations may be used for the successful introduction and implementation of the developed strategy.

The analyzed network is constituted of high number of separate cohesive groups, which can be challenging for joint management of natural resources, because of the distinctive features and ties each cohesive group possesses. It is essential that the existent communication with other stakeholders is maintained, in order for them to express their challenges and actively participate in the future strategies. In this respect, it is also necessary that this informal information channels are formalized through regular information and feedback meetings for communicating the current and relevant needs of the primary pasture users.

Higher density of the networks usually indicates increased levels of collective action, and is directly connected to the communication possibilities, an improved communication alternatively increases the level of social capital, reciprocity of relation and information exchange and trust. It can also increase the levels of knowledge and understanding, thus increase the level of power and influence for the farmers that are in the primary focus. In agriculture, it is very common that information are transferred through informal social ties. The analysis confirms that this circumstance for the farmers, users of the Shara Mountain pastures, whose relations with other farmers, but also representatives of the institutions are mostly based on close, informal relations.

Proper pasture management strategy should provide the basis for priorities setting, resource allocation, capability requirements and budgeting activities. As demonstrated through the SNA, PEMP representatives are the nucleuses that can influence and are expected to be the main creator, driver and pursuer of the strategies for capacity building and improved pasture management of the mountain grazing areas of Shara. PEMP should be actively involved in the communication, education and raising of the awareness for the importance of proper pasture management.

Acknowledgment. The research was conducted within the project: Conservation of biodiversity in mountain grazing areas of Shara, report "Stakeholder analysis and networking capabilities" on the request of "Mileucontact Macedonia". The results are used with full knowledge and approval of the commissioning party.

References

1. SSO: State Statistical Office, Republic of North Macedonia (2019)
2. Reed, M.S., Graves, A., Dandy, N., Posthumus, H., Hubacek, K., Morris, J., Prell, C., Quinn, C.H., Stringer, L.C.: Who's in and why? A typology of stakeholder analysis methods for natural resource management. J. Environ. Manag. **90**(5), 1933–1949 (2009)
3. Prell, C., Hubacek, K., Reed, M.: Stakeholder analysis and social network analysis in natural resource management. Soc. Nat. Res. **22**(6), 501–518 (2009)
4. Brugha, R., Varvasovszky, Z.: Stakeholder analysis: a review. Health Policy Plann. **15**(3), 239–246 (2000)
5. Medicamento, U., Degennaro, B.: Social Networks and Supply chain Management in rural Areas: A case Study Focusing on Organic Olive Oil. MPRA Paper No. 14558 (Munich Personal ReOPEc Archive), (2006). Accessed 12 Apr 2009
6. Borgatti, S.P., Everett, M.G., Freeman, L.C.: UCINET for Windows Software for Social Network Analysis. Anal Technologies, Harvard (2002)
7. Wasserman, S., Faust, K.: Social Network Analysis, Methods and Applications. Cambridge University Press, New York (1994)
8. Lin, N.: A network theory of social capital. In: Castiglione, D., van Deth, J.W., Wolleb, G. (eds.) The Handbook of Social Capital, pp. 50–69. Oxford University Press, London (2008)

Exploring the Links Between Farmers' Intention to Apply for Rural Development Program and Networks

Aleksandra Martinovska Stojcheska(✉), Ana Kotevska, Emelj Tuna, and Ana Simonovska

Faculty of Agricultural Sciences and Food, University Ss Cyril and Methodius in Skopje, 16-ta Makedonska Brigada 3, 1000 Skopje, North Macedonia
amartinovska@fznh.ukim.edu.mk

Abstract. The rural areas in North Macedonia encompass around half of the territory and population in the country, where agriculture is core activity and source of income, with lack of other employment opportunities. The rural development program (RDP) budget is not fully utilized. Informal networks among rural population are still the most valuable source of social capital and information exchange, whereas formal networks, such as associations or cooperatives, are still lacking to ensure their sustainable and functional role. This paper explores the factors influencing farmers' intention to apply for RDP, with an emphasis on network membership. The Theory of Planned Behavior and Social Capital Theory are used as framework. A direct survey on 296 family farms was carried out in 2014. The study is conducted using multivariate statistics, structural equation modelling and social network analysis. The model explains farmers' mid-term intention to use RDP. Farmers' attitudes and perceived behavioral controls, as well as network activity and participation, significantly influence the intention to apply for RDP funds. Network participation, as well as education levels and investment tendency are positively related to the intention to use RDP. These findings could be reflected in appropriate measures to encourage farmers' participation in institutionalized networks, thus in addition to their primary aim, to increase the RDP effectiveness as means of providing better opportunities and quality of life in rural areas.

Keywords: North Macedonia · Social Capital Theory · Theory of planned behaviour

1 Introduction

Networks play an important role in information dissemination, particularly in the otherwise scarce information flow in rural areas. Informal networks are still the most valuable source of social capital and information exchange in post-socialist countries such as North Macedonia. Formal networks, for instance cooperatives, remain an unappealing concept for Macedonian farmers, reminiscent to its socialistic definition, even after decades in the new political and economic system.

M. Brka et al. (Eds.): AgriConf 2019, IFMBE Proceedings 78, pp. 201–208, 2020.
https://doi.org/10.1007/978-3-030-40049-1_26

Rural areas accommodate nearly half of the population in the country (42% in 2017, [1]). Agriculture is the main source of income in rural areas with lack of other employment opportunities. Although there is rural development program (RDP) to support the improvement of life and conditions in rural areas, the budget is not fully utilized. The funds for structural and rural development measures are still at low level, although with an upward trend in recent years, taking up about 25% of the total support to agriculture and rural development [2].

Social capital is thought to influence rural development, networking, and decision-making in rural areas. It facilitates the utilization of local resources by creating social networks, trust, and civicness [3]. Because of the small size and low level of modernisation, restoring social capital potentials and changing farmers' attitudes towards cooperation is an important determinant for the economic growth of farms and rural areas. Social capital includes formal or informal social structures of cooperation for mutual benefit that offer a suitable arrangement for better information flow [4, 5].

Social psychology provides suitable conceptual frameworks and methodological tools to understand farmer behaviour and design interventions to support a behavioural change. TPB argues that attitudes towards a certain behaviour, subjective norm i.e. approval and acceptance by the referential environment and perceived controls over the behaviour affect the intention and consequently performance of the behavior in question [6]. Once an intention is formed, next is to close the intention-behaviour gap, by helping overcome the obstacles to perform the behaviour (increasing behavioural control) [7]. For doing so, it is important to identify the internal and external factors that form the control, hence design an assistance to provide individuals with the required self-esteem, tools and resources (trainings, networks, etc.) to encounter unanticipated difficulties and gain actual control over the behaviour performance.

The aim of this paper is to determine the factors influencing farmers' intention to apply for Rural Development Program (RDP), with an emphasis on the influence of social capital and network membership. The theory of planned behavior and social capital theory enable deeper understanding of the influencing factors. This assessment is necessary for improvement of the RDP funds' absorption capacity among farmers.

2 Method and Data

Structured questionnaire was designed to obtain measures of the TPB constructs, social capital/networking measures and additional relevant farmer background information. Eliciting accessible TPB beliefs was done first by using open-ended questions to agricultural experts, supplemented with modal accessible beliefs from the literature review [6, 8, 9]. Then, the questions were additionally simplified when testing the questionnaire with farmers. The TPB statements were assessed on a 5-point disagree-agree scale. Social network analysis (SNA) as a methodological approach provided presentation and quantification of social capital by analysing social structures. Using SNA extended name generating table [10], we gathered information on the farmer's informal cooperation and its effect on the level of application for RDP.

A direct survey on 296 family farms was carried out in 2014 in two regions in the country. The selection of the sample considered previously defined criteria: rural household type, and economically and demographically viable farm households (with at least two members, one being younger than 50).

Following the descriptive statistics to profile the sample, a two-step structural equation modeling (SEM) approach is used to analyze farmers' intention to apply for RD support. In the first step, a measurement model based on the TPB conceptual framework is specified (Model 1), with four groups of latent factors: attitude toward the behavior (ATT), subjective norm (SN), perceived behavioral control (PBC) and behavioral intention (BI), which demonstrated adequate convergent and discriminant validity, as well as reliability. In the second step, a structural model is defined to assess the impact of the behavioral latent factors on the farmers' intention to use RD support.

SNA measures were applied to reflect on the network structure: average degree of centrality as a measure of the quantity of each ego's relations by depicting the number of ties established with its alters (in-degree i.e. number of ties received by a node and out-degree i.e. number of ties initiated by a node) and normalized betweenness centrality referring to the share of each tie passing through a certain node (betacentre). These second level variables were added to the model, to estimate their influence on the latent constructs. Furthermore, the model is controlled by region (with existing network organisation, or not), and membership (Model 2).

Finally, since past behaviour (PB) can strengthen the intention to achieve the behaviour in question [11], the model is enhanced by adding an indication whether the farmer has applied for RD support in the last 3 years (Model 3).

3 Results

3.1 Sample Profile

The profile of the farmers included in the survey is presented in Table 1, through the prism of active membership in an organization (cooperative or professional farmers' organization). There is significant difference between the age of non-members and members; those belonging to an organisation are younger, which is reflected in the shorter faming experience. The level of education is similar throughout the sample, with average 10 years of formal education, the median being on high-school level.

In terms of TPB mean constructs, there is significant difference in all elements except the perceived controls. Namely, all respondents show high intentions to use RD support, are very positive about the benefits of RDP and recognise support from others to engage, though these findings are much more accentuated in the case of active members. The level of perceived control i.e. knowledge, experience and funding, is assessed as equally low by most farmers in the sample.

Members had significantly higher RDP application rate in the last three years (60.7% as compared to 22.9 for non-members), and half of them actually received support (compared to only 15% of non-members).

Table 1. Farmers' profile, TBP constructs and social capital measures, mean ± st.dev. (median).

	Non-members (n = 271)	Members (n = 28)	KW asym.sig.
Age (years)	47.1 ± 12.4 (46.5)	39.8 ± 7.3 (38.0)	.002
Formal education (years)	10.1 ± 2.9 (12.0)	10.2 ± 2.6 (12.0)	.719
Farming experience (years)	25.8 ± 11.8 (25.0)	20.5 ± 7.5 (20.0)	.029
Intention to apply for RDP	3.4 ± 1.0 (3.5)	4.0 ± 1.0 (4.5)	.002
Attitude towards use of RDP	3.9 ± 0.8 (4.0)	4.5 ± 0.7 (5.0)	.000
Social norm	3.3 ± 1.0 (3.0)	4.1 ± 0.8 (4.0)	.000
Perceived behavioural control	2.3 ± 1.0 (2.0)	2.5 ± 1.0 (2.7)	.134
OutDegree	1.5 ± 1.0 (1.0)	1.3 ± 0.5 (1.0)	.257
Indegree	1.3 ± 1.6 (1.0)	3.2 ± 6.0 (1.0)	.169
Betacentrality	69.5 ± 143.8 (8.8)	116.3 ± 119.3 (123.4)	.026
Past RD application (%)	22.9	60.7	.000
Past RD received support (%)	15.5	53.6	.000

3.2 SEM Models

Due to space limitation, we focus on presenting the final model (Model 3), although Tables 1 and 2 presents the model fit and results of the two preceding models.

Table 2. Model fit indicators.

	Ref. value [12]	Base Model 1 TPB	Model 2 TPB+SNA	Model 3 TPB+SNA+PB
CMIN		54.617	13.994	11.502
DF		28	8	9
P	>0.05	0.002	0.082	0.243
CMIN/DF	<5 (<3)	1.951	1.749	1.278
SRMR	<.09	0.0602	0.0264	0.0264
GFI	>0.9 (0.95)	0.966	0.990	0.992
AGFI	>.80	0.933	0.943	0.953
RFI	>0.9	0.941	0.938	0.951
CFI	>0.9 (0.95)	0.981	0.994	0.998
RMSEA	<0.08 (0.06)	0.056	0.050	0.031
LO 90		0.034	0.000	0.000
HI 90		0.079	0.093	0.076
PCLOSE	>0.05	0.293	0.443	0.709
SQ. CORR.		0.630	0.815	0.828

SEM results show the hypothesized model fits the data well. The overall model chi-square ($\chi2$) is with value of 11.5 and 9 degrees of freedom, i.e. the $\chi2$/df ratio is 1.3 (p = .243). Other supplementary fit statistics also indicate good model fit [1]. The model explains 83% of the intention to apply for RDP.

In the model specification, 10 variables enter into the latent constructs explaining the interrelations between the farmers' behavioural determinants. Two statements measure the intention to use RDP, in short-term and mid-term prospect (BI1: "I intend to apply for the RDP in one of the next calls", BI2: "I intend to apply for the RDP in the next 3–5 years"). The included statements demonstrate farmers' attitudes toward the use of RDP (ATT1: "RDP supports the survival of small family farms", ATT2: "RDP increases the income of farms and rural households", ATT3: "The principle of co-finance in RD projects is good motivator for farmers",), subjective norms (SN1: "People I respect approve my application for RDP" and SN2: "Many people I know pursue me to apply for RDP") and perceived behavioural control (PBC1: "I have enough information to independently prepare the application and procedure", PBC2: "My knowledge and experience is enough to independently prepare the application and procedure" and PBC3: "I have enough own means to co-finance RDP investment") (Fig. 1).

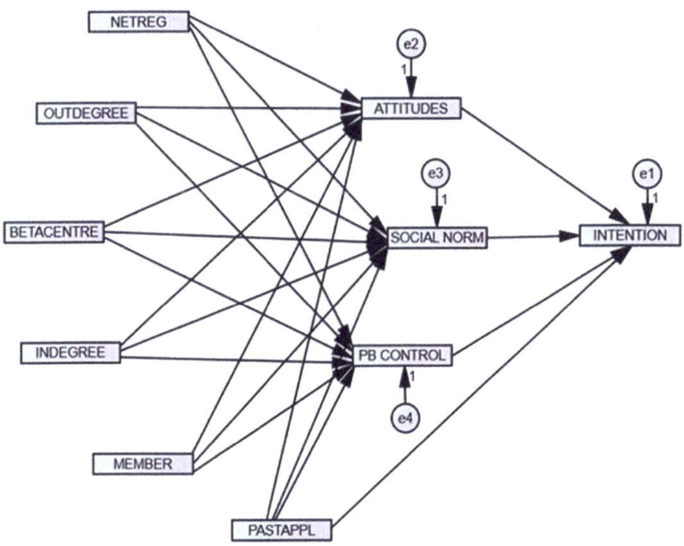

Fig. 1. Illustration of Model 3 (TPB + SNA + PB).

Most paths and relations in the model are statistically significant (Table 3). On TPB level, farmers' positive attitude towards RDP influences farmers' intention to apply most pronouncedly. In addition, stronger perception of controls contribute to stronger intention to apply for RD support. The subjective norms are statistically significantly influencing the intention to apply, but in a negative direction, hence the need for stronger encouragement.

Table 3. TPB and social network determinants of farmers' intention to apply for RDS, standardised estimate (standard error).

	TBP	TBP + SNA	TBP + SNA + PB
Intention <- Attitude	0.891*** (0.152)	1.052*** (0.045)	1.007*** (0.046)
Intention <- Socialnorm	−0.294*** (0.086)	−0.403*** (0.033)	−0.368*** (0.033)
Intention <- PBControl	0.280*** (0.064)	0.350*** (0.026)	0.325*** (0.025)
Intention <- Pastappl			0.109*** (0.052)
Attitude <- Netreg		0.297*** (0.077)	0.353*** (0.077)
Attitude <- OutDeg		0.207** (0.044)	0.143* (0.044)
Attitude <- Betacentre		−0.160** (0.000)	−0.162** (0.000)
Attitude <- Indegree		0.039 (0.015)	0.032 (0.014)
Attitude <- Member		0.259*** (0.113)	0.186** (0.116)
Attitude <- Pastappl			0.237*** (0.080)
Socialnorm <- Netreg		0.294*** (0.093)	0.309*** (0.095)
Socialnorm <- OutDeg		−0.268*** (0.053)	−0.283*** (0.054)
Socialnorm <- Betacentre		0.103* (0.000)	0.102* (0.000)
Socialnorm <- Indegree		0.043 (0.018)	0.041 (0.018)
Socialnorm <- Member		0.221*** (0.137)	0.203*** (0.144)
Socialnorm <- Pastappl			0.062 (0.099)
PBControl <- Netreg		0.231*** (0.125)	0.296*** (0.124)
PBControl <- OutDeg		−0.149* (0.071)	−0.218** (0.071)
PBControl <- Betacentre		0.135* (0.000)	0.132* (0.000)
PBControl <- Indegree		0.076 (0.024)	0.068 (0.023)
PBControl <- Member		−0.047 (0.184)	−0.128* (0.188)
PBControl <- Pastappl			0.269*** (0.130)

The second stage of the analysis reveals the impact of the region, networks, and past behavior. Region positively and significantly influences all latent factors. Membership in an organization is a significant predictor, positively influencing the attitudes and social norm towards the use of RDP, but is negatively influencing the controls. Membership compensates for farmers lower level of perceived behavioral control.

The number initiated ties (OutDegree) has significant positive influence on attitudes, but significant negative influence on social norms and controls. Conversely, the InDegree, i.e. the number of ties received by the node, has no significant influence on either of the factors. In general, the larger the number of links, the wider and more efficient the network use. Betacentrality, as a measure of the effect of one's neighbor's connections on ego's power, has significant positive influence on social norms and controls, but significant negative influence on attitudes.

The past application variable is statistically significant and positively influences the attitudes and controls of farmers towards the behaviour; moreover, in direct relation, it indicates that farmers that have already applied for RDP are more likely to apply again.

4 Discussion

The success of RDP depends upon the intention and ultimately actual behavior of farmers in effectively using RDP funds. Farmers in the survey with more positive attitude and stronger perceived behavioral control (abilities and self-confidence) have stronger behavioral intention to use the available support. This finding can be used when building RD policy and strategies to increase the absorption. The better the abilities of the farmers, their control over information, knowledge and experience to independently prepare the RDP application, the higher the intention to request support. Past experience can increase the confidence and strengthen the possibility of actual future behaviour, whereas absence of similar experience can strengthen the skepticism.

RDPs often require implementation through strong community coordination and efficient knowledge and information sharing. To overcome the potential barriers, networks can play an important role in information dissemination, particularly in the otherwise scarce information flow in rural areas. Since dispersion of information at local level is very important for improving the consumption of RD funds, it is necessary for rural population to achieve higher levels of social capital. In this regard, formal and informal cooperative actions in rural areas have a recognized importance.

Simplifying the procedure and rules, coupled with intensified extension support, would make applying more accessible and successful. Appropriately defined guidance and trainings can contribute to boosting farmer's self-confidence. Another important aspect is the ability and preparedness to co-finance investments, thus easier access to finance would additionally advance the use of RD support.

The literature suggests that the subjective norm construct is generally a weaker predictor of intentions and behavior [13]. In agricultural related studies, this can be to some extent explained by the importance of independence to farmers [14], making it understandable that farmers would not be very willing to declare in a survey that they are under social influence by the respected others [15].

5 Conclusion

Networks do support an increase of application for the RDP funds. The model explains that farmers from a region with an existing network organization are more likely to positively assess the attitudes, social norms and controls towards the use of RDP. Networks support knowledge and information sharing, thus improving farmers' attitudes. Farmers participating in networks have stronger social support. They have higher position in the information flow and increased opportunities to receive information and access resources related to RD support, thus they are regarded as referential for their knowledge and information about RDP. In addition, they have more positive attitude towards the use of RDP and are more inclined to apply for RDP funds. Additional investments in the social capital and trust among population can further contribute to increased use of RDP funds.

Acknowledgement. The research is done within the regional project "The impact of socio-economic structure of rural population on success of rural development policy", within the

Regional Research Promotion Programme in the Western Balkans (RRPP), run by the University of Fribourg upon a mandate of the Swiss Agency for Development and Cooperation, SDC, Federal Department of Foreign Affairs. The views expressed are those of the authors and do not necessarily represent opinions of SDC and University of Fribourg.

References

1. WBD, World Bank Database. https://data.worldbank.org/indicator/sp.rur.totl.zs. Accessed 30 May 2019
2. Dimitrievski, D., Kotevska, A., Stamenkovska, I.J., Tuna, E., Nacka, M.: FYR of Macedonia: agricultural policy brief. In: Volk, T., Erjavec, E., Ciaian, P., Gomez y Paloma, S. (eds.) Analysis of the Agricultural and Rural Development Policies of the Western Balkan Countries, pp. 80–90. Joint Research Centre, European Commission (2016)
3. Wiesinger, G.: The importance of social capital in rural development, networking and decision-making in rural areas. J. Alp. Res. **95**(4), 43–56 (2007)
4. Coleman, J.S.: Social capital in the creation of human capital. Am. J. Soc. **94**, 95–120 (1988)
5. Woolcock, M.: Social capital. In: Christensen, K., Levinson, D. (eds.) Encyclopedia of Community: From the Village to the Virtual World, pp. 1258–1262. Sage, Thousand Oaks (2003)
6. Ajzen, I.: The theory of planned behavior. Organ. Behav. Hum. Decis. Process. **50**, 179–211 (1991)
7. Ajzen, I.: The theory of planned behaviour: reactions and reflections. Psychol. Health **26**(9), 1113–1127 (2011)
8. Willock, J., Deary, I.J., Edwards-Jones, G., Gibson, G.J., McGregor, M.J., Sutherland, A., Dent, J.B., Morgan, O., Grieve, R.: The role of attitudes and objectives in farmer decision making: business and environmentally-oriented behaviour in Scotland. J. Agric. Econ. **50**(2), 286–303 (1999)
9. Bergevoet, R.H., Ondersteijn, C.J.M., Saatkamp, H.W., Van Woerkum, C.M.J., Huirne, R.B. M.: Entrepreneurial behaviour of Dutch dairy farmers under a milk quota system: goals, objectives and attitudes. Agric. Syst. **80**(1), 1–21 (2004)
10. Lin, N.: A network theory of social capital. In: Castiglione, D., van Deth, J.W., Wolleb, G. (eds.) The Handbook of Social Capital, pp. 50–69. Oxford University Press, London (2008)
11. Ajzen, I.: Residual effects of past on later behavior: habituation and reasoned action perspectives. Pers. Soc. Psychol. Rev. **6**(2), 107–122 (2002)
12. Hu, L., Bentler, P.M.: Cutoff criteria for fit indexes in covariance structure analysis: conventional criteria versus new alternatives. Struct. Eqn. Model.: Multi. J. **6**(1), 1–55 (1999)
13. Armitage, C.J., Conner, M.: Efficacy of the theory of planned behaviour: a meta-analytic review. Br. J. Soc. Psychol. **40**(4), 471–499 (2001)
14. Gasson, R.: Goals and values of farmers. J. Agric. Econ. **24**(3), 521–542 (1973)
15. Burton, R.J.: Reconceptualising the 'behavioural approach' in agricultural studies: a socio-psychological perspective. J. Rural Stud. **20**(3), 359–371 (2004)

Food Technologies

Content and Nutritional Value of Selected Biogenic Elements in Monofloral Sunflower Bee-Collected Pollen from Serbia

Aleksandar Ž. Kostić[1(✉)], Biljana P. Dojčinović[2], Nebojša Nedić[3], Danijel D. Milinčić[1], Nebojša Đ. Pantelić[1], Sladjana P. Stanojević[1], Miroljub B. Barać[1], Živoslav Lj. Tešić[4], and Mirjana B. Pešić[1]

[1] Faculty of Agriculture, Chair of Chemistry and Biochemistry, University of Belgrade, Nemanjina 6, 11080 Belgrade, Serbia
akostic@agrif.bg.ac.rs
[2] Institute of Chemistry, Technology and Metallurgy, University of Belgrade, Njegoševa 12, 11000 Belgrade, Serbia
[3] Faculty of Agriculture, Chair for Breeding and Reproduction of Domestic and Bred Animals, University of Belgrade, Nemanjina 6, 11080 Belgrade, Serbia
[4] Faculty of Chemistry, Chair of Analytical Chemistry, University of Belgrade, Studentski Trg 12-16, 11000 Belgrade, Serbia

Abstract. Pollen (floral or bee-collected) can be used as a good source of different nutrients including several important biogenic elements (minerals). The aim of this study was to determine content of selected biogenic elements (phosphorous, potassium, sulphur, calcium, magnesium, iron and zinc) in monofloral bee-collected pollen originated from sunflower (*Helianthus annuus* L.) plants grown in Serbia and to estimate its nutritional value regarding to elements content. In that purpose ICP-OES analytical method was applied. According to obtained results the examined pollen sample contains above mentioned elements in following quantities: P – 3523.98 mg/kg; K – 2869.13 mg/kg; S – 1951.74 mg/kg; Ca – 1919.20 mg/kg; Mg – 654.36 mg/kg; Fe –83.55 mg/kg; Zn – 45.92 mg/kg of dry weight. Based on recommended daily intakes for iron, phosphorus and zinc application of pollen, as food or food ingredient, will cover meaningful part of human needs for these three important elements.

Keywords: Bee-collected pollen · Sunflower · ICP-OES · Biogenic elements

1 Introduction

Unlike honey, which is a source of energy for bees, pollen supplies these insects with different important nutrients such as proteins, sugars, lipids, etc. [1]. Although pollen has been used as folklore medicine or as supplement for a long time in human history, its significant role in nutrition as food ingredient has been established recently. There are several articles with pollen application in food industry as enhancer of some chemical or techno-functional properties of different products [2–4]. In addition, it has been proved that floral or bee-collected pollen represents excellent source of different macro-, micro- or trace elements, especially regarding iron (Fe) and zinc (Zn) content

M. Brka et al. (Eds.): AgriConf 2019, IFMBE Proceedings 78, pp. 211–217, 2020.
https://doi.org/10.1007/978-3-030-40049-1_27

[5–7] for human. Depending on plants, pollen production during pollination process can be significant. Sunflower (*Helianthus annuus* L.) is one of the most important industrial crops growing all around the world mostly due to its oil production. As an entomophilous plant, normal and satisfactory development and yield of sunflower is impossible without bees as pollinators. According to literature data an average number of pollen grains per gram for sunflower is 5516 [8] with average number of pollen grains per plant between 21924 and 40448 [9]. Bearing in mind proximate number of sunflower plants (50000–80000) grown per hectare [10] of field it can be calculated that, during one season, pollen yield for sunflower is between 198.5 and 586.4 kg/ha. Since no significant additional financial investment is needed for the collection considerable amount of pollen can be easily collected as useful accompanying by-product from sunflower plants. In that sense the aim of this work was to determine quantity of selected biogenic elements (phosphorus, calcium, potassium, magnesium sulphur, iron and zinc) in sunflower pollen from Serbia by using ICP-OES (Inductively Coupled Plasma-Optical Emission Spectroscopy) analytical method and to estimate its nutritional potential if pollen will be used as food ingredient.

2 Materials and Methods

Pollen collection was conducted at private apiary located in Uljma village (Vojvodina province, Serbia) during intensive sunflower inflorescence (June 2017). Pollen samples were collected with pollen traps application placed at the entrance of several hives. One, representative and unique, sample was prepared by mixing pollen grains from different hives. After assortment, gathering and homogenization pollen sample was placed in plastic vacuum bag protected by aluminum foil from sunlight. Plastic bag was transported to the laboratory and stored in refrigerator until further analysis.

2.1 Palynological Analysis

Determination of botanical origin of collected pollen sample was conducted according to previously described method [11].

2.2 ICP-OES Analysis

Determination of content of selected biogenic elements (phosphorus, calcium, potassium, magnesium sulphur, iron and zinc) was performed with application of ICP-OES method given and described in our previous publication [6].

2.3 Statistical Analysis

All measurements were run in triplicate ($n = 3$). The results are given as mean values ± standard deviations. In order to determine significant differences ($p < 0.05$)

Statistics 8.0 software (StatSoft. Inc., Tulsa, OK, USA) was used and t-test was applied.

3 Results and Discussion

3.1 Palynological Analysis

Depending on botanical origin bee-collected pollen can be monofloral (predominantly contains pollen grains originate from one plant species/genus) or polyfloral (mixed pollen came from more than one plant species/genus). Since the possibility that bees will collect only one pollen type is a quite rare collection of monofloral pollen can be difficult if we try to prepare 100% "clean" sample. In that sense, Campos et al. [12] proposed that any pollen which has more than 80% of one pollen grains type should be characterized as monofloral sample. After conducted palynological analysis, based on numbering and microscopic examination of 500 pollen grains, the examined pollen sample can be classified as monofloral sample since it contains 80.4% of sunflower pollen grains.

3.2 ICP-OES Analysis

Biogenic macro- and/or microelements (the term minerals is also used in nutrition sciences) present important and inevitable part of healthy and balanced human diet and can be provided from plant or animal sources [13]. Due to different roles in our body and several important functions phosphorus, potassium, sulphur, calcium, magnesium, iron and zinc stood out. Bees use pollen as important basis of all nutrients, including diverse minerals. Accordingly, in this study the examination of above-mentioned selected biogenic elements in monofloral sunflower pollen from Serbia was conducted. The results of ICP-OES analysis, expressed as mg/kg of dry weight of pollen sample, are given in Table 1.

Table 1. The content (mg/kg of dry weight) of selected biogenic elements in monofloral bee-collected pollen sample.

Element	Content [mg/kg]
P	3523.98 ± 102.10
K	2869.13 ± 114.92
S	1951.74 ± 64.28[a]
Ca	1919.20 ± 93.85[a]
Mg	654.36 ± 38.14
Fe	83.55 ± 4.96
Zn	45.92 ± 2.71

[a]the results in the same column with same letter did not differ significantly ($p < 0.05$).

According to obtained results the predominant macroelement was phosphorus (3523.98 mg/kg) followed by potassium with determined content at 2869.13 mg/kg. Based on results of statistical analysis data (Table 1) amounts of sulphur and calcium were without significant differences. The quantities of the rest of three examined elements (magnesium, iron and zinc) were significantly different in range between 45 and 660 mg/kg of dry weight of pollen sample. Compared to some literature data it can be found that phosphorous content in sunflower monofloral pollen from Serbia is similar to result for sample of Turkish bee-collected pollen from Eastern Anatolia (unknown botanical origin) – 3518.33 mg/kg [14]. In case of potassium the given result is similar to polyfloral and buckwheat pollen samples from Turkey [15] as well as to results from our previous study [6] for samples collected from Vojvodina province (Serbia). The amount of calcium is similar to content of this element in polyfloral pollen sample from Turkey [15]. In addition, magnesium content in examined bee-collected pollen sample is in range with results for different pollen samples (diverse palynological origin) from Turkey [14, 15]. Compared to results for sulphur content in floral maize pollen samples (2450–3040 mg/kg) from Serbia [7] it can be concluded that sunflower bee-collected pollen contains a slightly lower amount of this element – 1951.74 mg/kg. Regarding to content of iron result from present study is almost identical with mean value for iron content in one study of Turkish bee pollen samples [15] while zinc content was higher to some extent. Contrary, in the study of Altunaltmaz et al. [14] the determined mean value for amount of iron in pollen samples from Turkey was significantly higher (203.16 mg/kg) i.e. lower (29.15 mg/kg) in case of zinc. All these observations confirm, once again, the importance of botanical and geographical origin of pollen for its chemical composition.

3.3 Nutritional Value of Sunflower Pollen Sample Regarding to Selected Biogenic Elements Content

Although quantity of minerals for human nutrition is quite lower compared to other nutrients such as proteins, carbohydrates and lipids they are essential for normal functioning of our body. According to literature, currently, there are 23 macro, micro and/or trace elements involved in different biochemical and physiological processes [16]. Compared with recommended (RDI) or adequate daily intake (ADI) values given for investigated elements in this section calculations for % of daily intake for phosphorus, potassium, sulphur, calcium, magnesium, iron and zinc are made and presented on Figs. 1 and 2.

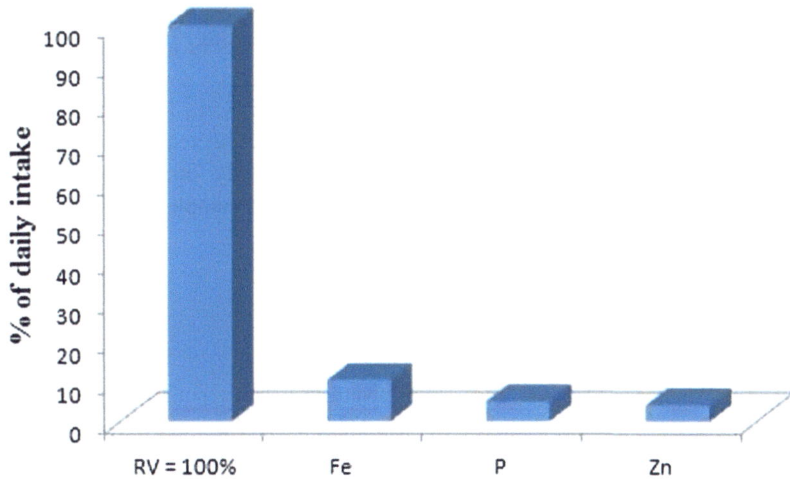

Fig. 1. Calculated daily intake of P, Zn and Fe based on 10 g (one cup) of bee pollen consumption (RV – reference value).

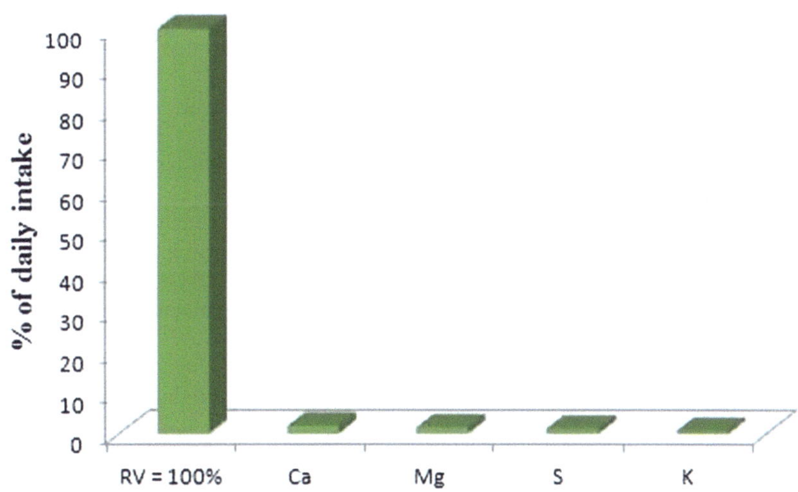

Fig. 2. Calculated daily intake of K, S, Mg and Ca based on 10 (one cup) g of bee pollen consumption (RV – reference value).

All estimates are built on presumption of 10 g (one cup) of bee pollen used as food or as food ingredient. Based on recommendation of Food and Nutrition board of USA Institute of Medicine (NBSC) dietary intake values for selected elements are: 700 mg/day (RDI for phosphorus), 4300 mg/day (ADI for potassium), 2800 mg/day (RDI for sulphur), 1000 mg/day (RDI for calcium), 420 mg/day (RDI for magnesium), 11 mg/day (RDI for zinc) and 8 mg/day (RDI for iron) [17–19].

It can be seen from results presented in Figures that 10 g of monofloral sunflower bee-collected pollen is potential source of meaningful quantity of iron (10.44% of daily need) and zinc (4.175%) as addition to some regular food.

This is in accordance with previous observations of Serra-Bonvehí and Escolá Jordá [5] and Kostić et al. [6, 7] who confirmed that bee-collected or floral pollen is a good source of these elements for human. Besides, daily intake for phosphorus (5.034%) originated from presumed amount of examined sample should not also be ignored.

References

1. Li, Q.-Q., Wang, K., Marcucci, M.C., et al.: Nutrient-rich bee pollen: a treasure trove of active natural metabolites. J. Funct. Foods **49**, 472–484 (2018)
2. Krystyjan, M., Gumul, D., Ziobro, R., et al.: The fortification of biscuits with bee pollen and its effect on physicochemical and antioxidant properties in biscuits. LWT-Food Sci. Technol. **63**, 640–646 (2015)
3. de Florio Almeida, J., dos Reis, A.S., Heldt, L.F.S., et al.: Lyophilized bee pollen extract: a natural antioxidant source to prevent lipid oxidation in refrigerated sausages. LWT-Food Sci. Technol. **76**, 299–305 (2017)
4. Conte, P., Del Caro, A., Balestra, F., et al.: Bee pollen as functional ingredient in gluten-free bread: a physical-chemical, technological and sensory approaches. LWT-Food Sci. Technol. **90**, 1–7 (2018)
5. Serra-Bonvehí, J., Escolá Jordá, R.: Nutrient composition and microbiological quality of honeybee collected pollen in Spain. J. Agric. Food Chem. **45**, 725–732 (1997)
6. Kostić, A.Ž., Pešić, M.B., Mosić, M.D., et al.: Mineral content of bee pollen from Serbia. Arh. Hig. Rada. Toksikol. **66**, 251–258 (2015)
7. Kostić, A.Ž., Kaluđerović, L.M., Dojčinović, B.P., et al.: Preliminary investigations of mineral content of pollen collected from different Serbian maize hybrids–is there any potential nutritional value? J. Sci. Food Agric. **97**, 2803–2809 (2017)
8. Buchmann, S.L., O'rourke, M.K.: Importance of pollen grain volumes for calculating bee diets. Grana **30**, 591–595 (1991)
9. Astiz, V., Hernandez, L.F.: Pollen production in sunflower (*Helianthus annuus* L.) is affected by air temperature and relative humidity during early reproductive growth. Phyton-Int. J. Exp. Bot. **82**, 297–302 (2013)
10. Fernández-Luqueño, F., López-Valdez, F., Miranda-Arámbula, M., et al.: An introduction to the sunflower crop. In: Arribas, J.A. (ed.) Sunflowers: Growth and Development, Environmental Influences and Pests/Diseases, 1st edn. Nova Science Publishers Inc., New York (2014)
11. Kostić, A.Ž., Barać, M.B., Stanojević, S.P., et al.: Physicochemical composition and techno-functional properties of bee pollen collected in Serbia. LWT-Food Sci. Technol. **62**, 301–309 (2015)
12. Campos, M.G.R., Bogdanov, S., Almeida-Muradian, L.B., et al.: Pollen composition and standardisation of analytical methods. J. Apic. Res. **47**, 154–161 (2008)
13. Gharibzahedi, S.M.T., Jafari, S.M.: The importance of minerals in human nutrition: Bioavailability, food fortification, processing effects and nanoencapsulation. Trend Food Sci. Technol. **62**, 119–132 (2017)
14. Altunaltmaz, S.S., Tarhan, D., Aksu, F., et al.: Mineral element and heavy metal (Cadmium, Lead and Arsenic) levels of bee pollen in Turkey. Food Sci. Technol. **37**, 136–141 (2017)

15. Kalaycioğlu, Z., Kaygusuz, H., Döker, S., et al.: Characterization of Turkish honeybee pollens by principal component analysis based on their individual organic acids, sugars, minerals and antioxidant activities. LWT-Food Sci. Technol. **84**, 402–408 (2017)
16. Quintaes, K.D., Diez-Garcia, R.W.: The importance of minerals in the human diet. In: de la Guardia, A., Garrigues, S. (eds.) Handbook of Mineral Elements in Food, 1st edn, pp. 1–18. Wiley, West Sussex (2015)
17. NBSC: Dietary Reference Intakes for Water, Potassium, Sodium, Chloride, and Sulfate. Standing Committee on the Scientific Evaluation of Dietary Reference Intakes, Food and Nutrition Board. The National Academies Press, Washington, DC (2005)
18. NBSC: Dietary Reference Intakes for Calcium, Phosphorus, Magnesium, Vitamin D, and Fluoride. Standing Committee on the Scientific Evaluation of Dietary Reference Intakes, Food and Nutrition Board. The National Academies Press, Washington, DC (1997)
19. NBSC: Dietary Reference Intakes for Vitamin A, Vitamin K, Arsenic, Boron, Chromium, Copper, Iodine, Iron, Manganese, Molybdenum, Nickel, Silicon, Vanadium and Zinc. Standing Committee on the Scientific Evaluation of Dietary Reference Intakes, Food and Nutrition Board. The National Academies Press, Washington, DC (2001)

Metal Content in Honey with Different Coloring from Industrial Areas of Bosnia and Herzegovina

Josip Jurković(✉), Mejrima Ploskić, Senad Murtić, Teofil Gavrić, Lejla Biber, Enisa Omanović Mikličanin, and Anera Kazlagić

Faculty of Agriculture and Food Science, University of Sarajevo, Sarajevo, Bosnia and Herzegovina
j.jurkovic@ppf.unsa.ba

Abstract. Honey is the food which is used by humans since ancient times and is highly valued due to its nutritional value. Most metals which are found in soil - plants system can be found in honey also. The aim of this research is to see could some metals from industrial areas of Bosnia and Herzegovina be found in honey samples and to see is there a difference in amount of metals in dark and light versions of honey. In honey samples some basic quality parameters were measured: water activity, pH, electrical conductivity, moisture content, dry matter content. Also, concentrations of five metals were determined: K, Na, Ca, Mg and Fe. Concentration of metals were determined by means of FAAS (Flame Atomic Absorption Spectrophotometry) and FP (flame photometry). For statistical evaluation, descriptive statistics was used, one – way ANOVA, and correlation. The results are showing differences between honeys from different industrial areas, and among honeys of different coloring. High correlations (above 0.900) were found between concentrations of K and pH (0.962) and between K and EC (0.975).

Keywords: Honey · Metals · Industry · Color · Correlation

1 Introduction

Honey is a natural supersaturated sugar solution, which is consumed as a food with high nutritive value [1]. It is mainly composed of carbohydrates (glucose and fructose) around 65%. Second component in high percentage is water, around 18%. Honey also contains lipids, proteins and other bioactive compounds in a small percentage. Among some other component honey contains metals. It is estimated that honeybees travel around 7 km^2, and during that time they are in contact with air, water and soil. Because of that honey is the result of a bio – accumulative process that is useful for the collecting of information about environmental aspects and it could be used as bio – indicator of environmental pollution [2]. Mineral component of honey is quite variable. Honey can contain between 0.04 to 0.2% of minerals. The mayor elements which are present in honey are K, P, Mg, Al, Ca, Na, Fe, Mn, Cu, Zn, Cl and S [3]. The composition of honey regarding minerals and heavy metals is largely dependent on the soil composition, a s well on the type of floral plant, because minerals are transported in

M. Brka et al. (Eds.): AgriConf 2019, IFMBE Proceedings 78, pp. 218–228, 2020.
https://doi.org/10.1007/978-3-030-40049-1_28

plants trough the roots and are present in nectar which is collected by the honeybees [2]. Some heavy metals are essential for normal growth and development of plants, such as: Fe, Co, Mn, Ni, Zn and Cu [4]. Metals that are present in soil and plants (honey) can originate from natural resources or from anthropogenic resources. Anthropogenic resources are: industrial smelter pollution, industrial emissions, improper procedures in honey processing, usage of agrochemicals, mining, ore processing, improperly stored mine wastes etc. [1]. According to Pfund's scale honey can be divided due to coloring into two main groups: light honey (Pfund' scale value from <9 mm to 34 mm) and dark honey (Pfund's scale value from 35 mm to >114 mm). Darker honeys usually contain more trace metals than lighter variations [5]. Because of all of this fact, the aim of this research is to investigate metals content in honey samples from industrial parts of Bosnia and Herzegovina, and to see do darker honeys variations contain more metals. In this research the main metals in honey (potassium, sodium, calcium, magnesium and iron) will me researched and some basic quality parameters. Potassium will be investigated, since potassium is metal that is the most abundant in honey samples, and it would be interesting to investigate correlations between this metal and some other parameters of honey quality. Sodium, is also in rather high concentrations in honey, and it is expected that in the area of sodium chloride mining we get higher concentration of sodium. Calcium and magnesium content could lead us to some conclusions about honey origin. Iron is heavy metal which is common in honey samples. Iron and steel industry are important part of economy of Bosnia and Herzegovina. It is possible that iron in honey except naturally comes from anthropogenic sources.

2 Materials and Methods

2.1 Samples

In this investigation different honey samples were used. They originate from different areas of Bosnia and Herzegovina. Three samples of honey were from industrial parts of Bosnia and Herzegovina (Tuzla, Vareš, Zenica), and one was from mountain area (Treskavica), Fig. 1.

Area of Tuzla. This area is well known for its salt (NaCl) mine, which is still very active.

Area of Vareš. Second area under investigation was city of Vareš, which is known for its iron ore deposits (hematite and magnetite).

Area of Zenica is third area under the research. Zenica is well known for its steel industry - still very active.

Area of Treskavica. As a clean area, mountain of Treskavica was chosen.

In total four honeys from four different parts of Bosnia and Herzegovina were researched. All honey samples can be classified as meadow honeys. From these four parts of Bosnia and Herzegovina two types regarding color of honey were collected: light and dark honey. At the end we got eight different honey samples:

- Light and dark honey from area of Tuzla (T)
- Light and dark honey from area of Vareš (V)
- Light and dark honey from area of Zenica (Z)
- Light and dark honey from mountain of Treskavica (M)

Fig. 1. Sampling areas

Only one honey was bought from honey fair. Others were bought directly from beekeepers which do not produce large amount of honey.

2.2 Methods

After the collection of honey samples, the analysis was started. In the beginning some basic quality parameters were measured: water activity, pH value, electrical conductivity, moisture content and color.

Water activity was measured with activity meter (Hygrolab, rotronic). Calibration of an instrument was made with $Ca(NO_3)_2$. Measurement of pH was made with pH – meter (Mettler Toledo MP230). For calibration of instrument buffers with known pH value were used. Before the measurement the honey samples had to be prepared. 5 g of honey was transferred in volumetric flask, dissolved in deionized water and filled until the mark of 50 mL. Measurement of electrical conductivity (EC) was conducted with usage of Mettler Toledo MC 126 instrument, as follows: amount of honey which contain 20 g of honey's dry matter was transferred to volumetric flask of 100 mL. 40 mL of this solution was tempered on 20 °C in water bath. After tempering the measurement of EC was done. Determination of water content was made by refractometric method. Sample was introduced in Abbe's refractometer where index of refraction was measured. Index of refraction depends on temperature, and this method is set for 20 °C. If there is a difference in temperature, it must be taken in account. Color of honey samples was determined with honey color photometer (HI 96785). Results from colorimeter are compared with Pfund's scale.

Sample Preparation for Metals Analysis

Honey sample preparation is very important and it can be difficult, since it contains high amount of carbohydrates and it has complex biological matrix. For this investigation wet digestion of samples was used. 2.000–2.5000 g of sample was introduced in laboratory glass. Two different strong acids were added: Concentrated nitric (HNO_3) and concentrated perchloric ($HClO_4$) acid in the 1:1 ratio. In total 20 mL of acids was added. After the addition of the acids, the digestion starts on a hot plate. Duration of digestion was three hours. After digestion, glassware was washed with 1 M HCl, and transferred in 50 mL volumetric flask [6].

Methods for Metal Analysis

Methods of analytical chemistry that are most widely used to determine metal content in honey are based on spectroscopy, such as inductively coupled plasma atomic emission spectrometry (ICP-AES) inductively coupled plasma mass spectrometry (ICP – MS), atomic absorption spectrometry wit atomization in flame or graphite furnace (FAAS, GFAAS), Flame photometry (FP) is used for analysis of alkaline metals and some alkaline earth metals [7]. For current study, determination of metal content in honey samples was measured with two different methods: Atomic absorption spectrometry with atomization in flame, and flame photometry.

Atomic absorption spectrophotometry. This method for metal analysis was used for measurement of calcium, magnesium and iron. Measurements were conducted on Atomic absorption spectrophotometer, Shimadzu AA-7000. For measurement of every metal different concentrations of standards were used. Addition of lanthanum solution (La) was necessary for measurement of calcium and magnesium.

Flame photometry. For determination of sodium and potassium content this method was used. Flame Photometer, FP902, PG instruments was used. The procedure with standards was similar to procedure with atomic absorption spectrophotometry.

Every measurement was made in three replications, and the average value was calculated. At the 24 samples were analyzed (8 × 3).

Statistical Evaluation

Results of this research are statistically evaluated with usage of statistical program SPSS 20.0 (SPSS Inc., Chicago, IL, USA) through next analysis:

- Descriptive statistics (mean, standard deviation).
- For testing of honey origin and coloring influence two – way ANOVA was used. Samples were analyzed to average values for all tested quality parameters. When p – value was lower than 0.05, difference among tested groups are considered statistically significant.
- Tukey test was used when ANOVA testing showed influence of average values. When p – value was lower than 0.05, the difference between groups are having statistically significant difference [8].

- Correlation is used to see similarity between two groups of results. Description of correlation coefficients which describe strength of connection between two variables is as follows: 0.0–0.19 very weak connection, 0.2–0.39 weak connection, 0.40–0.59 moderate connection, 0.60–0.79 strong connection, 0.80–1.0 very strong connection. Correlation can be uphill (coefficient of correlation (r) is positive), or downhill (coefficient of correlation (r) is negative).

3 Results and Discussion

In this section of paper, the results of research are shown. Results are divided in two parts. First part is dealing with selected basic quality parameters (water activity, pH, electrical conductivity, moisture content and dry matter), and the second one with analysis of metals content (K, Na, Ca, Mg and Fe).

3.1 Results of Selected Basic Quality Parameters

The results of basic quality parameters are shown with Table 1.

Table 1. Results of basic quality parameters

Parameter/sample	Color	T	Z	V	M
Water activity (aw)	Dark	0.634 ± 0.089	0.709 ± 0.026	0.633 ± 0.069	0.746 ± 0.009
	Light	0.684 ± 0.097	0.725 ± 0.021	0.593 ± 0.071	0.740 ± 0.012
pH	Dark	3.06 ± 0.00	3.94 ± 0.03	4.61 ± 0.01	4.26 ± 0.02
	Light	3.78 ± 0.02	3.75 ± 0.03	4.02 ± 0.02	4.02 ± 0.03
EC (mS/cm)	Dark	1.03 ± 0.00	5.62 ± 0.03	11.96 ± 0.22	8.99 ± 0.11
	Light	5.57 ± 0.35	4.03 ± 0.15	7.74 ± 0.00	5.25 ± 0.00
Moisture content (%)	Dark	17.80	16.20	15.40	16.80
	Light	18.20	18.60	13.20	17.40
Dry matter	Dark	82.2	83.8	84.6	83.2
	Light	81.8	81.4	86.8	82.6

T – honey from area of Tuzla, Z – honey from area of Zenica, V – honey from area of Vareš, M – honey from mountain of Treskavica.

Water activity. All of honey samples had low water activity. The highest water activity had mountain honey (0.740). Lowest water activity showed honey from the area of Vareš (0.593). Water activity is similar in value but results are statistically different from one honey to another. Water activity, as expected, shows strong correlation with water content (r = 0.672)

pH value. Results of pH measurement of all samples expect one were within the accepted range (3.2–4-5) [9]. Highest pH had dark mountain honey (4.26), and lowest pH had dark honey from Tuzla (3.06). Values were very similar, but there are some statistically significant differences which is proved with Tukey test. All off honeys samples were acidic. pH value shows very strong correlation (very high correlation, $r = 0.954$, Table 3, Fig. 9). This correlation means that with higher pH we have higher electrical conductivity, and it could be explained by the fact that there are some basic compounds that contribute to electrical conductivity.

Electrical conductivity. Highest electrical conductivity had honey from the area of Vareš (11.96 mS/cm). Honey samples from Tuzla showed the lowest electrical conductivity (1.03 mS/cm). It is significant that in all measured honeys the higher electrical conductivity showed darker variations of honey. Higher electrical conductivity means higher number of dissolved ions in honey. The higher concentration of ions (K^+, Na^+, Ca^{2+}, H^+, OH^- etc.) is known to be a characteristic of darker variations of honey compared to lighter variations [2]. Electrical conductivity shows very strong correlation to pH value ($r = 0.954$, Table 3). Higher pH value gives higher conductivity. Explanation is similar to explanation for pH value description.

Moisture content. Light honey from area of Zenica had the highest value for moisture content (18.60%). Honey samples from area of Vareš had the lowest amount of moisture (13.50%). Honey samples should not have amount of moisture bigger than 20% [10], so this result is inside the allowed values.

Color determination. According to Pfund's scale, honeys had white, extra light brown, light brown, brown and dark brown color. None of the samples had water white and extra white color.

3.2 Results of Metal Content

The results of metals analysis are shown in the Table 2.

Table 2. Results of metal analysis

Parameter/sample	Color	T	Z	V	M
K (mg/kg)	Dark	47.19	1382.48	2506.24	2078.37
	Light	1464.46	1009.73	1521.38	1154.37
Na (mg/kg)	Dark	88.62	52.24	38.11	37.90
	Light	142.18	35.29	45.91	37.77
Ca (mg/kg)	Dark	60.90	76.26	60.00	157.63
	Light	139.53	109.00	102.57	118.00
Mg (mg/kg)	Dark	46.87	61.56	65.65	76.87
	Light	49.72	49.85	96.71	42.62
Fe (mg/kg)	Dark	2.2	0.1	7.63	6.2
	Light	–	–	10.16	–

T – honey from area of Tuzla, Z – honey from area of Zenica, V – honey from area of Vareš, M – honey from mountain of Treskavica.

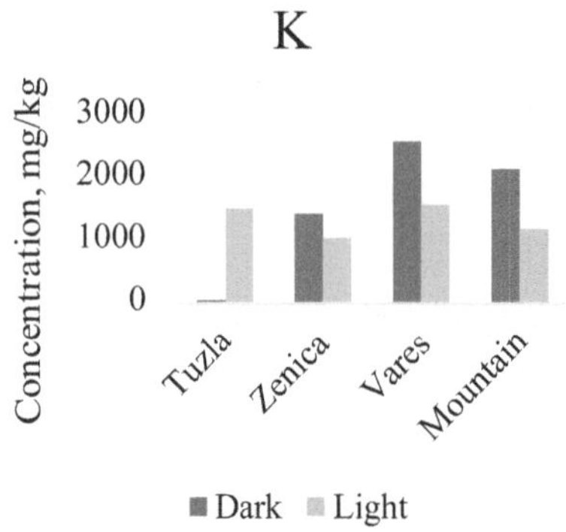

Fig. 2. Concentration of potassium

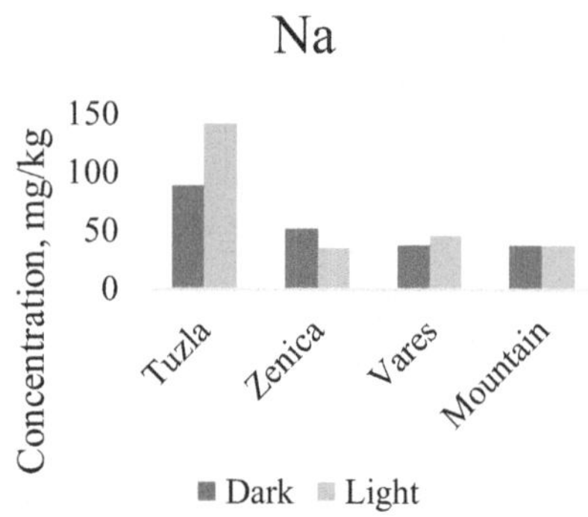

Fig. 3. Concentration of sodium

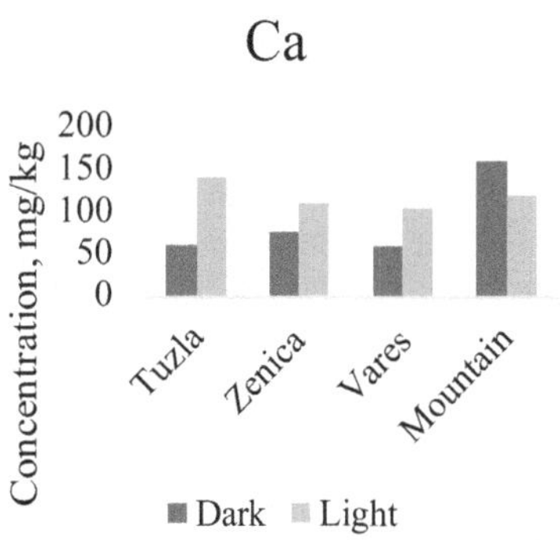

Fig. 4. Concentration of calcium

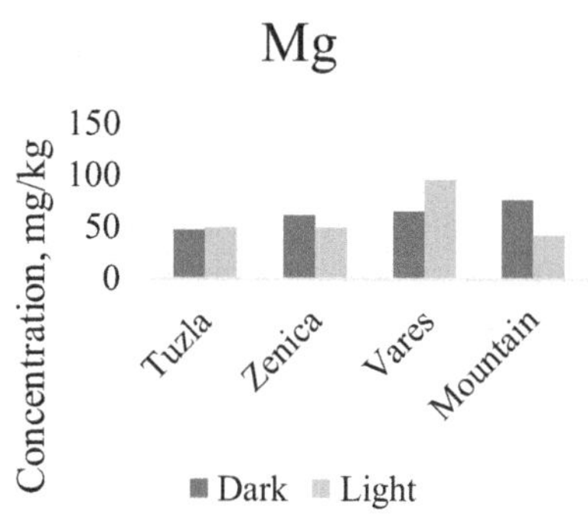

Fig. 5. Concentration of magnesium

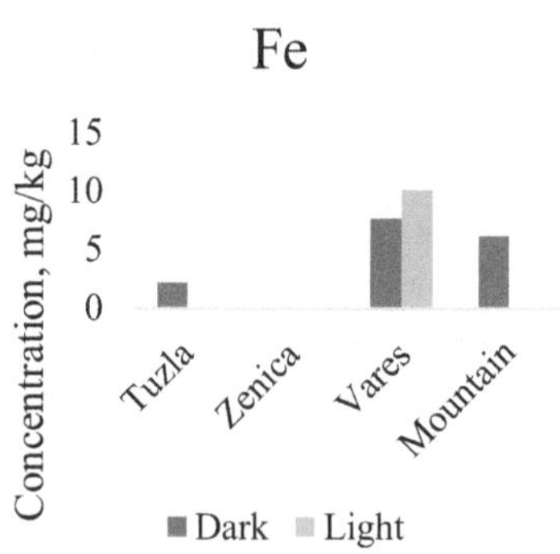

Fig. 6. Concentration of iron

Correlation Between Metal Content and Selected Basic Parameters
In this section the results of correlations are shown. Correlation was calculated for every measured parameter. The correlations above 0.500 are bolded, and correlations above 0.900 are shown also graphically.

Table 3. Correlation between metal content and selected basic parameters

	K	Na	Ca	Mg	aw	pH	EC	Moisture	Dry m.
K	1								
Na	−0.307	1							
Ca	0.236	0.134	1						
Mg	**0.507**	−0.334	0.104	1					
Aw	0.055	**0.531**	**0.687**	−0.487	1				
pH	**0.962**	−0.429	0.178	0.452	−0.03	1			
EC	**0.975**	−0.379	0.105	**0.585**	−0.144	**0.954**	1		
Moisture	−0.543	0.369	0.242	**−0.870**	**0.672**	−0.477	**−0.582**	1	
Dry m.	0.453	−0.369	−0.242	**0.870**	**−0.672**	0.477	**−0.582**	−1	1

Very high correlations (above 0.950) are shown in the next graphics, those correlations are between potassium and pH, potassium and electrical conductivity and pH and electrical conductivity.

Potassium. Results of analysis of potassium are shown in Table 2, and Fig. 2. Concentrations of potassium are in some cases higher than in other investigation papers [11–13]. Honey is naturally very rich in potassium, because it is part of the pollen which is used by the bees (potassium is the most abundant metal in honey) [14]. All of the samples had high concentrations of potassium (above 1000 mg/kg), only one sample had very low concentration (47.19 mg/kg) that was dark honey from area of Tuzla. Regarding this result it is possible that this honey with low concentration of potassium is not real, natural product. The highest amount of potassium was found in honey sample from area of Vareš and in honey sample from mountain. Analysis of variance showed that there is a significant difference in potassium content regarding honey origin. Potassium showed two very high correlations (it has very strong correlation) (Table 3), one was with pH ($r = 0.962$, Fig. 7) and other was with electrical conductivity ($r = 0.975$, Fig. 8). The first correlation (correlation is positive, so it means that higher amount of potassium gives higher pH value) can be explained with fact that potassium is part of salts that come from reaction of strong base (KOH) and weak acid (for example H_2CO_3), this kind of salts give basic reaction in water solutions. Because of that, higher concentrations of potassium give higher pH (more basic). Another high correlation can be explained by the fact that this high concentration of potassium contributes in large manner to overall conductivity of honey sample, because potassium ion is very mobile.

Sodium. Amounts of sodium that were found in this research are shown in Table 2, and Fig. 3. Concentrations of sodium in honey samples were not high, it is inside of the

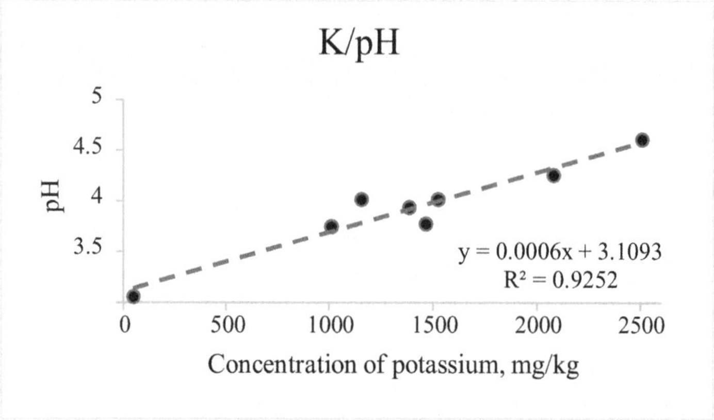

Fig. 7. Correlation between potassium and pH (r = 0.962)

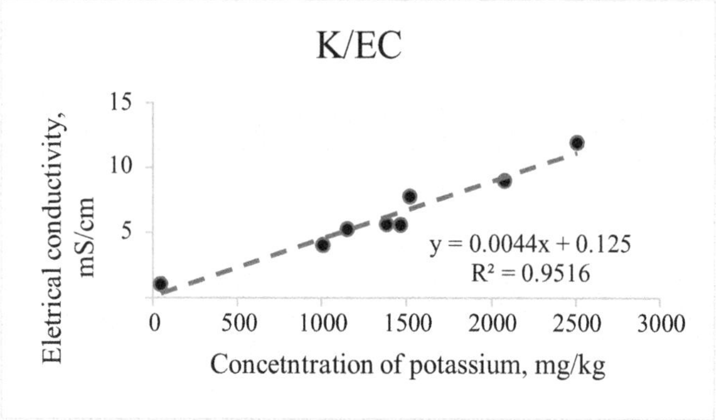

Fig. 8. Correlation between potassium and EC (r = 0.975)

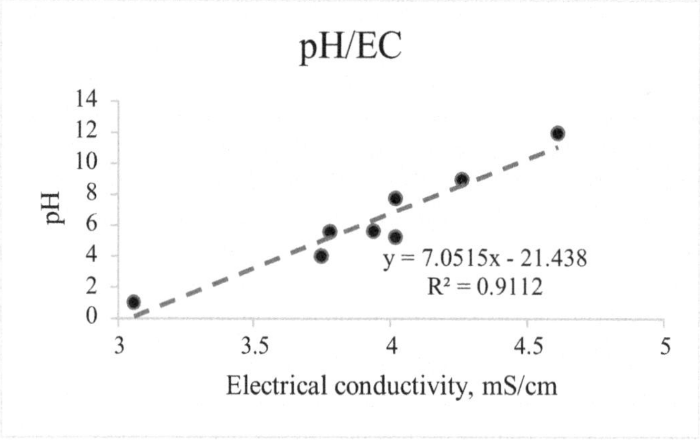

Fig. 9. Correlation between pH and EC (r = 0.954)

range reported by [6]. The highest amount of sodium was found in samples of honey from area of Tuzla (142.18 mg/kg for light honey and 88.62 for dark honey). This result can be explained by the fact that Tuzla has history in halite - sodium chloride (NaCl) mining, and sodium from soil comes to plants trough root system. Sodium is a constituent of pollen which is used by bees. Minimal concentration of sodium (35.29 mg/kg) was found in honey sample from area of Zenica. Analysis of variance showed that there is significant difference among dark and light honey specimens.

Calcium. Results of calcium gained from honey samples were statistically different (Table 2, Fig. 4). Highest value had mountain honey (157.63 mg/kg), and lowest honey from area of Vareš (60.00 mg/kg). Concentrations of calcium are in according to other research papers [3, 15]. Statistically significant difference was also found among honeys with different origin, and different color. Amounts of calcium in samples of honey are higher in light honeys.

Magnesium. Results of magnesium concentration are in the similar amount as reported by [2]. The highest concentration of magnesium was found in honey from area of Vareš (96.71 mg/kg). Lowest concentration of magnesium was found in mountain honey (42.62 mg/kg). (Table 2, Fig. 5). Statistically different results are gained between honeys of different origin. In case of magnesium, darker honeys had bigger amount. Magnesium shows very high correlation (very strong correlation) with dry matter/moisture content ($r = 0{,}870$, $r = -0.870$, Table 3.). This correlation means that with higher amount of dry matter, content of magnesium gets higher. Magnesium is mainly in dry matter of honey, not in soluble fraction.

Iron. Iron is heavy metal and in honey it is described as trace element which is in low concentrations. The results of this research also confirmed this fact. Iron could not be detected in some samples with atomic absorption spectrophotometry with flame atomization. Darker honeys are richer in iron. The highest amounts of iron were found in honeys from area of Vareš (10.16 mg/kg), which is expected since area of Vareš is rich in iron ores (Table 2, Fig. 6). This amount of iron is in many cases higher or similar to other papers [16, 17].

4 Conclusions

- Darker honeys are in almost every sample richer in metal content.
- Honeys with higher electrical conductivity had higher pH.
- Areas with larger production of some metals (Na, Fe) have larger amounts of this metals in honey samples.
- Industry (mining) contributes to metal content of honeys.
- Potassium in honey samples largely contributes to pH value of honey and to its conductivity.
- Low amount of potassium could mean that honey is not honeybee product.
- Magnesium is mostly found in dry matter of honey

References

1. Aghamirlou, H.M., Khadem, M., Rahmani, A., Sadeghian, M., Mahvi, A.H., Akbarazadeh, A., Nazmara, S.: Heavy metals determination in honey samples using inductively coupled plasma-optical emission spectrometry. J. Environ. Health Sci. Eng. **13**, 39 (2015)
2. Solayman, Md., Islam, Md.A., Paul, S., Ali, Y., Khalil, MdI, Alam, N., Gan, S.H.: Physicochemical properties, minerals, trace elements, and heavy metals in honey of different origins: a comprehensive review. Food Sci. Food Saf. **15**(1), 219–233 (2015)
3. Saadiyah, A.D., Ameera, H.H.: Determination of heavy metals and trace element levels in honey samples from different regions of Iraq and compared with other kind. Am. J. Appl. Chem. **3**(3), 83–92 (2015)
4. Mejías, E., Garrido, T.: Analytical procedures for determining heavy metal content in honey: a bioindicator of environmental pollution (2017)
5. Shah, A., Skinadar, F., Ullah, I., Shah, A., Ud-Din Khan, S., Ali Rana, U., McCoy, T.: Spectrophotometric determination of trace elements in various honey samples, collected from different environments. J. Food Nutr. Res. **2**(9), 532–538 (2014)
6. Pohl, P.: Determination of metal content in honey by atomic absorption and emission spectrometries. Trends Anal. Chem. **28**(1), 117–128 (2009)
7. Vincevica-Gaile, Z., Klavins, M., Rudovica, V., Viksna, A.: Geographical dissemination of trace and major elements in honey. Trans. Ecol. Environ. **167**, 211–218 (2011)
8. Formicki, G., Greń, A., Stawarz, R., Zyśk, B., Gal, A.: Metal content in honey, propolis, wax, and bee pollen and Implications for metal pollution monitoring. Pol. J. Environ. Stud. **22**, 99–106 (2013)
9. da Silva, P.M., Gauche, C., Gonzaga, L.V., Costa, A.C.O., Fett, R.: Honey: chemical composition, stability and authenticity. Food Chem. **196**, 309–323 (2016)
10. Thrasyvoulou, A., Tananaki, C., Goras, G., Karazafiris, E., Dimou, M., Liolios, V., Kanelis, D., Gounari, S.: Legislation of honey criteria and standards. J. Apic. Res. **57**, 88–96 (2018)
11. Alvarez-Suarez, J.M., Giempieri, F., Damiani, E., Regoli, F., Quiles, J.L., Battino, M.: Radical-scavenging activity, protective effect against lipid peroxidation and mineral contents of monofloral Cuban honeys. Plant Foods Hum. Nutr. **67**, 31–38 (2012)
12. Baroni, M.V., Arrua, C., Nores, M.L., Fayé, P., del Pilar Diaz, M., Chiabrando, G.A., Wonderlin, D.A.: Composition of honey from Còrdoba (Argentina): assessment of North/South provenance by chemometrics. Food Chem. **114**, 727–733 (2009)
13. Corredera, L., Bayarri, S., Pérez-Arquillué, C., Lázaro, R., Molino, F., Herrera, A.: Evaluation of heavy metals and polycyclic aromatic hydrocarbons in honeys from different origins. J. Food Prot. **77**, 505–509 (2014)
14. Liberato, M.C.T.C., Morais, S.M., Magalhães, C.E.C., Magalhães, I.L., Cavalcanti, D.B., Silva, M.M.O.: Physicochemical properties and mineral and protein content of honey samples from Ceará state, Northeastern Brazil. Food Sci. Technol. **33**(1), 38–46 (2013)
15. Ajibola, A., Chamunorwa, J.P., Kennedy, H.E.: Nutraceutical values of natural honey and its contribution to human health and wealth. Nutr. Metab **9**, 1–12 (2012)
16. Kılıç Altun, S., et al.: Analyses of mineral content and heavy metal of honey samples from South and East Region of Turkey by using ICP-MS. Int. J. Anal. Chem. **2017**, 1–6 (2017)
17. Altunatmaz, S.S., Tarhan, D., Aksu, F., Ozsobaci, N.P., Or, M.E., Barutçu, U.B.: Levels of chromium, copper, iron, magnesium, manganese, selenium, zinc, cadmium, lead and aluminium of honey varieties produced in Turkey. Food Sci. Technol. **39**, 392–397 (2018)

Quantification of Selected Toxic and Potentially Toxic Elements in Vegetables, and Health Risk Assessment

Irena Miljković[1], Jelena Popović-Djordjević[1(✉)], Miloš B. Rajković[1], Aleksandar Ž. Kostić[1(✉)], Blaženka Popović[2], and Ilija Brčeski[3]

[1] Faculty of Agriculture, Chair of Chemistry and Biochemistry, University of Belgrade, 11080 Belgrade, Serbia
{jelenadj,akostic}@agrif.bg.ac.rs
[2] Faculty of Agriculture, Chair of Agribusiness Management, University of Belgrade, 11080 Belgrade, Serbia
[3] Faculty of Chemistry, University of Belgrade, 11000 Belgrade, Serbia

Abstract. Vegetables play an important role in human nutrition and their importance in diet is determined by its chemical composition. Numerous studies indicate a high degree of contamination of soil and plants produced in certain agro-ecological conditions, especially near urban and industrial areas. Accordingly, toxic and potentially toxic elements (aluminum, chromium, copper, manganese, nickel, lead and zinc) were quantified in edible parts of four vegetable species (potato, cabbage, carrot and broccoli) by means of inductively coupled plasma–optical emission spectrometry (ICP–OES). In addition, health risk assessment was done and expressed through estimated weekly intake of elements (EWI) and hazard quotient (HQ). Samples of each species originating from different localities were collected from three green markets in the city of Belgrade (Serbia). The content of studied elements ranged from 1.32 to 4.00, 0.31 to 5.77, 3.50 to 23.11, 1.82 to 11.17 and 2.11 to 10.62 mg/kg for aluminum, cooper, manganese, nickel and zinc, respectively. Chromium was detected in carrot sample (1.76 mg/kg), whereas concentration of lead in two samples of broccoli exceeded the maximum allowed concentration (>0.3 mg/kg) set by the national regulations. The average estimated weekly intakes for Al, Cu, and Zn were within the safe limits set by FAO/WHO, while for Ni and Pb in some vegetables were considerably higher than proposed values. In respect to the obtained results, a great weekly health risk was indicated for chromium, nickel and lead with the highest HQ values: 2.44, 10.32 and 2.02, respectively. The results of the study impose the necessity for continuous monitoring of harmful elements content in individual vegetable crops as well as strict regulative guide-lines in order to diminish possibility of contamination.

Keywords: Vegetables · Toxic elements · ICP–OES · Health risk assessment

I. Miljković and J. Popović-Đorđević—These authors equally contributed to the work.

M. Brka et al. (Eds.): AgriConf 2019, IFMBE Proceedings 78, pp. 229–236, 2020.
https://doi.org/10.1007/978-3-030-40049-1_29

1 Introduction

Vegetables are fresh and edible parts of green plants and they represent an important portion of human diet especially significant for good health and prevention of diseases. Vegetables contain important nutrients and they are a source of various constituents which enable normal body functioning. Furthermore, they are highly appreciated mainly because of the high content of carbohydrates, vitamins and minerals. Depending on the vegetable species, roots, stalks, leaves, fruits or seeds are used as food. Each group contributes to diet in its own way. Different elements (most oftenly named minerals) are present in the organism in low concentrations, but they have a wide variety of roles that they play. Out of twenty-two minerals are naturally present in the living organisms [1]. However, vegetables are not only a source of essential macro- and microelements, but they can also contain toxic elements. Elevated concentration of toxic elements in vegetables can be the consequence of their natural presence (soil composition) or anthropogenic influence (industrial and air pollution) [2, 3].

Besides the fact that contamination of agricultural soil and products (fruits, vegetables and cereals) by elements such as arsenic (As), cadmium (Cd), chromium (Cr), manganese (Mn), nickel (Ni), mercury (Hg) and lead (Pb) pose a serious ecological problem, exposure to toxic elements can cause damage to different organs. In that respect, United States Environment Protection Agency (US EPA) and International Agency for Research on Cancer (IARC) classified As, Cd, Cr, Pb and Hg as 'known' or 'probable' human carcinogens based on epidemiological and experimental researches which show that there is a connection between exposure to their impact and frequency of cancer in people and animals. On the other hand, iron, copper and zinc are considered to be important nutrients whereas deficiency and excessive intake of these essential elements can lead to adverse acute and chronic disorders [4, 5].

Vegetables play an important role in everyday diet of the inhabitants of Serbia, noting the biggest average consumption in Central Serbia compared to the other parts of national territory [6]. Local agricultural production near the big cities is extremely important because green markets are mostly supplied by fresh products from agricultural production sites (villages) nearest to the city. Since the public is concerned about food safety, it has become necessary to monitor the quality of different vegetable species in order to ensure the benefits of their consumption. There are limited available data on the content of toxic elements in vegetables in Serbia [5].

The aim of this study was to analyze the content of toxic (Cr and Pb) and potentially toxic elements (Al, Cu, Mn, Ni, and Zn), in selected vegetables from green markets in Belgrade, and to assess potential health risk for the consumers.

2 Materials and Methods

2.1 Samples Collection

Samples of four vegetable species, potato (*Solanum tuberosum* L.), cabbage (*Brassica oleracea* L. var capitata), carrot (*Daucus carota* L. ssp. sativus Hoffm.) and broccoli (*Brassica oleracea* var. silvestris), were collected during January 2018 from three localities - city's green markets in Belgrade: locality I – "Zemun", locality II – "Zeleni venac" and locality III – "Blok 44 (Novi Beograd)-organic".

2.2 Preparation of Samples for Analysis

First, the fresh vegetables were washed with distilled and then with de-ionized water and afterwards dried in the air. Air-dried fruits were kept for few hours at 4 °C before analysis. Composite samples were made from each vegetable species and they comprised 5–10 individual samples. Only the parts of the plants that are used in human nutrition were subjected to the further analysis. The samples were first ground (in a porcelain mortar using a pestle) and then were subjected to the process of "*wet digestion*". Briefly, the composite samples (0.5 g) of each vegetable species was measured on the analytical scale and placed into 25 ml glasses. Nitric acid and hydrogen-peroxide (7/1 ml, respectively) were added, and mixtures were heated on the water bath for 6 h ($\sim$80 °C). After the digestion of samples was completed, mixtures were transferred into volumetric flasks and diluted with deionized water. The same procedure was applied for a blank [7].

Chemicals and reagents (of analytical purity) were procured from Merck (Darmstadt, Germany). Ultrapure de-ionized water was used to prevent samples contamination. Quantitative paper (pore size 2–4 µm) was procured from FILTER–LAB, FILTROS ANOIA, S.A. Barcelona, and the cuvettes from ISOLAB Laborgeräte GmbH, Germany.

Analytical method ICP - OES (Inductively coupled plasma - optic emission spectrometry) was used to quantify seven elements: aluminium (Al), chromium (Cr), copper (Cu), manganese (Mn), nickel (Ni), lead (Pb) and zinc (Zn) in the vegetable samples. Analysis was done on the instrument Spectroblue (SPECTRO Analytical Instruments GmbH, Germany) which is equipped with Spectro Smart Analyzer program for data processing. The method US EPA 200.7 was applied [8]. The obtained results from three measurements are presented as mean values expressed as mg/kg.

2.3 Health Risk Assessment

Intake of Al, Cr, Cu, Mn, Ni, Pb and Zn by consuming selected vegetables was calculated on a weekly basis and expressed as EWI (estimated weekly intake) by using the equation $EWI = ADC \times c \times 7/ABW$ [5]. where ADC is average daily vegetable consumption (0.345 kg), c is the concentration of the analyzed element (mg/kg) and ABW is the average body weight (74.7 kg) of the Serbian population calculated according to literature (84.3 for an adult man and 65.0 for an adult woman) [9]. Average daily vegetables consumption was calculated using the annual vegetable consumption in Central Serbia [6]. The calculation of EWI was based on the assumption that an adult daily consumes 0.345 kg of examined vegetables.

Human health risk caused by consuming examined vegetables is set at a weekly level (short-term risk) and expressed as hazard quotient (HQ) according to the equation $HQ = EWI/PTWI$; where EWI is estimated weekly intake of toxic elements in mg/kg of body weight (bw) and PTWI is provisional tolerable weekly intake of toxic and potentially toxic elements in mg/kg of body weight [10].

3 Results and Discussion

Many data bases about food do not include data on potentially toxic elements and they are mainly focused on nutritionally important elements. Data about potentially toxic elements are more often available in reviewed publications. Although it is not expected that certain types of fruit and vegetables present a significant food source of potentially toxic elements, local agricultural production, including the choice of pesticides and fertilizers, irrigation water and other factors may be responsible for letting these elements enter to the food chain. Besides this, there are certain species such as leaf and root vegetables which can adopt and accumulate toxic and potentially toxic elements from the soil in the areas with high concentrations and in this way they can greatly contribute to the intake of these elements depending on the local consumption. [3]. In Central Serbia, potato and cabbage are the most consumed vegetables with 32.6 and 18.7 kg per capita, respectively. The consumption of root vegetables is 6.8 kg per capita and other vegetables 17.6 kg per capita [6]. Figure 1 shows the content of the examined elements in potato, cabbage, carrot and broccoli from three city green markets. Two kinds of vegetables had the highest content of the analyzed elements: broccoli - Mn, Pb and Zn, and carrot - Al, Cu and Ni. The minimum and maximum concentrations of toxic elements in the examined types of vegetables are shown in Table 1. Aluminum was detected at concentrations (mg/kg) from 1.32 (potato from locality I) to 4.41 (carrot-locality II). Chromium was detected only in carrot from the organic green market (1.76 mg/kg, locality III). The lowest concentration of Cu (0.31 mg/kg) was measured in the cabbage from the organic green market, and the highest content was measured in the carrot from the green market at locality II. Manganese wasn't detected in vegetable samples from the organic green market or in

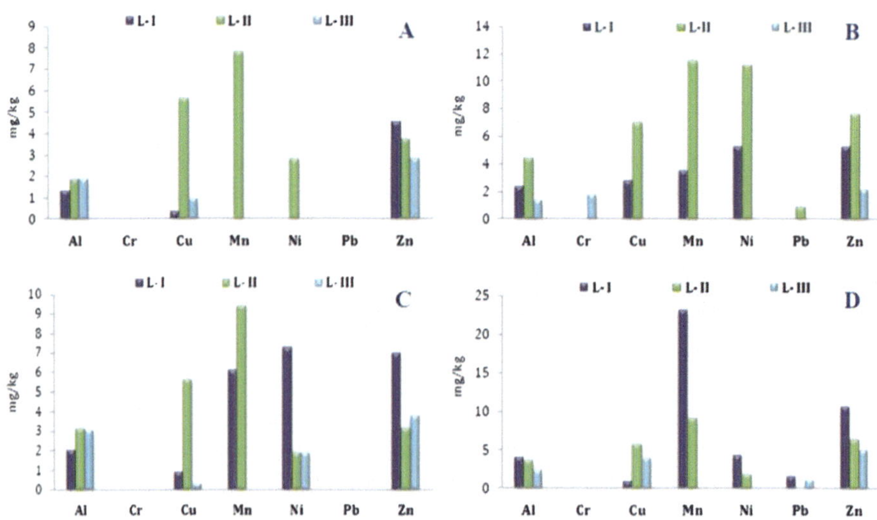

Fig. 1. The content of Al, Cr, Cu, Mn, Ni, Pb and Zn in potato (A), carrot (B), cabbage (C) and broccoli (D) from city's green markets I-III

the potato from the green market at locality I. The lowest (3.50 mg/kg) and the highest (23.11 mg/kg) concentration of Mn was detected in the carrot and broccoli, respectively, from the green market at locality I.

Except in cabbage (1.87 mg/kg) Ni was not detected in vegetables collected from the organic green market. In vegetable samples from the other two green markets, the content of nickel varied from 1.82 mg/kg in broccoli up to 11.17 mg/kg in carrot (both kinds of vegetables taken from locality II). Lead was detected only in broccoli (locality I and III) and potato (locality II). Concentrations of Pb in broccoli were above the maximum allowed concentration (MAC = 0.1 mg/kg) prescribed by the national regulations [11]. Zinc was detected in all tested samples at concentrations from 2.11 mg/kg in carrot from locality III to 10.62 mg/kg in broccoli from locality I.

Table 1. The minimum and maximum concentrations (mg/kg) of the examined elements in vegetables

c (mg/kg) elements	Potato		Cabbage		Carrot		Broccoli	
	Min	Max	Min	Max	Min	Max	Min	Max
Al	1.32	1.87	2.06	3.13	1.34	4.41	2.42	4.00
Cr	/*	/	/	/	/	1.76	/	/
Cu	0.35	5.61	0.31	5.63	2.80	7.05	0.94	5.77
Mn	/	7.79	6.07	9.38	3.50	11.46	9.12	23.11
Ni	/	2.80	1.87	7.28	5.25	11.17	1.82	4.37
Pb	/	/	/	/	/	0.88	1.07	1.56
Zn	2.81	4.52	3.13	6.97	2.11	7.64	4.97	10.62

*/ – not detected

The Provisional Tolerable Weekly Intake PTWI was established by the Joint FAO/WHO Expert Committee on Food Additives JECFA and refers to the assessment of the safety factor for establishing the acceptable or tolerable intake of substances that demonstrate toxicity [10].

In order to assess the safety of consumption of the examined vegetables in relation to the content of the selected elements, the estimated weekly intake (EWI) for Al, Cr, Cu, Mn, Ni, Pb and Zn was calculated and the obtained values were compared to the provisional tolerable weekly intake of the elements (PTWI, Table 2). The results indicate that the EWI for Cr, Mn, Ni and Pb is higher than PTWI if the daily consumption is 345 g of vegetables, which is the average daily consumption of vegetables per capita in Central Serbia. Observed by vegetable kinds, the maximum/minimum estimated weekly intake (EWI) of the elements was: Mn/Ni in potato, Mn/Al in cabbage, Ni/Pb in carrot and Mn/Pb in broccoli.

In assessing the short-term risk for human health, it is considered that high risk exists if the hazard quotient (HQ) for an element is higher than 1 [13]. According to the minimum and maximum concentrations of the analyzed elements (mg/kg) presented in Table 1, it can be noticed that the examined vegetables: potato, cabbage, carrot and broccoli contain concentrations of Al, Cu and Zn that do not present a short-term risk for the health of people consuming them with HQ < 1 (Table 2). The consumption of carrot and broccoli, which contain the highest concentrations of Cr, Mn and Pb, presents a short-term risk for human health (HQ > 1), since the intake of Cr, Mn and Pb is above the values of the provisional tolerable weekly intake (PTWI) in the indicated samples. The consumption of potato, cabbage, carrot and broccoli containing the minimum and maximum concentrations of Ni indicate an increased short-term health risk in sample from green market (L-II, Table 2). The excess of nickel may present a risk for human health, and the main pathway to Ni access the plant is the root uptake or foliage [14]. On the other hand, plant-available fraction of Ni in soil affects its uptake by plants [15].

The examinations of the Al content is especially important because its role in the development of dementia and Alzheimer's disease-AD has been extensively examined in recent years.

Considering that in the tested potato samples Cr and Pb were no detected there is no short-term health risk for the consumers (HQ = 0). In case of nickel, with HQ > 1 (2.59), potato consummation, in estimated quantity, presents significant health risk. In addition, presence of Ni corresponds to significant health risk for consumers due to the increased HQ values (1.73–10.32) in all examined samples, with the exception of samples from green markets L-I and L-III.

Cabbage, as well as potato, had only in the case of Ni an increased value of the hazard quotient (HQ = 1.73–6.72). This may be attributed to the soil naturally rich with nickel or fertilizers with microelements which were applied foliarly [14].

Carrot appears as a plant species with an increased short-term health risk due to the presence of Cr and Pb (in one sample) and Ni (in all samples). The concentration of Mn detected in this vegetable is such that the hazard quotient (HQ) is ≈1. Since it is a root vegetable, whose edible part is directly used from the soil, the effects of soil and/or micro-fertilizers on the increased content of Ni are possible [14].

Table 2. Estimated weekly intake (EWI) and short-term health risk expressed through the hazard quotient (HQ) for selected elements

	Potato			Cabbage			Carrot			Broccoli		
	EWI	HQ		EWI	HQ		EWI	HQ		EWI	HQ	
		Min	Max		Min	Max		Min	Max		Min	Max
Al	0.05	0.02	0.03	0.09	0.03	0.05	0.09	0.02	0.07	0.11	0.04	0.07
Cr	/	/	/	/	/	/	0.03	/	2.44	/	/	/
Cu	0.01	0.02	0.36	0.12	0.02	0.36	0.16	0.18	0.46	0.13	0.06	0.37
Mn*	0.13	0	0.65	0.27	0.050	0.79	0.24	0.29	0.96	0.60	0.77	1.94
Ni	0.05	0	2.59	0.18	1.73	6.72	0.27	4.85	10.32	0.11	1.68	4.04
Pb	/	/	/	/	/	/	0.01	/	1.14	0.04	1.38	2.02
Zn	0.12	0.22	0.35	0.18	0.24	0.54	0.16	0.16	0.59	0.28	0.38	0.82

PTWI - provisional tolerable weekly intake (mg/kg of body weight) for Al-2.00; Cr-0.023; Cu-0.50; Mn-0.42*; Ni-0.035; Pb-0.025; Zn-0.42; *calculated through tolerable daily intake (TDI = 0.06 mg/kg) [12].

In the case of broccoli, where the aboveground parts of the plant are used for the human consumption, there is also a noticeable contamination of Ni and Pb in samples from all studied localities, and Mn in sample from green market L-I. The assumption is that the contamination comes from the resources used for the fertilization and protection of the plants or from the soil pollution (as in the case of carrot where the root is used in human diet).

4 Conclusion

Two plant species were examined where the root (potato and carrot) is used in the diet and two where the above-ground parts is used (cabbage and broccoli). The results of the study showed that no regularity was observed in the analyzed elements, nor in the different types of vegetables, but the individual contamination of the examined plant species was noticed. The results obtained for the content of Pb in the analyzed vegetable samples indicated its increased concentration in broccoli from two green markets, in comparison to the permitted values prescribed by the national regulations. According to the obtained results, EWI for Cr (carrot), Mn (broccoli), Ni (all vegetables) and Pb (broccoli) in individual vegetables was above the safe values established by the FAO/WHO. In accordance with the above, a short-term health risk related to the following elements was observed: nickel by consuming all the tested vegetable types; chromium by the consumption of carrot in which Cr is detected; manganese and lead by consuming carrot and broccoli which contain the highest concentrations of these two elements. The results obtained may indicate the contamination of vegetables through soil, water and air, as well as the use of plant fertilizers and protection products.

Acknowledgments. This work was supported by the Ministry of Education, Science and Technological Development of the Republic of Serbia (Research grants Nos. 46009, 43009, and 31003). The vegetable samples studied in this work were collected under the Permission No. 1436/1, issued by the Public Utility Company "JKP Gradske pijace" – Beograd.

References

1. Hanif, R., Iqbal, Z., Iqbal, M., et al.: Use of vegetables as nutritional food: role in human health. AJABS **1**, 18–22 (2006)
2. Khan, K., Lu, Y., Khan, H., Ishtiaq, M., et al.: Heavy metals in agricultural soils and crops and their health risks in Swat District, Northern Pakistan. Food Chem. Toxicol. **58**, 449–458 (2013)
3. de la Guardia, M., Garrigues, S. (eds.): Handbook of Mineral Elements in Food. Wiley, Hoboken (2015)
4. Tchounwou, P.B., Yedjou, C.G., Patlolla, A.K., et al.: Heavy metals toxicity and the environment. EXS **101**, 133–164 (2012)
5. Popović-Dordevic, J., Bokan, N., Dramicanin, A., et al.: Content and weekly intake of essential and toxic elements in Serbian vegetables. J. Environ. Prot. Ecol. **18**, 889–898 (2017)

6. Vlahović, B., Tomić, D., Andrić, N.: Consumption of vegetables in Serbia-a comparative aproach. 45. Savetovanje agronoma Srbije, Zlatibor, Srbija (2011). (in Serbian)
7. Popović Djordjević, J., Marjanović, Ž.S., Gršić, N., et al.: Essential elements as a distinguishing factor between mycorrhizal potentials of two cohabiting truffle species in riparian forest habitat in Serbia. Chem. Biodivers. **16**, e1800693 (2019)
8. U.S. Environmental Protection Agency, Method 200.7.: Determination of metals and trace elements in water and wastes by inductively coupled plasma-atomic emission spectrometry, revision 4.4 (1994)
9. Kos, J., Lević, J., Duragić, O., et al.: Occurrence and estimation of aflatoxin M1 exposure in milk in Serbia. Food Control **38**, 41–46 (2014)
10. Herrman, J.L., Younes, M.: Background to the ADI/TDI/PTWI. Regul. Toxicol. Pharmacol. **30**, S109 (1999)
11. Official Gazette of RS No. 22/18 (2018). (in Serbian)
12. Concha, G., Eneroth, H., Hallström, H., et al.: Contaminants and minerals in foods for infants and young children. Part 2: risk and benefit assessment. National Food Agency, Sweden (2013). https://www.livsmedelsverket.se/globalassets/publikationsdatabas/rapporter/2013/2103_livsmedelsverket_1_part_2_contaminants_and_minerals_in_foods_for_infants_and_young_children_risk_and_benefit_assessment.pdf
13. Leung, A.O., Duzgoren-Aydin, N.S., Cheung, K.C., et al.: Heavy metals concentrations of surface dust from e-waste recycling and its human health implications in southern China. Environ. Sci. Technol. **42**, 2674–2680 (2008)
14. Harasim, P., Filipek, T.: Nickel in the environment. J. Elem. **20**, 525–534 (2015)
15. Shaheen, S.M., Antić-Mladenović, S., Wang, S.L., Niazi, N.K., Tsadilas, C.D., Ok, Y.S., Rinklebe, J.: Nickel mobilization/immobilization and phytoavailability in soils as affected by organic and inorganic amendments. In: Tsadilas, C., Rinklebe, J., Selim, M. (eds.) Nickel in Soils and Plants, pp. 283–310. CRC Press (2017)

Development and *In-House* Validation of HACH Spectrophotometry Method for Determination of Phosphoric Acid in Cola Beverages

Lejla Jesenković-Habul(✉), Amira Čustović, and Selma Dizdarević

Institute for Public Health of the Federation of Bosnia and Herzegovina, Sarajevo, Bosnia and Herzegovina
l.habul@zzjzfbih.ba

Abstract. A cheap and a rapid HACH spectrophotometric method 8048 was developed and validated for the quantitative estimation of phosphoric acid in soft drinks. The developed method resulted in phosphoric acid exhibiting linearity in the range from 0.02 to 2.5 mg/L. The precision is exemplified by relative standard deviation of 3.45%. Percentage Mean recovery was found to be in the range of 90–107%, during accuracy studies. The limit of detection (LOD) and limit of quantitation (LOQ) were found to be 0.0075 mg/L and 0.02 mg/L respectively. Accordingly, it is concluded that the developed HACH spectrophotometric method 8048 is accurate, precise, linear and rugged, and therefore the method can be used for the routine analysis of phosphoric acid in soft drinks.

Keywords: Soft drinks · Phosphoric acid · HACH method

1 Introduction

The overall consumption of soft drinks including colas, diet sodas and sports drinks is increasing drastically worldwide, especially among young adults. These drinks contain numerous ingredients which have implications on human health [1].

Phosphoric acid is commonly added to beverages to improve flavor and taste. Use of phosphoric acid has been associated with adverse health effects. Higher concentration of phosphorus in blood, referred as “hyperphosphatemia”, may lead to organ damage, most notably kidneys. Furthermore, phosphoric acid interferes with calcium absorption and contributes to imbalances that lead to additional loss of calcium, increasing the risk of brittle bone disease. The phosphoric content of beverages may also contribute to vascular calcification and cardiovascular disease [2, 3].

The National Legislative allows up to 700 mg/L of phosphoric acid in soft drinks, measured as PO_4^{3-} and expressed as mg/L P_2O_5 [4].

A number of methods for determination of phosphate concentration has been reported, including titrimetry, ion chromatography [5, 6] spectrophotometry and potentiometry [7, 8].

M. Brka et al. (Eds.): AgriConf 2019, IFMBE Proceedings 78, pp. 237–243, 2020.
https://doi.org/10.1007/978-3-030-40049-1_30

UV/VIS spectrophotometric method is usually used in our laboratory for determination of phosphoric acid.

In this study, we have developed HACH spectrophotometric method for determining phosphate ions (PO_4^{3-}) and compared with conventional spectrophotometric method (UV/VIS). HACH spectrophotometric method 8048 has been successfully applied to monitoring and ensuring safety of beverages samples from a local market in terms of phosphate content.

2 Materials and Methods

2.1 Sample Collection and Storage

Samples of beverages were collected from different markets in Sarajevo, Bosnia and Herzegovina. The samples used in this study were soft drinks like *Colas* (13 samples). Soft drinks like *colas* are considered to be the most popular drinks which contain phosphoric acid. In this study, soft drinks *Cockta* and *Schweppes Tonic water*, which do not contain phosphoric acid, were used as blank samples. *Cockta* blank sample was treated in the same manner as the samples to be tested, due to its color and composition that are very similar to Cola juices, while *Schweppes Tonic water* blank sample was measured directly.

2.2 Preparation Sample

To remove the carbon (IV) oxide, approximately 100 mL of each sample was placed into sample in ultrasonic bath (15 min) before the analysis. All glasses were washed with 6 N hydrochloric acid (1:1) and deionized water prior to the analysis.

2.3 Determination of Phosphoric Acid by UV/VIS Spectrophotometric Method

The method for the determination of phosphoric acid in beverages is based on spectrophotometric determination at 830 nm of the color intensity of the bluish-purple colored heteropolymolybdenum (V) complex. In an acidic medium, phosphates bond with ammonium molybdate to form ammonium phosphomolybdate. With the reducing agent (ascorbic acid), ammonium phosphomolybdate is reduced to form a bluish-purple colored heteropolymolybdenum (V) complex also known as 'molybdenum blue'. The intensity of the blue coloration is proportional to the concentration of phosphates (9).

A stock solution containing 1000 ppm P_2O_5 was prepared by dissolving 1.9412 g of dry potassium di-hydrogen phosphate (KH_2PO_4) in 1 L of distilled water. A working standard solution was prepared by diluting 2.5 mL of stock solution to 50 mL, containing 4 ppm P_2O_5.

Ammonium molybdate solution was prepared by dissolving 1.00 g of ammonium molybdate tetrahydrate ($(NH_4)_6Mo_7O_{24}$ •$4H_2O$) in 50 mL of deionized water. For preparation of the ascorbic acid solution, 1.76 g of ascorbic acid ($C_6H_8O_6$) was dissolved in 100 mL of deionized water. Diluted sulfuric acid was prepared diluting 17 mL of concentrated sulfuric acid (H_2SO_4) in 200 mL of deionized water.

A reducing solution was prepared by dissolving 39 mL of 60 mL of the ascorbic acid solution and 125 mL of diluted sulfuric acid in a 250 mL deionized water. The solution is stable for 24 h.

Standard solutions containing 0.4, 0.8, 1.2, 1.6 and 2 ppm P_2O_5 were prepared by diluting 0.5, 1.0, 1.5, 2.0 and 2.5 mL of the working standard and 2 ml reducing solution to 5 mL respectively. The obtained linear equation is equal to $y = 0.0475x + 0.0092$, and the correlation coefficient amounts to $R^2 = 0.9937$.

After degasis, 2.5 mL of the sample was transferred into a 50-mL volumetric flask. Flask was fill up with deionized water up to the mark. Each solution was dispensed in test tubes, as indicated: 0.25 mL volume from diluted sample, 2.0 mL reducing solution, 2.75 mL deionized water which made a final volume of 5 mL.

After 10 min, the absorbance of the solution was measured at 830 nm using a standard quartz cell in a UV-Visible spectrophotometer. The reagent blank was determined in the same manner using sample solution without reducing the solution.

2.4 Determination of Phosphoric Acid by HACH Spectrophotometric Method

The method for the determination of total phosphates in beverages samples using HACH spectrophotometric method is also based on the reaction between orthophosphate with molybdate in an acid medium to produce a phophomolybdate complex, giving an intense molybdenum blue color. Method which we used was Phosphorus, Reactive (Orthophosphate) Method 8048 (0.02 to 2.50 mg/L PO_4^{3-}) [9].

After degassing, a 0.2 ml sample was transferred to a volumetric flask of 100 ml and filled with deionized water up to the mark. The sample cell was filled with 10 ml of preprepared sample and PhosVer 3 phosphate reagent added to the pillow. The sample cell was sealed and shaken vigorously for 20–30 s. Absorbance was measured after 2 min. The wavelength of measurement was 880 nm.

2.5 Method Validation

The HACH spectrophotometric method was validated for the parameters like linearity, accuracy, precision, intra-day precision, inter-day precision/intermediate precision/ ruggedness, limit of detection (LOD) and limit of quantitation (LOQ) and finally uncertainty measurement to describe the quality of results.

Accuracy was determined by means of recovery experiments, by the determination of % mean recovery of sample at three different levels (50–100%) in relation to the maximum allowed concentration (700 mg/L). The recovery experiments were performed by adding known amounts of the phosphoric acid in the prepared Cola sample. At each level, three determinations were performed. This step also included review of linearity of the calibration curve.

Method precision was determined by performing assay of internal control material (PT) under the tests of repeatability (Intra-day precision) and Intermediate precision (Inter-day precision). Intra-day precision study was carried out by preparing internal control material as sample and analyzing it at three different times within a day. The same procedure was followed during 3 days to determine inter-day precision. The results were reported as % RSD.

Ruggedness was determined by carrying out analysis by two different analyst and also by carrying out the analysis on two different instruments and method. Results was indicated as SD (n = 3).

The sensitivity of measurements of phosphoric acid by the use of the proposed method was estimated in terms of the limit of quantification (LOQ) and limit of detection (LOD). Calculation of LOD was based on determinations for blank samples. A more labor-consuming method is using measurement for a series of blank samples. It involves 10 independent measurements for 10 independently prepared blank samples. For the thus obtained 10 results, one calculates the mean value and the standard deviation (SD). LOD is equal to the mean value magnified by three times the standard deviation in this instance: LOD = Xm + 3*SD. Limit of quantification (LOQ) is equal to LOQ = 3*LOD.

Intermediate precision of the method was checked by monitoring values obtained from the analysis of the internal control sample (PT which we froze in little portion after receiving), by plotting them on a control chart. Limits are set for the values on the chart is conventionally 'warning limits' are set at ±2SD about the mean value, and 'action limits' are set at ±3SD about the mean value. Mean value for our material is 380.00 mg/L and it was central line in control chart.

Nordtest QuaLab Excel program was used for estimating measurement uncertainty. The results obtained from control charts (n = 20) and the results of comparison between measurement values and the "accurate value" for PT were used for estimating measurement uncertainty. Value of measured uncertainty amounts to ±12.3% of the obtained measurement result.

3 Results and Discussion

With the purpose of showing that the developed analytical method is appropriate for the determination of phosphates in "Cola" juices samples, in this paper we conducted method validation by checking the following parameters: linearity, detection and quantification limit, precision, and trueness.

Analytical method validation is a process that is established by laboratory studies in which the performance characteristics of the method meet the requirements for the intended analytical application according to International Conference on Harmonization (ICH) guidelines [10] and Quality Assurance and Quality Control in the Analytical Chemical Laboratory: A Pratical Approach [11].

In this study, we used Phosphate Standard Solution, 1 mg/L as PO_4^{3-} for control of calibration curve and Carbonated Drink (degassed) Sample 516 proficiency test (PT) Scheme: Soft Drinks & Fruit Juice (QBS) (laboratory involve in Dec. 2018) and we used it as internal control material which we froze in little portion after finishing the initial control. "Accurate" value for this parameter in PT was 370.1 mg/L PO_4^{3-} and it is expressed as mean Satisfactory Range as 332.7 to 406.7 mg/L PO_4^{3-}.

Proficiency testing (PT) is also an essential element of laboratory quality assurance. Regular proficiency testing provides independent feedback on the quality of analytical results, enabling laboratories to monitor and improve performance over time. It also enables the analyst to compare the performance of their method with that of other methods used and can be use for validation method if no other control material exists or if laboratory does not have it.

Recovery percentage was calculated as shown in Table 1. The accepted limits of recovery are 90%–107% (for concentration analite ≥ 100 mg/L) and all observed data are within the required range which indicates good recovery values and suggests that the proposed method is accurate.

Table 1. Accuracy

Concentration level (%)	*% Mean recovery
50	99,2
75	102,4
100	106,2

*Mean of three replicates

The precision result showed a good repeatability with relative standard deviation percentage below 5.3% (for concentration analite ≥ 100 mg/L), which indicates that the method developed is method precise by the test of repeatability and inter day precise (Table 2).

Table 2. Intra-day precision (day 1) and Inter-day precision results (day 3)

n	Results (mg/L P_2O_5)		
	Day 1	Day 2	Day 3
1	380.00	395	400
2	385.00	390	415
3	385.00	395	405
4	390.00	395	415
5	385.00	400	405
6	390.00	390	415
7	385.00	400	405
Average	385.71	395.00	408.57
Standard deviation (SD)	3.45	4.08	6.26
Relative standard deviation (%RSD)	0.89	1.03	1.53

Table 3. LOD and LOQ

LOD	LOQ
0.0075 mg/L	0.02 mg/L
Eligibility criterion is met	10*LOD > c_{min} LOD < c_{min} $c_{min} \geq$ LOQ

Concentration of phosphoric acid in beverage samples obtained from two spectrophotometric methods and analysis by two different analysts varied from 274.80 to 691.12 mg/L P_2O_5 for UV/VIS and from 337.50 to 723.15 mg/L P_2O_5 for HACH spectrophotometric method. Apart from chemical composition of the sample and possible interference, consequences of such variations are the steps of the analysis as well (major dilutions). Considering that in this paper we used *Cockta* as blank, which has composition similar to that of *Cola*, it is easy to notice in analysis results the contribution (interference) of other present components, which is not the case with blank sample of *Schweppes Tonic Water* (Table 3).

Various types of blank are available to the analysts, Reagent blanks or Sample blanks. In this study we decided to use sample blanks because these are essential sample matrices with no analyte present, like *Cockta* sample without a specific phosphoric acid. Sample blanks may be difficult to obtain but such materials are necessary to give a realistic estimate of interferences that would be encountered in the analysis of test samples (Table 4).

Table 4. Ruggedness – concentration of phosphoric acid in beverage samples obtained from two spectrophotometric methods and analysis by two different analysts

Sample	Lab No.	Method 1 (UV/VIS)	Method 2 (HACH)
		Mean ± SD (mg/L P_2O_5)	Mean ± SD (mg/L P_2O_5)
Cola Spektar drink	1	336.80 ± 15.20	416.25 ± 13.28
Coca cola original teste	3	360.40 ± 7.53	431.25 ± 6.34
Pepsi	4	376.00 ± 10.75	393.75 ± 10.81
Cola stevia Vitinka	5	311.60 ± 12.32	356.25 ± 12.58
Coca cola zero sugar limun	6	333.20 ± 8.82	375.00 ± 7.38
Pepsi max	7	330.80 ± 9.21	382.50 ± 5.11
Sky cola	8	395.20 ± 12.45	475.00 ± 10.46
Cola Sinalco	9	317.20 ± 10.67	404.50 ± 9.87
RC Cola	10	274.80 ± 13.05	337.50 ± 8.87
coca cola zero sugar	13	312.80 ± 9.21	386.25 ± 7.34
Cola plus -zmajići	13	432.61 ± 7.47	491.25 ± 12.04
Panonska ledena Cola	14	475.34 ± 10.42	506.25 ± 9.21
Joys Cola	15	691.12 ± 12.33	723.15 ± 11.45
Cockta finest herbal	16	22.00 ± 9.38	15.00 ± 7.21
Schweppes Tonic water	17	0.0053 ± 0	0.0075 ± 0

4 Conclusion

Analyses and control of presence of phosphoric acid in the beverage samples is highly significant both for consumers and producers, taking into account the influence of phosphoric acid to human health and the increasing production of such product type. Modern approach to samples control instructs that methods used in the laboratory should be quick, accurate, and reliable, which facilitates analyst's work and provides the end user with a quick and reliable result. In order for the laboratory to provide analysis and its control, a validated HACH method 8048 is considered as quick, accurate, and reliable, which is the reason it will be used in future activities.

Unlike other methods used for phosphates analysis in juices, HACH spectroscopic method 8048 quickly gives a reliable result (five minutes per sample) without preparation of reagents in advance for the conduct of the analysis.

References

1. Vasanti, S.M., Schulze, M.B., Hu, F.B.: Intake of sugar-sweetened beverages and weight gain: a systematic review. Am. J. Clin. Nutr. **84**(2), 274–288 (2006)
2. Tucker, K.L., Morita, K., Qiao, N., Hannan, M.G., Cupples, A., Kiel, D.P.: Colas, but not other carbonated beverages, are associated with low bone mineral density in older women: The Framingham Osteoporosis Study. Am. J. Clin. Nutr. **84**(4), 936–942 (2006)
3. Kregiel, D.: Health safety of soft drinks: contents, containers, and microorganisms. Biomed. Res. Int. **2015**, 15 (2015). https://doi.org/10.1155/2015/128697. Article ID 128697
4. Pravilnik o prehrambenim aditivima. Službeni glasnik BiH, broj 33/18
5. Drossman, H.: Chemical speciation analysis of sports drinks by acid-base titrimetry and ion chromatography: a challenging beverage formulation project. J. Chem. Educ. **84**(1), 124 (2007)
6. Bello, M.A., Gonzalez, A.G.: Determination of phosphate in cola beverages using nonsuppressed ion chromatography: an experiment introducing ion chromatography for quantitative analysis. J. Chem. Educ. (1996). https://doi.org/10.1021/ed073p1174
7. Utange, P.I., Wuana, R.A., Akpoghol, T.V.: Potentiometric and spectrophotometric determination of phosphoric acid in some beverages. Pelagia Res. Libr. Chem. Sin. **6**(3), 93–99 (2015)
8. Murphy, J., Riley, J.P.: A modified single solution method for the determination of phosphate in natural waters. Anal. Chem. Acta **27**, 31–36 (1962)
9. Phosphorus, reactive (orthophosphate) method 8048 (0.02 to 2.50 mg/L PO43-). https://www.hach.com/quick.search-download.search.jsa?keywords=DOC316.53.01119
10. International Conference on Harmonization (ICH) guidelines
11. Konieczka, P., Namiesnik, J.: Quality Assurance and Quality Control in the Analytical Chemical Laboratory A Pratical Approach. CRC, Boca Raton (2009)

Analysis of the Fat Substitution in Biscuit Type Cookies by the Addition of Inulin

Mirsad Veladžić, Suzana Jahić(✉), and Aida Džaferović

Biotechnical Faculty, University of Bihać,
Luke Marjanovića bb, 77 000 Bihać, Bosnia and Herzegovina
suzanajahic2002@gmail.com

Abstract. Cookies are the products with a high content of sugar and fat, and one of the methods of improv- ing their nutritional profile is the addition of the functional products in the recipe for making cakes, such as inulin. Addition of inulin, reduce the fat content in the final product, without a significant change in the water content and sensory properties. This study aimed at developing formulations of cookies with partial replacement of fat by inulin as well as evaluating physical, chemical and sensory characteristics. In the recipe for the cookies 2%, 4% and 6% of inulin was added with the reduction in fat content. Cookies with inulin content of 2% had reduction of the total fat 44.6%, while in the cookies with 4% of inulin, the reduction of the fat content was 64.8%, and the cookies with 6% of inulin had the reduction of the fat 75.8%. In the samples with 2% and 6% of inulin, the percent of the moisture was higher with regard to a standard, but, in the sample with 2% of inulin, the percent of the moisture was lower ($p < 0.01$) regard to a standard. Once content of inulin was increased, the size of the cookies after baking also increased; all samples had a higher thickness regard to a standard, and the diameter of the cookies was higher regard to a standard in the all samples, except in case of sample with 2% of inulin. According to the results of the sensory analysis, the Commision classified the samples with 2% and 6% of inulin into the category very good and the samples with 4% of inulin in the category good.

Keywords: Fat substitution · Cookies · Inulin

1 Introduction

Biscuits type cookies are mainly made of wheat flour and have a high content of sugars and fat [1]. These ingredients are responsible for the distinctive characteristics of this kind of product [2]. Due to consumers' demand for healthier products, such products have been increasingly required reformulation of the previous recipes. Studies have shown that there is a high correlation between the excess of fat in diets and the risk of cardiovascular diseases [3]. Vegetal hydrogenated fat used in biscuit production has a high content of *trans* - fatty acids, which affects the factors of cardiovascular risks: increase in cholesterolemia, increasing of LDL cholesterol and reduction of HDL cholesterol [4]. One of the strategies used to improve the nutritional profile of these highly consumed worldwide products is the incorporation of dietary fibers to its

M. Brka et al. (Eds.): AgriConf 2019, IFMBE Proceedings 78, pp. 244–250, 2020.
https://doi.org/10.1007/978-3-030-40049-1_31

formulation. Dietary fibers are of the plant origin, those are not hydrolyzed by digestive enzymes of the human upper gastrointestinal tract and undergoes to partial or total fermentation by colon bacteria. One enzyme, the inulinase is able to hydrolyze inulin. This enzyme is not present in the human gastrointestinal tract, and therefore inulin reaches the colon intact, where is fermented by anaerobic bacteria producing short-chain fatty acids [5]. These characteristics generate beneficial health effects such as risk reduction of diabetes, cardiovascular disease, some cancers and obesity [6]. According to [7], inulin is present as a reserve carbohydrate found in roots and tubers of plants such as the Jerusalem artichoke (*Helianthus tuberosus),* yacon (*Smallanthus sonchifolius),* chicory (*Cichorium intybus),* dahlia (*Dahlia sp.).* Inulin is a polymer of linear D-fructose bonded by β - (1–2) linkages with or without an end glucose residue, with a degree of polymerization ranging between 10 and 25. Mostly, inulin is used as a fat replacement in some kind of food products [8]. The use of inulin in a biscuit production is technologically justified; compared to standard samples, the samples with addition of inulin don't have these a significant change in the sensory properties, and no changes in the softness or the content of water [9]. The aim of this paper was to produce biscuits type cookies with a reduced percentage of fat, partly replacing fat with the inulin and estimation the physical, chemical and sensory properties of biscuits.

2 Materials and Methods

2.1 Samples

The ingredients used to produce the cookies were: wheat flour, starch, sodium bicarbonate, sugar, hydrogenated vegetal fat, salt, vanilla essence, water and inulin extract (Table 2). All items were acquired on local stores in the city Bihać. Extract of inulin was obtained from the Jerusalem artichoke (*Helianthus tuberosus*) in the Laboratory of the Biotechnical Faculty. For the production of the extract, fresh tubers of Jerusalem artichokes were used, which were grown on the experimental plots of the Biotechnical Faculty in Bihać. After washing, the tubers were cut into slices and then placed in a double-coated device (SP 100 L EH, Križevci). Distilled water was used as an extraction agent (1:6) and the extraction lasted 60 min at a temperature of 80 °C. The obtained extract was discharged from the duplicate, and the residue in the duplicate was again distilled with water at the same temperature for the same time period. Obtained resulting extract was coupled with the first extract, and then evaporated at a temperature 90 °C to a concentration of 30° Brix measured with a refractometer at a temperature of 20° (Optech, Optical Technology, Germani). The evaporated extract is filtered by a funnel and a wrinkled paper filter (Filter Disc, 125 mm diameter, 84 g/m^2) using a vacuum pump (Conecta). The resulting extract was dried on Mini Spray Dryer B-290 (BUCHI). The temperature of the inlet air was 190 °C and the output temperature was 90 °C. Water content was determined by drying to a constant weight at 105 °C, protein content by the Kjeldahl method, fat content by the Soxhlet method, mineral matter by burning and sampling the sample at 550 °C. The total acids were determined by titration of the sample with sodium hydroxide, the content of crude fibers by the Kurschner-Hanak method, starch is determined by the Grossfeld method, and natural invert is determined by the method of Luff-Shorl [10]. The amount of inulin was obtained mathematically as a difference in the amount of water, protein, fat, mineral matter, acids, natural invert, crude fibers and starch to 100%. Table 1 shows chemical composition of inulin extract.

Table 1. Chemical composition of inulin extract

Parameters	Content %
Water	0.5
Protein	11.0
Fat	1.0
Mineral matter	6.5
Total acids	1.0
Natural invert	4.0
Crude fibers	8.0
Starch	11.0
Inulin	57.0

2.2 Biscuits Preparation

Four samples of cookies were prepared with a partial replacement of fat with inulin: the sample without inulin – standard (Sample 1), sample with 2% of inulin (Sample 2), sample with 4% of inulin (Sample 3) and sample with 6% of inulin (Sample 4).

Table 2. Formulation of the cookies

Ingredients %	Sample 1	Sample 2	Sample 3	Sample 4
Wheat flour type 500	43.4	43.4	43.4	43.4
Starch	7.2	7.2	7.2	7.2
$NaHCO_3$	0.4	0.4	0.4	0.4
Hydrogenated vegetal fat	18.1	13.6	9.1	4.6
Sodium chloride	0.7	0.7	0.7	0.7
Sugar	21.7	21.7	21.7	21.7
Vanilla essence	0.6	0.6	0.6	0.6
Water	7.9	8.9	9.9	10.9
Inulin extract	0.0	3.5	7.0	10.5

Hydrogenated vegetal fat mixed with water, vanilla essence, sugar and starch in a mixer for 2 min. at low speed and then 4 min. at high speed until a cream was formed. Flour, sodium bicarbonate, sodium chloride and inulin extract were sifted together and added to the cream. The dough was manually sheeted to a 9 mm thickness and cut in circles of 45 mm of diameter and placed on aluminium sheet. The cookies were baked in an air circulation oven at 160 °C for 30 min. The samples were allowed to cool at room temperature for 2 h.

2.3 Chemical and Physical Analysis of Biscuits

All analytical determinations were performed at least in triplicate. Values of different parameters were expressed as the mean ± standard deviation. The composition and cookies quality were determined through physical-chemical assays: moisture [11], ash [12] and crude fat [13]. The cookies physical evaluation was carried out through setting their mass, diameter, and thickness before and after baking. The cookies were measured through a caliper rule and weighted on a digital semi-analytical scale. All analytical determinations were performed in triplicate.

2.4 Sensory Analysis of Biscuits

Sensory evaluation of cookies for the appearance, color, structure and breaking, chewability, odor and taste, and overall acceptability were done in order to determine consumer acceptability. The produced cookies were evaluated by a five-member sensory evaluation panel at the Laboratory of the Biotechnical Faculty in Bihać, using the scoring method [14]. The procedure of scoring was based on an assessment of quality attributes chosen. A score list with 5 points (1 to 5) was applied. Each quality level expressed with the corresponding score was precisely defined. The assessors were instructed to rinse their mouths with water after every sample. The individual parameters were corrected by the correction factor, whereby weighted points were obtained, the collection of which gives the total number of points for the sensory quality of the biscuits. Depending on the sum of points it was determined in which category the cookies belong to by sensor quality [14].

Table 3. Category of sensor quality

Quality category	Excellent	Very good	Good	Frail	Not match
The total sum of weighted points	19.1–20.0	16.1–19.0	13.1–16.0	11.1–13.0	<11.1

2.5 Statistics Analysis

The given results in this paper were analyzed by applying the analysis of variance (ANOVA) and the significance of the obtained differences was evaluated using the Tukey test. The level of significance $p < 0.05$ was used for all comparisons and for the discussion of the obtained results.

3 Results and Discussion

The results show (Table 4) a significant reduction in fat content in relation to the particular recipe. For Sample 2, reduction of total fat was 44.6% in relation to the standard recipe. For Sample 3, reduction of total fat was 64.8% and for Sample 4 was 75.8%. Similar results obtained by [1] using inulin for the cookies in addition from 1.1 to 3.3%. The significant difference ($p < 0.01$) of the fat was found between the standard (Sample 1) and other inulin-containing samples, as well as between Sample 2 and Samples 3 and 4. The significant difference ($p < 0.01$) of moisture was found between the standard (Sample 1) and Samples 2 and 3, as well between the Sample 2 and Samples 3 and 4. The most content of ash was in Sample 4 (2.44%) and the smallest content of ash had a standard (Sample 1 with 1.36%). According to literature data [1] there is no significant differences in moisture and ash content for the cookies samples.

Table 4. Chemical analysis of cookies

Content %	Sample 1 Standard	Sample 2 2% inulin	Sample 3 4% inulin	Sample 4 6% inulin
Crude fat	$30.55 \pm 1.955^{AB,AC}$	$16.91 \pm 11.45^{BC,BD}$	10.74 ± 0.66	7.39 ± 0.84
Moisture	$10.89 \pm 0.45^{AB,AC}$	$6.94 \pm 0.17^{BC,BD}$	12.25 ± 0.39	11.31 ± 0.44
Ash	$1.36 \pm 0.10^{AC,AD}$	$1.58 \pm 0.04^{BC,BD}$	2.22 ± 0.19	2.44 ± 0.22

* Results are expressed as the mean ± standard deviation of three replicates. Means followed by different uppercase letters in the same line are significantly different through the Tukey test ($p < 0.01$).

Table 5. Values of weight, thickness, and diameter of the cookies before baking

Sample parameters	Sample 1 Standard	Sample 2 2% inulin	Sample 3 4% inulin	Sample 4 6% inulin
Weight/g	17.64 ± 1.01	19.01 ± 1.12	20.83 ± 0.68	20.24 ± 0.87
Thickness/mm	8.90 ± 0.38	9.39 ± 0.35	10.98 ± 0.36	9.56 ± 0.41
Diameter/mm	46.17 ± 1.29	45.77 ± 0.81	45.53 ± 0.75	47.41 ± 1.56

Table 6. Values of weight, thickness, and diameter of the cookies after baking

Sample parameters	Sample 1 Standard	Sample 2 2% inulin	Sample 3 4% inulin	Sample 4 6% inulin
Weight/g	14.49 ± 0.78	14.89 ± 1.46	17.03 ± 1.77	15.95 ± 1.30
Thickness/mm	11.78 ± 3.748	15.45 ± 1.07	16.65 ± 0.99	15.84 ± 0.79
Diameter/mm	50.40 ± 1.06	48.27 ± 1.41	50.72 ± 1.39	53.08 ± 1.34

In the relation to the results shown in the Tables 5 and 6 (results are expressed as the mean ± standard deviation of three replicates), it was found a change in the thickness and diameter of the cookies before and after baking; in the all samples, higher values for these parameters recorded after baking compared to the values before baking. As the percentage of inulin extract increases, a considerable changes of the thickness of the cookies compared to standard is noticed. After baking, the thickness of standard increased by 24.45%; for Sample 2 the thickness increased by 39.23%, for Sample 3 - 30.05% and for Sample 4 - 39.65%. After baking, diameter of standard increased by 8.39%; for Sample 3 diameter increased by 10.24%, for Sample 4 - 10.68%. For Sample 2, after baking, diameter increased by 5.18% (less then standard). After baking, the weight of samples decreased; for Sample 1 (standard) by 17.86%, for Sample 2 - 21.60%, for Sample 3 - 18.24 and for Sample 4 - 21.30%. During baking, inulin progressively dissolves resulting in a rise of the spread rate and maximum diameter, which improved cookies quality [2]. This effect is similar to that achieved by an increase of fat level since fat melts during baking and contributes to the system mobility [15] (Table 7).

Table 7. The obtained values for the sensory analysis of cookies by the scoring method

Sensory properties	Sample 1 Standard	Sample 2 2% inulin	Sample 3 4% inulin	Sample 4 6% inulin
Shape and surface	3.76 ± 0.32	3.66 ± 0.59	3.32 ± 0.99	4.26 ± 0.43
Color, baking	2.96 ± 0.37	3.09 ± 0.52	2.99 ± 0.57	3.46 ± 0.43
Structure	3.67 ± 0.33	3.57 ± 0.51	3.14 ± 0.32	3.12 ± 0.69
Odor and taste	6.75 ± 0.30	6.64 ± 0.52	5.49 ± 0.65a,c	6.02 ± 0.71b,c
Total	17.14	16.96	14.94	16.86

According to the results of sensory analysis, standard and the samples with addition of 2% and 6% of inulin belong to the category very good (Table 3). The sample with 4% of inulin belongs to the category good. For the odor and taste, the highest score there was in Sample 1 (standard) and the smallest score there was in the sample with the addition of 4% of inulin. The smallest total number of points recorded in the sample with the addition of 4% of inulin (14.94 points). In standard recorded the highest total number of points (17.14). According to the results of [1], the sample with 75% partial replacement of fat by inulin had smaller acceptability in relation to the standard.

4 Conclusion

With the partial replacement of fat by inulin in the cookies, achieved a positive reduction in the percentage of the total fat with a smaller change in the moisture and ash content in the final product. Addition of 4% and 6% of inulin in the cookies caused the increase in the moisture content, relation to the standard recipe. Addition of 2% of inulin in the cookies caused the reduction in the moisture content, relation to the standard. The cookies with inulin addition had an increase in the thickness and diameter after baking in relation to the standard; only the sample with 2% of inulin had less increasing in diameter relation to the standard. According to the sensory analysis of the cookies, the samples with addition of 2% and 6% of inulin belong to the category very good, and the sample with the addition of 4% of inulin belongs to the category good. Inulin added as an ingredient in cookies and biscuit recipes can be considered as an effective substitute for fat, whereby the obtained products possess a good technological and acceptable sensory properties.

References

1. Estevam, L.E., Benossi, L., Rodrigues, M.D., Miguel, J.B., Giriboni Monteiro, A.R.: Development of biscuit type cookie with partial replacement of fat by inulin. Int. J. Nutr. Food Sci. **2**(5), 261–265 (2013)
2. Blanco Canalis, M.S., Leona, A.E., Ribotta, P.D.: Effect of inulin on dough and biscuit quality produced from dfferent flours. Int. J. Food Stud. **6**, 13–23 (2017)
3. Viana, R.F., Silva, M.D.V., Carvalho, G.M., Oliveira, L.A., Silvestre, C.P.M.: Efeito da substituicao parcial de gordura pela globina e plasma bovinos em pate de presunto. Ac. Sci. Bio. Sci. **25**(1), 233–240 (2003)
4. Rique, R.B.A., Soares, A.E., MEirelles, M.C.: Nutricao e exercicio na prevencao e controle das doencas cardiovasculares. Rev Bras Med Esp. **8**(6), 244–254 (2002)
5. Lunn, J., Buttriss, J.L.: Carbohydrates and dietary fibre. Nut. Bull. **32**(1), 21–64 (2007)
6. Topping, D.: Cereal complex carbohydrates and their contribution to human health. J. Cereal Sci. **46**(3), 220–229 (2007)
7. Chi, Z.M., Zhang, T., Cao, T.S., Liu, X.Y., Cui, W., Zhao, C.H.: Biotechnological potential of inulin for bioprocesses. Bioresour. Technol. **102**, 4295–4303 (2011)
8. Meyer, D., Bayarri, S., Tarrega, A., Costell, E.: Inulin as texture modifier in dairy products. Food Hydrocolloids **25**(8), 1881–1890 (2011)
9. Franck, A.: Technological functionality of inulin and oligofructose. Br. J. Nutr. **87**(2), 287–291 (2002)
10. Trajković, M., Mirić, M., Baras, J., Šiler, S.: Analize životnih namirnica. Tehnološko-metalurški fakultet Beograd, Beograd (1983)
11. AOAC: Official Methods of Analysis, 15th edn. Association of Official Analytical Chemists, Washington, DC, USA (2000). (AOAC 925.09)
12. AOAC: Official Methods of Analysis, 5th edn. Association of Official Analytical Chemists, Washington, DC, USA (2000). (AOAC 923.03)
13. AOAC: Official Methods of Analysis, 15th edn. Association of Official Analytical Chemists, Washington, DC, USA (2000). (AOAC 922.06)
14. Pajin, B.: Handbook of Confectionary. Faculty of Technology, Novi Sad (2009)
15. Pareyt, B., Talhaoui, F., Kerckhofs, G., Brijs, K., Goesaert, H., Wevers, M., Delcour, J.A.: The role of sugar and fat in sugarsnap cookies: Structural and textural properties. J. Food Eng. **90**(3), 400–408 (2009)

Fermentation of Cow's Milk and Soy Milk Mixture with *L. acidophilus* Probiotic Bacteria with Yoghurt Culture

Edina Šertović[1(✉)], Zlatan Sarić[2], Rajka Božanić[3], Miroljub Barać[4], Irena Barukčić[3], and Aleksandar Kostić[4]

[1] Biotechnical Faculty, University of Bihac, Luke Marjanovica bb, 77000 Bihac, Bosnia and Herzegovina
edina.sertovic@gmail.com
[2] Faculty of Agriculture and Food Sciences, University of Sarajevo, Zmaja od Bosne, 71000 Sarajevo, Bosnia and Herzegovina
[3] Faculty of Food Technology and Biotechnology, University of Zagreb, Pierottijeva 6, 10000 Zagreb, Croatia
[4] University of Belgrade, Faculty of Agriculture, Nemanjina 6, 11080 Belgrade, Serbia

Abstract. The aim of the paper was to produce a functional product and to determine the physico - chemical, microbiological and sensory properties of probiotic beverages produced from different mixtures of cow's milk and soy milk. The ratios of cow's milk and soy milk were 100:0%, 25:75%, 50:50%, 75:25% and 0:100%. The samples were fermented with probiotic bacteria *Lactobacillus acidophilus* (La5) with the addition of yogurt culture. The fermentation of samples was performed at +43 °C until coagulation and pH value of 4.6. Characteristics of the obtained beverages were monitored during storage at 1st, 7th, 14th and 21st day of storage at +4 °C. Produced probiotic beverages were observed and change in active acidity, titration acidity, change in the number od probiotic bacteria, and sensory properties and acceptability of products were tracked. After fermentation, the number of lactobacilli in the samples produced gradually decreased during 21 days, but was still above the probiotic minimum (CFU 10^6 mL^{-1}) so the samples had probiotic properties. Samples of fermented beverages got the best score of their sensory properties on the seventh day of preservation. The sensory properties of the samples were mainly influenced by the type and ratio of the used milk. Mixing cow's milk with soy milk significantly improved the sensory properties of the product. The acceptance test showed good acceptance of fermented beverages samples by potential consumers, apart from the sample that was 100% soy milk.

Keywords: Cow's milk · Soy milk · Probiotics · *Lactobacillus acidophilus*

1 Introduction

Soy milk contains various oligosaccharides, including raffinose and stachyose, which can lead to digestive disorders [1]. Fermentation of soy milk gives the ability to transform and improve the taste and texture [2, 3]. Also, fermentation of soy milk is

M. Brka et al. (Eds.): AgriConf 2019, IFMBE Proceedings 78, pp. 251–259, 2020.
https://doi.org/10.1007/978-3-030-40049-1_32

considered as a good basis for the functional food development and production. Fermentation of soy milk with probiotic bacteria reduces the level of oligosaccharide and raises the level of free isoflavones [4]. Fermentation of soy milk with probiotic bacteria has numerous health benefits, including antimicrobial, antimutagenic, anticancerogenic, hypertensive properties, and it is also reducing serum cholesterol and mitigating lactose intolerance [5, 6]. For fermented probiotic products, the ability of culture to make the positive influence on the sensory properties of the product is essential. The aim of this study was to produce a functional milk beverage (probiotic yoghurt) that has useful characteristics of probiotic and soy milk. For this purpose, the physico-chemical, microbiological and sensory properties of fermented probiotic beverages produced from different ratios of cow's milk and soy milk were estimated using *L. acidophilus* probiotic bacteria during 21 days of storage.

2 Materials and Methods

Five different proportions of homogenized milk in soy beverage (100:0, 75:25, 50:50, 25:75, 0:100) were prepared by mixing UHT sterilised milk and soy milk. Samples were inoculated with probiotic starter culture (La5) and yoghurt culture (YF-L811). Probiotic monocultures for inoculation were in a lyophilized DVS (Direct Vat Set) form. After incubation, the inoculum was cleaved into milk samples intended for the production of probiotic beverages. Fermentation was carried out at 43 °C until pH reached a value of 4.6. Characteristics of the obtained probiotic dairy beverages are monitored during and at the end of fermentation. Three different repeated-batch fermentations were performed.

2.1 Microbiological Analysis

The viable cell counts of the probiotic strain in the produced samples were determined by a standard pour plating method using the MRS agar (Merck, Germany) supplemented by clindamycin (Sigma-Aldrich, Germany). Clindamycin was added to a sterilized MRS agar cooled to 43 °C just before pouring into the Petri dish in order to prevent the growth of the used yoghurt culture during the fermentation process (ISO/IDF (2006). Subsequently, Petri dishes were incubated at 37 °C for 72 h after which colonies of probiotic bacteria (La5) were enumerated. The obtained data represent the arithmetic average of the enumerated colonies expressed as CFU mL^{-1}.

2.2 Physical, Chemical and Sensory Analysis

The active acidity (pH) of probiotic beverages was determined by the pH meter pH3110 (Portable meter ProfiLine) and the titratable acidity by the Soxhlet - Hankel method and expressed as % of lactic acid. During the 21 days of produced probiotic fermented beverage samples storage, changes in acidity (pH value and lactic acid content) were observed, as well as the microbiological and sensory properties of the product. Sensory properties are rated by a weighted scoring method (ISO,1985) by a group of 5 trained sensory analysts. Eligibility of probiotic beverages was performed by testing 30 younger consumers (students around 20 years old) using the verbal 9-point Hedonic Scale (Periam) [7].

2.3 Statistical Analysis of Results

The results of the analyzed samples are shown as the average repetition value ± standard deviation. One-way analysis of variance (ANOVA) and multiple comparisons (Duncan's post-hoc test) were used to estimate significant difference in data at the significance level of $p < 0.05$. Statistics were implemented using Microsoft Office 2014 and demo versions of the MS Office XLSTAT-Pro 2014 statistical package.

3 Results and Discussion

The fermentation of milk and soy milk samples (100:0, 0:100, 25:75, 50:50, 75:25) with probiotic bacteria La5 and the starter culture addition, lasted 5 to 7 h (Table 1). Table 1 shows changes in pH value during fermentation in all analyzed samples.

Table 1. Fermentation time and acidity of produced yoghurt samples

Para\s	Yogurt samples				
	1A	2A	3A	4A	5A
Fermentation time (minutes)	420	360	360	360	360
pH	4.57 ± 0.08^{ab}	4.49 ± 0.05^{bc}	4.47 ± 0.02^{c}	4.57 ± 0.04^{ab}	4.59 ± 0.02^{a}
% lactic acid	0.67 ± 0.002^{a}	0.67 ± 0.007^{a}	0.58 ± 0.003^{b}	0.53 ± 0.01^{c}	0.47 ± 0.001^{d}

1A–100% cow's milk; 2A–75% cow's milk + 25% soy milk; 3A–50% cow's milk + 50% soy milk; 4A–25% cow's milk + 75% soy milk; 5A–100% soy milk; Data represent mean values (±SD) of three repetitions. ^{abc} Duncan's Test ($p < 0.05$) confirmed the statistically significant difference between mean values (±SD).

A decrease in pH value of Sample 1A (100% cow's milk) at the beginning of fermentation was rather slow, probably due to the presence of cow milk proteins with higher buffer capacity than that of soy protein [8, 9]. In general, it could be noticed that increasing soy milk share was accompanied by a decrease in pH and increased acidity, which could be attributed to the significantly reduced buffer capacity [10, 11]. This justifies the fact that the fermentation of sample 1 (100% cow's milk) lasted longer in relation to other samples with equal fermentation time. The inclusion of soy milk in cow's milk resulted in an increase in the rate of pH drop during fermentation as well as increase in acidity. Acidity analysis results of the samples during the storage days are shown in Figs. 1 and 2. During 21 days of storage at +4 °C, the dynamics of the various decreases in the pH value in the time sequence were observed. Namely, the decrease in pH values is caused by organic acids due to fermentation of carbohydrates with probiotic bacteria and starter cultures [8, 12]. The greatest decrease in pH value happened between day 1 and day 7 - with sample produced from 100% cow's milk (1A) and with samples produced from different ratios of cow's milk and soy milk, while in soy milk samples (5A) – very small decrease in pH values throughout the storage period was observed. Also, in samples 4A, between day 14 and day 21 almost no pH change occurred. It was found that the titration acidity increase was proportional to the pH decrease (Fig. 2). Titration acidity is expressed as a percentage of lactic acid.

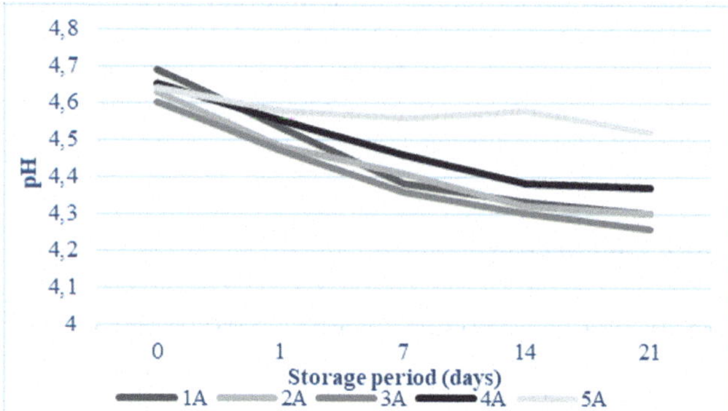

Legend: 1A-100% cow's milk; 2A-75% cow's milk + 25% soy milk; 3A-50% cow's milk + 50% soy milk; 4A-25% cow's milk + 75% soy milk ; 5A-100% soy milk . The data represents the mean of three repetitions.

Fig. 1. Change in pH values of cow's milk and soy milk yoghurt fermented with probiotic bacteria La5 during the storage

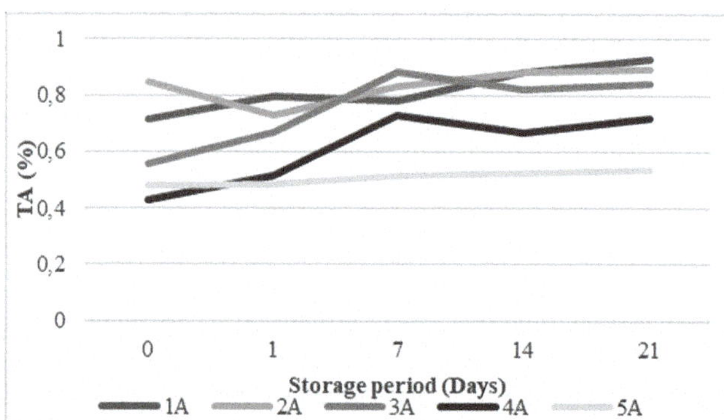

Legend: 1A-100% cow's milk; 2A-75% cow's milk + 25% soy milk; 3A-50% cow's milk + 50% soy milk; 4A-25% cow's milk + 75% soy milk ; 5A-100% soy milk . The data represents the mean of three repetitions. The titration acidity (TA) is expressed as the percentage of lactic acid.

Fig. 2. Change of TA (%) during the storage of cow and soy milk yoghurt fermented with probiotic bacteria *Lactobacillus acidophilus*

The results of the change in the number of bacteria (La5) in the probiotic beverages of soy and cow's milk during the storage are shown in Table 2. According to the results from Table 2, during 21 days of storage in the refrigerator in all samples, except for the

4A sample, the number of live lactobacillus cells increased. Although a significant decrease in pH values between the first and the seventh day occurred, this did not significantly affect the survival of lactobacilli. A sudden increase in acidity in yoghurt reduces the probiotic bacteria survival ability [13]. Mortazavian et al. [14], stated in their paper that the survival of lactobacilli in yoghurt is related to storage temperature, and the ideal temperature for storing this product is 2 °C. Lowering the temperature during the preservation of yoghurt reduces the activity of the yoghurt culture (*L. delbrueckii* subsp. *bulgaricus*) and increases the conditions suitable for growth of lactobacilli that have the ability to adapt.

Table 2. Change in the number of growing colonies (log CFU mL^{-1}) *Lactobacillus acidophilus* (La5) during the cow's milk and soy milk fermented samples storage

Sample	Number of bacteria (log CFU/ml)				
	0	1	7	14	21
1A	8.29 ± 0.16^{a}	7.98 ± 1.13^{a}	8.19 ± 0.12^{a}	7.91 ± 1.14^{ab}	7.11 ± 0.83^{b}
2A	7.64 ± 0.20^{a}	7.52 ± 0.29^{a}	7.75 ± 0.35^{a}	8.22 ± 0.00^{a}	6.53 ± 0.00^{b}
3A	8.41 ± 0.83^{a}	7.97 ± 0.30^{a}	8.25 ± 1.05^{a}	6.79 ± 0.08^{c}	7.11 ± 0.00^{b}
4A	8.03 ± 0.07^{a}	8.2 ± 1.13^{a}	7.54 ± 0.79^{a}	7.02 ± 0.00^{bc}	n. d.
5A	7.97 ± 0.27^{a}	8.15 ± 1.12^{a}	7.29 ± 1.06^{a}	n. d.	8.49 ± 0.00^{a}

Legend: A - *L. acidophilus;* 1A–100% cow's milk; 2A–75% cow's milk + 25% soy milk; 3A–50% cow's milk + 50% soy milk; 4A–25% cow's milk + 75% soy milk; 5A–100% soy milk. Data represent mean values (±SD) of three repetitions. [abc] Duncan's Test ($p < 0.05$) confirmed the statistically significant difference between mean values (±SD).

Considering the results obtained, the best substrate for lactobacillus growth might be the 5A sample (100% soy milk) which is equal with the results from Farnworth et al. [8]. Presence of oligosaccharides that also function as prebiotics in soy beverage, contributes to lactobacillus growth. Also, the soy milk contains amino acids and peptides that promote the growth of probiotic bacteria [15]. During 21 days of storage, the least-increasing lactobacillus was found in sample 2A (75% cow's milk + 25% soy milk). According to Champagne and Gardner [16], La5 bacteria is growing better in soy milk than in cow's milk.

Sensory analysis was performed by a group of trained panelists who evaluated the scent, consistency, color, appearance and taste of samples (sample 1–5) of probiotic beverages during the 21 days of storage in the fridge at +4 °C. The evaluation results are shown in Table 3 – there is a clear comparison between the average scores of different samples (sample 1A-5A) of probiotic beverages produced by La5 and yoghurt culture.

Table 3. Sensory scores of probiotic beverages during storage.

Properties	Storage days	1A	2A	3A	4A	5A
Flavour (max 12)	1	10.5 ± 0.41[a]	8.6 ± 2.35[b]	8.08 ± 2,13[b]	7.36 ± 2.75[bc]	6.25 ± 2.78[c]
	7	10.4 ± 1.08[a]	9.3 ± 1.28[ab]	8.89 ± 1,61[bc]	8.17 ± 2.11[bc]	7.67 ± 2.63[c]
	14	9.4 ± 1.34[a]	8.5 ± 1.41[ab]	8.42 ± 1,71[ab]	8.45 ± 1.82[ab]	7.98 ± 2.29[b]
	21	9.2 ± 1.19[a]	8.0 ± 2.31[b]	7.75 ± 2,91[c]	7.75 ± 2.38[c]	9.33 ± 1.91[a]
Odour (max 2)	1	1.9 ± 0.18[a]	1.52 ± 0.52[b]	1.45 ± 0.59[b]	1.38 ± 0.58[b]	1.39 ± 0.72[b]
	7	1.8 ± 0.39[a]	1.74 ± 0.37[a]	1.65 ± 0.47[a]	1.68 ± 0.47[a]	1.65 ± 0.56[a]
	14	1.8 ± 0.30[a]	1.70 ± 0.30[a]	1.66 ± 0.38[a]	1.65 ± 0.36[a]	1.58 ± 0.43[a]
	21	1.7 ± 0.43[a]	1.61 ± 0.35[a]	1.71 ± 0.32[a]	1.69 ± 0.58[a]	1.70 ± 0.38[a]
Appearance (max 1)	1	0.9 ± 0.03[a]	0.92 ± 0.18[a]	0.90 ± 0.18[a]	0.89 ± 0.19[a]	0.85 ± 0.24[a]
	7	0.9 ± 0.04[a]	0.98 ± 0.04[a]	0.92 ± 0.14[ab]	0.88 ± 0.19[ab]	0.83 ± 0.26[b]
	14	0.9 ± 0.02[a]	0.96 ± 0.07[ab]	0.90 ± 0.21[ab]	0.85 ± 0.20[b]	0.87 ± 0.19[b]
	21	0.9 ± 0.04[a]	0.95 ± 0.09[a]	0.89 ± 0.19[a]	0.83 ± 0.29[a]	0.91 ± 0.14[a]
Color (max 1)	1	1.0 ± 0,00[a]	0.90 ± 0.20[a]	0.90 ± 0.17[a]	0.85 ± 0.23[a]	0.85 ± 0.24[a]
	7	1.0 ± 0.00[a]	0.97 ± 0.04[a]	0.88 ± 0.15[ab]	0.85 ± 0.18[b]	0.83 ± 0.19[b]
	14	1.0 ± 0.00[a]	0.97 ± 0.05[a]	0.92 ± 0.14[ab]	0.88 ± 0.16[b]	0.86 ± 0.17[b]
	21	0.9 ± 0.08[a]	0.96 ± 0.06[a]	0.90 ± 0.16[ab]	0.80 ± 0.31[b]	0.94 ± 0.09[ab]
Consistency (max 4)	1	**3.7 ± 0.40[a]**	3.4 ± 0.67[ab]	3.45 ± 0.60[ab]	**3.29 ± 0.70[ab]**	3.31 ± 0.73[b]
	7	3.6 ± 0.53[a]	3.5 ± 0.58[a]	3.43 ± 0.91[a]	3.30 ± 0.81[a]	3.18 ± 0.98[a]
	14	3.6 ± 0.41[a]	3.5 ± 0.66[a]	3.36 ± 0.92[a]	3.24 ± 0.91[a]	3.18 ± 0.78[a]
	21	3.7 ± 0.34[a]	3.03 ± 0.82[a]	3.22 ± 0.38[a]	**3.02 ± 1.18[a]**	3.63 ± 0.47[a]
Total (max 20)	1	18.2[I] ± 1.7[a]	15.3[II] ± 3.6[b]	14.6[III] ± 03.3[b]	13.6[III] ± 3.8[b]	12.7[IV] ± 4.0[b]
	7	18.0[I] ± 1.5[a]	16.7[II] ± 1.6[ab]	16.3[II] ± 1.7[abc]	15.30[II] ± 2.5[bc]	14.2[III] ± 3.2[c]
	14	16.8[II] ± 1.6[a]	15.8[II] ± 1.8[ab]	15.2[II] ± 2.7[ab]	15.02[III] ± 3.0[ab]	14.5[III] ± 3.2[b]
	21	**16.6[II] ± 1.6[a]**	14.6[II] ± 2.8[bc]	14.4[III] ± 2.4[c]	**15.4[II] ± 1.7[abc]**	16.4[II] ± 1.9[ab]

1A–100% cow's milk; 2A–75% cow's milk + 25% soy milk; 3A–50% cow's milk + 50% soy milk; 4A–25% cow's milk + 75% soy milk; 5A–100% soy milk. Data represent mean values (±SD) of three repetitions. [abc] Duncan's Test ($p < 0.05$) confirmed the statistically significant difference between mean values (±SD).

After the first day of storage, the highest sensory scores were obtained with samples 1A (100% cow's milk) and 2A (75% cow's milk + 25% soy milk), while sample 5A (100% soy milk) was rated with the lowest scores. Such ratings clearly point to poorer sensory properties of soy milk, as the total grade of fermented beverage is almost proportional to the share of soy milk. However, seven days later, the ratings have changed significantly. In sample 1A the overall rating decreased, while in other samples - the overall rating increased. These changes are likely to be related to the acidity change, which negatively affects the taste of fermented cow's milk – as proven by the results of other authors [17]. In sample 5A, the sensory taste rating increased with longer storage time. On the first day - it was 6.25, while on the 21st day 9.33, which contributed to an increase in the overall 5A sample rating. Also, in sample 5A (100% soy milk) there was a small decrease in pH value throughout the storage period, so acidification could not influence the evaluation. Although fermentation gives better flavor to a soy, beny taste is still present, proven by the lowest overall ratings, which is consistent with the results of the research conducted by Park et al. [18]. Sensory ratings of the consistency on the first day ranged from 3.29 to 3.74 compared to the highest possible value 4. According to the results of the sensory analysis, it is evident that the best consistency ratings have samples 1 and 3.

Based on data determined by the hedonic scale, the basic statistical parameters (x, s, Cv) and desirability percentage are calculated (Table 4).

Table 4. Scores obtained of probiotic beverages after 1st day of storage at +4°C using hedonic scale

Scores	Samples of fermented beverages				
	1A	2A	3A	4A	5A
9	5	1	2	1	0
8	17	14	13	3	2
7	7	9	6	5	2
6	1	2	3	6	2
5	0	2	0	3	3
4	0	1	4	5	4
3	0	0	2	3	4
2	0	1	0	3	8
1	0	0	0	1	5
Total	30	30	30	30	30
X	7.9^{a}	7.1^{ab}	6.7^{b}	5.2^{c}	3.5^{d}
S	0.7	1.5	1.5	2.1	2.2
Desirability (%)	100^{a}	93^{ab}	75^{b}	33.3^{c}	13.3^{d}
Cv	9.3	20.7	22.7	40.7	61.2

1A–100% cow's milk; 2A–75% cow's milk + 25% soy milk; 3A–50% cow's milk + 50% soy milk; 4A–25% cow's milk + 75% soy milk; 5A–100% soy milk. Data represent mean values (±SD) of three repetitions.[abc] Duncan's Test ($p < 0.05$) confirmed the statistically significant difference between mean values (±SD). x = mean; s = standard deviation; Cv = variability coefficient.

The results are showing the opinion of potential customers (30) of new products. Samples 1A (100% cow's milk), 2A (75% cow's milk + 25% soy milk) and 3A (50% cow's milk +50% soy milk) were rated as "I like it very much" by most of the respondents, using hedonic scale, which indicates that such type of products would be well accepted by consumers. Sample 4A (25% cow's milk + 75% soy milk) for most potential consumers was rated as acceptable, or as "I like it" on hedonic scale, meaning they would consume. From the other side, sample 5A (100% soy milk) received the lowest ratings. Most respondents rated it as "I do not like it" on the hedonic scale or "I extremely don't like it" and they would not consume. Similar research was conducted by Beherens et al. [19], for fermented soy beverage with the addition of various types of fruit and nuts.

4 Conclusions

Functional probiotic beverages were successfully produced from different ratios of cow's milk and soy milk mixture and the probiotic strain of *L. acidophilus* with yogurt culture. Adding soy milk to cow's milk resulted in probiotic beverages with high nutritional value. The number of probiotic bacteria in all samples of the produced beverages were in the range of 7.52–8.20 log CFU mL^{-1}, which was above the generally accepted probiotic minimum (10^6 CFU mL^{-1}), so the produced samples could be considered as probiotic products. Cow's and soy milk fermentation during the 21 days samples storage at +4 °C did not show a significant pH value decrease in all samples compared to typical pH value (pH = 4.5). During the 21 days of sample storage in the refrigerator at +4 °C, the number of lactobacilli gradually decreased, however, the number of lactobacilli live cells is above the probiotic minimum (10^6 CFU mL^{-1}) so the samples could be considered as a probiotic product. The sensory properties of the samples were mainly influenced by the type and ratio of the milk used in the mixture. Mixing cow's milk with soy milk improved the sensory properties of the product, especially the odor, flavor and color. Samples of fermented beverages are best evaluated on the 7th day of storage, and with the further storage time, ratings became lower. The reason for bad rating is the increase in acidity, which is expressed with low sensory ratings of fermented beverage flavors. Acceptability Test has shown that the combination of cow's milk with soy milk up to 50% does not change the sensory properties of cow's milk, so this ratio of cow's milk and soy milk in fermented beverage production is acceptable to consumers. Such fermented products have high nutritional value and satisfy all requirements of functional foods, and could also be intended for different age groups of consumers.

Acknowledgements. The authors have no conflict of interest to state regarding the results presented within this manuscript.

References

1. Scalabrini, P., Rossi, M., Spettoli, P., Matteuzzi, D.: Charaterization of Bifidobacterium strains for use in soymilk fermentation. Int. J. Food Microbiol. **39**, 213–219 (2016). https://doi.org/10.1016/S0168-1605(98)00005-1
2. Liu, J.R., Lin, C.W.: Production of kefir from soymilk with or without added glucose, lactose or sucrose. J. Food Sci. **65**, 716–719 (2000). https://doi.org/10.1111/j.1365-2621.2000.tb16078.x
3. Tsangalis, D., Ashton, J.F., McGill, A.E.J., Shah, N.P.: Biotrasformation of isoflavones by bifidobacteria in fermented soymilk supplemented with D-glucose and L-cystein. J. Food Sci. **68**, 623–631 (2003)
4. Wei, Q.K., Chen, T.R., Chen, J.T.: Using of Lactobacillus and Bifidobacterium to product the isoflavone aglycones in fermented soymilk. Int. J. Food Microbiol. **117**, 121–124 (2007)
5. Lourens-Hattingh, A., Viljoen, B.: Yoghurt as a probiotic carrier food. Int. Dairy J. **11**, 1–17 (2001)
6. Schrezenmeir, J., De vrese, M.: Probiotics, prebiotics, and symbiotics approaching a definition. Am. J. Clin. Nutr. **73**, 361S–364S (2001)

7. Božanić, R., Jeličić, I., Bilušić, T.: Analiza mlijeka i mliječnih proizvoda. Priručnik, Plejada, Zagreb (2010)
8. Farnworth, E.R., Mainville, I., Desjardins, M.P., Gardner, N., Fliss, I., Champagne, C.: Growth of probiotic bacteria and bifidobacteria in a soy yogurt formulation. Int. J. Food Microbiol. **116**, 174–181 (2007). https://doi.org/10.1016/j.ijfoodmicro.2006.12.015
9. Silva, C.F.G., Santos, F.L., Santana, L.R.R., Silva, M.V.L., Conceição, T.A.: Development and characterization of a soymilk Kefir-based functional beverage. Food Sci. Technol. **38**(3), 543–550 (2018). https://doi.org/10.1590/1678-457x.10617
10. Heydari, S., Mortazavian, A.M., Ehsani, M.R., Mohammadifar, M.A., Ezzatpanah, H., Sohrabvandi, S.: Biochemical, microbiological and sensory characteristics of probiotic yogurt containing various prebiotic or fiber compounds. Ital. J. Food Sci. **23**, 153–163 (2011)
11. Sadaghdar, Y.: Topic: Probiotic fermented milk-juice. MSc. thesis. Tehran, Iran: Azad Islamic University, Azad Islamic University, Science and Research branch (2012)
12. Wang, J., Guo, Z., Zhang, Q., Yau, L., Chen, W., Liu, X.M., Zhang, H.P.: Fermentation characteristics and transit tolerance of probiotic Lactobacillus casei Zhang in soymilk and bovine milk during storage. J. Diary Sci. **92**, 2468–2476 (2009)
13. Shah, N.P.: Probiotic bacteria: selective enumeration and survival in dairy foods. J. Diary Sci. **83**(4), 894–907 (2000)
14. Mortazavian, A.M., Ehsani, M.R., Mousavi, S.M., Rezaei, K., Sohrabvandi, S., Reinheimer, J.A.: Effect of refrigerated storage temperature on the viability of probiotic micro-organisms in yogurt. Int. J. Dairy Technol. **60**, 123–127 (2007)
15. Kazemi, A., Mazloomi, S.M., Hassanzadeh-Rostami, Z., Akhlaghi, M.: Effect of adding soymilk on physicochemical, microbial, and sensory characteristics of probiotic fermented milk containing Lactobacillus acidophilus. Iran. J. Vet. Res. **15**, 206–210 (2013)
16. Champagne, C.P., Gardner, N.J.: Challenges in the addition of probiotic cultures to foods. Food Sci. Nutr. **45**, 61–84 (2015)
17. Cakmakci, S., Cetin, B., Turgut, T., Gurses, M., Erdogan, A.: Probiotic properties, sensory qualities and storage stability of probiotic banana yogurts. Turkish J. Vet. Anim. Sci. **36**(3), 231–237 (2012)
18. Park, D.J., Oh, S., Ku, K.H., Mok, C., Kim, S.H., Imm, J.Y.: Characteristics of yogurt-like products prepared from the combination of skim milk and soymilk containing saccharified-rice solution. Int. J. Food Sci. Nutr. **56**(1), 23–34 (2005). http://dx.doi.org/10.1080/09637480500082181. PMid:16019312
19. Behrens, J.H., Roig, S.M., Da Silva, M.A.A.P.: Fermentation of soymilk by commercial lactic cultures: development of a product with market potential. Acta Aliment. Hung. **33**(2), 101–109 (2004)

Determination of Aflatoxin M1 in Raw Milk by the ELISA Method in the Una-Sana Canton

Huska Jukić[1(✉)], Samira Dedić[2(✉)], Miloš Rodić[3], Zlatko Jusufhodžić[3], and Dinko Demirović[3]

[1] Faculty of Health Studies, University of Bihać, 4, Nositelja hrvatskog trolista 4, 77000 Bihać, Bosnia and Herzegovina
huskaj037@gmail.com
[2] Faculty of Biotechnical Sciences, University of Bihać, Kulina Bana 2, Bihać, Bosnia and Herzegovina
samira.dedic@yahoo.com
[3] Public Institution "Veterinary Institute" Bihać, Omera Novljanina 6, Bihać, Bosnia and Herzegovina

Abstract. The presence of aflatoxin M1 (AFM1) in raw milk should be continuously monitored to protect the population from the risks associated with proven toxicity and carcinogenicity. The ELISA method offers many advantages, including shorter time of analysis, simultaneous analysis of a large number of samples, acceptable as a screening method, limited use of organic solvents, absence of complicated sample preparation steps and simplicity of analytical procedure compared to long-lasting and expensive chromatographic techniques. The test is highly sensitive with a detection limit of 0.01 μg/kg and a quantitative range of 0.01–0.1 μg/kg. In the Una-Sana Canton in 2018, 27 samples of raw cow milk were analysed and all the analyses were carried out twice. The highest concentration of AFM1 was in December and it was 0.032 μg/kg, which is a permissible concentration of AFM1 according to the Regulations on Maximum levels for certain contaminants in food, Official Gazette No. 37/09, which is in compliance with European Union regulations (EU) and the Codex Alimentarius Commission (CAC). In a case of a positive result, the samples are analysed several times and sent to a certified chemical laboratory for the super analysis for the confirmation of the results.

Keywords: Aflatoxin M1 · Raw milk · ELISA

1 Introduction

Mycotoxins (gr. myxes - mushroom, toxicon - poison) are secondary metabolic products that molds synthesize during growth on plant and animal substrates, but they are also pathogens of various diseases through food. These are chemical compounds of different s tructures and different biological effects, typically colourless and tasteless. They are stable and generally resistant to increased temperatures. Their biosynthesis depends on the type of toxic mold, climatic and environmental conditions, physico-chemical factors (temperature: from −5 to 60 °C, water content in food: 13% and higher, aw-values: above 0.65, pH - values: 3–9, and concentration of gases in the atmosphere and food composition [1].

M. Brka et al. (Eds.): AgriConf 2019, IFMBE Proceedings 78, pp. 260–267, 2020.
https://doi.org/10.1007/978-3-030-40049-1_33

The diseases they cause are mycotoxicosis. Since aflatoxins are thermostable and their amount is only slightly reduced during food processing, it is important to prevent their formation [2]. In addition to being acutely toxic, some mycotoxins are now linked with the incidence of certain types of cancer and this aspect has caused global concern over food safety, especially for milk and milk products [3]. Aflatoxin M1 is a hepatocarcinogenic agent found in the milk of animals that consumed food contaminated with aflatoxin B1, the major metabolite produced by mushrooms of the genus Aspergillus, in particular A. flavus, A. parasiticus and A. nomius [4]. The occurrence of aflatoxin M1 in milk depends on the species of the animal. AFM1 is the most commonly found in cow milk, and as a possible reason they suggest a different digestive tract in ruminants, as well as a higher amount of food intake, and a higher presence of nutrients (corn) that is more often contaminated with AFB1 [5, 6].

In order to evaluate the accuracy of the results of the different methods [7] have performed the AFM1 analysis of naturally contaminated milk samples using the enzyme immunological method (ELISA), high performance liquid chromatography with fluorescence detector (HPLC-FLD) and liquid chromatography-tandem high performance mass spectrometry (HPLC-MS/MS). The obtained validation parameters indicate that all three methods are suitable for the determination of AFM1 in milk samples. They found that the statistical analysis of the variance between the methods and the obtained coefficient of correlation indicates that there is a strong correlation between the methods. Also, all three methods are satisfactory in meeting the requirements for official control purposes.

Today, the enzyme immunological method (ELISA) is used as orientation test in routine analyzes for the determination of aflatoxins, mainly because of its simplicity, speed, and the sensitivity of less than 0.01 μg/L [8–10]. Development of an ELISA method enabled the execution of rapid, reproducible and sensitive analytical methods suitable for the determination of aflatoxin. Literature data suggest that when it comes to the determination of aflatoxin, ELISA has many advantages over other analytical methods. However, the quality of ELISA kits may vary from manufacturer to manufacturer. Moreover, there are differences in the different sets of the kits made by the same manufacturer. No matter which ELISA test kit is used to determine aflatoxin, the principle is the same. The basic principle of ELISA technique is shown in Fig. 1.

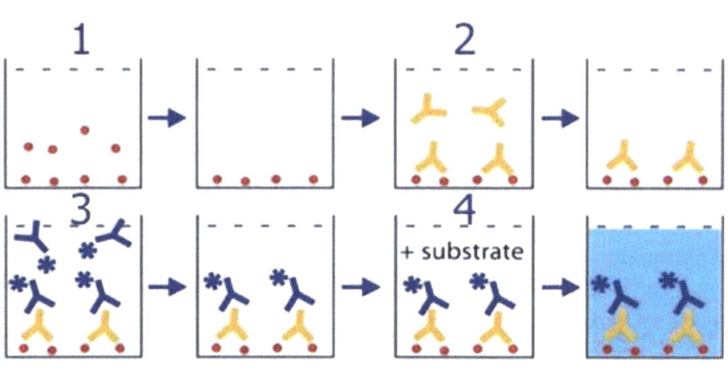

Fig. 1. Basic principle of ELISA technique (Kos, 2015)

Free aflatoxins from the sample and standard (Fig. 1. 1) in wells compete with enzyma tically bound aflatoxin (conjugate) for binding sites of the antibody (Fig. 1.2). After the incubation period (Fig. 1. 2), the wells are washed with a rinse aid, followed by the addition of a substrate which reacts with the conjugate bound and the color develops (Fig. 1. 3). After a specific incubation period (Fig. 1. 4), the stop reagent is added and the colour intensity is read on the ELISA reader. The concentration of aflatoxin is determined on the base of the standard's calibration curve and the concentration is inversely proportional to the intensity of the developed color.

The main advantages of ELISA technique, compared to other techniques for determining aflatoxins, are: minimal sample preparation, simple performance and low cost analysis. However, the main disadvantages of the ELISA technique are the emergence of "crossreactivity" and the possibility of obtaining false positive results. "Cross-reactivity" is defined as the ability of compounds with a similar chemical structure to react with antibodies. Due to the possible occurrence of "cross-reactivity" and obtaining false positive results, each aflatoxin detected by the ELISA method should be also confirmed by a chromatographic technique, especially if the concentration of aflatoxins is close to the MDK value [11, 12].

The European Commission and the Codex Alimentarius Commission have prescribed a maximum level of AFM1 in milk and milk products, and it should not exceed 50 μg/L in all countries, as well as in Bosnia and Herzegovina, according Official Gazette BiH, No. 37/09 and 39/12), while the maximum allowed concentration in the US is 0.5 μg/L [13].

Many studies have been carried out on the contamination of AFM1 in raw milk in the world and their results have shown that they have exceeded concentrations relative to the regulatory boundary of the European Community and Codex Alimentarius, and some of them are [14–21]. The aim of this study was to examine the presence of AFM1 in samples of raw cow milk produced in the Una-Sana Canton area by the ELISA method.

2 Material and Method of Work

In collaboration with the laboratory for the residue of the Public Institution "Veterinary Institute" Bihać, an accredited body (BAS EN ISO/IEC 17025: 2006), the determination of AFM1 in 27 samples of cow raw milk was performed by ELISA method. The concentrations of aflatoxin M1 are: 0 ppt, 5 ppt, 10 ppt, 25 ppt, 50 ppt, and 100 ppt.

According to data from kits producers, the detection limit for AFM1 in the specified matrix is 10–100 ppt (μg/kg). "Free" aflatoxin M1 in the sample or standard and the AFM1 enzyme conjugate is applied to the microtiter plate, which is covered with specific anti-aflatoxin M1 antibodies. "Free" AFM1 in the sample or standard and the enzymatically conjugated AFM1 compete for binding to these antibodies. After the wash, a chromogenic substrate, which forms a coloured complex with a conjugate, is added, so the colour intensity (absorbance) is inversely proportional to the concentration of "simple"AFM1 in the sample or standard.

An aliquot of untreated raw milk should be placed in the refrigerator overnight to allow separation of the fat globules on the surface, which will form a natural creamy effect. It is necessary to remove the upper fatty layer with spatula.

Pipette 200 µl of distilled water into a blank pit (a blank test), then pipet 200 µl of standard and sample into the appropriate wells. Cover the plate with adhesive and aluminium foil. Mix the plate manually and gently. Incubate for 30 min at room temperature, avoiding light. After incubation, discard the contents of the well in the appropriate container. Fill the well with the prepared washing solution (washing buffer) of 300 µl, and then discard the contents. Repeat the rinsing 2 times (total of 3 rinses in one series). After the last rinse, the plate is dried well with the absorbent paper. Add 200 ml of enzyme conjugate to each well, except blank holes. Cover the plate with adhesive and aluminium foil. Incubate for 15 min at room temperature, avoiding light. Repeat flushing 3 times. By multi-channel pipette, add 200 µl of chromogen solutio to each well, including a blank well. Cover the plate with adhesive and aluminium foil. Incubate for 10 min at room temperature. Stop the reaction by adding 50 µl of stop solution. The colour from the blue goes to yellow (Fig. 2), mix the plate manually.

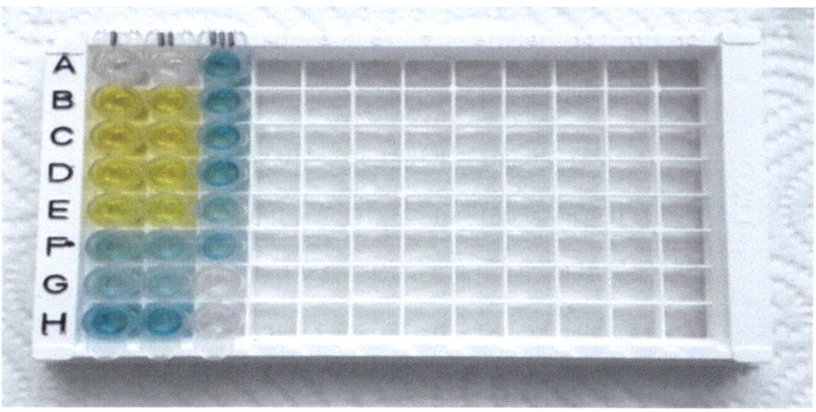

Fig. 2. Adding stop solution and the change of colour

Figure 2 shows a panel on which empty standards, sample, positive control and positive sample, are sown, in a duplicate to obtain a mean absorbance.

3 Results and Discussion

Using the absorbance read (standard solutions) presented in the table (Table 1), the percentages were calculated, based on which we can also display a diagram of the AFM1 concentration (Fig. 3), and calculate the AFM1 concentration in the samples (Table 2).

Table 1. Results (absorbance) of the ELISA reader of standard solutions and sample of the raw milk

Concentration	Well 1	Well 2	Well 3
A blank	0.061	0.056	1.264
B st 0 ppt	1.423	1.414	1.297
C st 5 ppt	1.299	1.291	1.311
D st 10 ppt	1.241	1.244	1.312
E st 25 ppt	0.986	0.963	0.620
F st 50 ppt	0.711	0.706	0.822
G st 100 ppt	0.494	0.505	0.046
H sample	1.326	1.303	0.050

$$\text{Calculate } \% \text{ Abs} = \text{Absaveragevalue} - \text{blank} / \text{Absneg.kontr.} - \text{Blank x } 100$$

$$\% \text{ Abs for negative control (p. 0 ppt)} = 100$$

$$\text{A. Blank} = 0.061$$

$$\text{B. } 1.423 - 0.061 = 1.362 / 1.362 \times 100 = 100\%$$

$$\text{C. } 1.299 - 0.061 = 1.238 / 1.362 \times 100 = 91\%$$

$$\text{D. } 1.241 - 0.061 = 1.18 / 1.362 \times 100 = 86.6\%$$

$$\text{E. } 0.986 - 0.061 = 0.925 / 1.362 \times 100 = 68\%$$

$$\text{F. } 0.711 - 0.061 = 0.65 / 1.362 \times 100 = 48\%$$

$$\text{G. } 0.494 - 0.061 = 0.437 / 1.362 \times 100 = 32\%$$

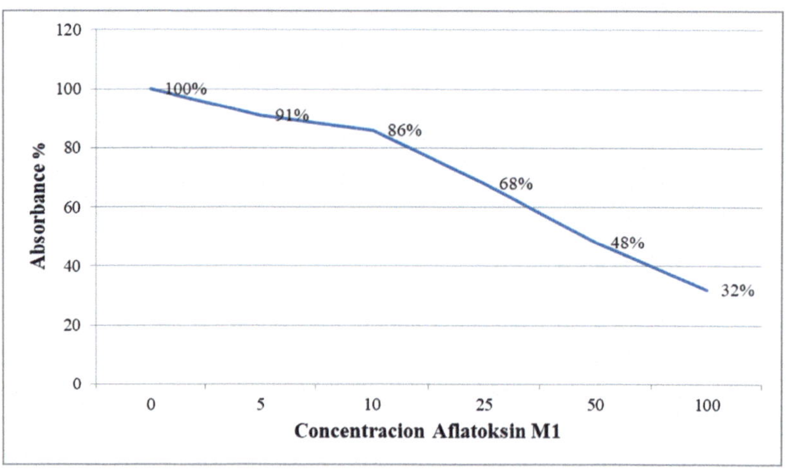

Fig. 3. Calibration curves of standard solutions of AFM1 with concentrations of 5, 10, 25, 50 and 100 ppt by ELISA method

Table 2. Results of AFM1 for the period 2018. in the USK area

Month	MU	MPC	NS	C (AV)	Results
January	μg/l	0.050	3	<0.010	Negative
February	μg/l	0.050	3	<0.019	Negative
March	μg/l	0.050	2	<0.010	Negative
April	μg/l	0.050	2	<0.015	Negative
May	μg/l	0.050	3	<0.010	Negative
June	μg/l	0.050	2	<0.020	Negative
July	μg/l	0.050	2	<0.010	Negative
August	μg/l	0.050	3	<0.010	Negative
September	μg/l	0.050	2	<0.015	Negative
October	μg/l	0.050	2	<0.014	Negative
November	μg/l	0.050	2	<0.010	Negative
December	μg/l	0.050	1	<0.032	Negative
Total			**27**		

MU - Measuring unit; MPC - Maximum permissible concentration;
NS - Number of samples; C (AV) Concentration μg/l average value.

According to the Ordinance on the Maximum Allowed Quantities of Certain Contaminants in Food [22], the maximum permitted limit for AFM1 content in milk and milk products is 0.05 μg/kg, which is in accordance with currently applicable regulations in the EU. On the basis of the results obtained in the twelfth month of 2018, from a sample of raw cow milk with a maximum concentration of 0.032 μg/kg, it can be concluded that it is in accordance with EU regula tions on the permitted concentration of AFM1. After laboratory analysis of raw cow's milk, negative results were obtained, which was proven by the absence of AFM1 in the samples submitted. If you get positive results, further testing is confirmed in the laboratory that is accredited and authorized for this region.

Sani et al. 2012 showed in their research that the incidence of AFM1 contamination of pasteurized milk samples was low [19]. Although 97.6% of the samples were contaminated with AFM1, the toxin concentration was lower than the Iranian national standard and the Food and Drug Administration (FDA) limit (500 ppt), and only in three (1.6%) samples the AFM1 concentration was greater than the maximum tolerance limit (50 ppt) which is adopted by the European Union and the Codex Alimentarius Commission. Using the ELISA method, Bellio et al., 2016 identified 36 (2.2%) positive milk samples that were subsequently confirmed by HPLC. The level of AFM1 in the positive samples ranged between 18 ± 2 and 208 ± 27 ppt. Of the total samples, only eight (0.5%) were found not to comply with the EU regulatory limit (50 ppt; range 74 ± 10 to 208 ± 27 ppt) [21]. Significantly increased concentrations of AFM1 (> 140 ng/L) in raw cow milk from Pakistan and Syria and extremely high concentrations of 492 ng/L in pasteurized milk from Syria were determined [17, 20]. The frequency of AFM1 observed in this study was lower than the frequency of AFM1 reported by the authors mentioned.

4 Conclusion

ELISA tests include wide application and are often used to determine different contaminants in food products. The ELISA method, which resulted in rapid results with high sensitivity, was used in this study. After laboratory analysis of raw cow's milk, the obtained results ranged from 0.010 to 0.032 μg/kg, which is less than the maximum allowable amount for AFM1 content in milk. If positive results are obtained, further testing is confirmed in a laboratory that is accredited and competent, with subsequent confirmation by one of the methods such as high-performance liquid chromatography-tandem mass spectrometry (HPLC-MS/MS) or high-performance liquid chromatography with a fluorescence detector (HPLC-FLD). The presence of AFM1 in milk should be constantly monitored in order to protect the population from the risks associated with proven toxicity and carcinogenicity.

Acknowledgements. The authors have no conflict of interest to state regarding the results presented within this manuscript.

References

1. Šarkanj, B., Kipčić, D., Vasić-Rački, Đ., Delaš, F., Galić, K., Katalenić, M., Dimitrov, N., Klapec, T.: Kemijske i fizikalne opasnosti u hrani., Osijek (2010)
2. Peraica, M., Rašić, D., Gluščić, V.: Utjecaj aflatoksina na zdravlje ljudi. Glasilo biljne zaštite **4**, 310–316 (2014)
3. Castegnaro, M., Mcgregor, D.: Carcinogenic risk assessment of mycotoxins. Rev. Med. Vet. **149**, 671 (1998)
4. Galvano, F., Galofaro, V., Ritieni, A., Bognanno, M., De Angelis, A., Galvano, G.: Survey of the occurrence of aflatoxin M1 in dairy products marketed in Italy: second year of observation. Food Addit. Contam. **18**(7), 644–646 (2001)
5. Barbiroli, A., Bonomi, F., Benedetti, S., Mannino, S., Monti, L., Cattaneo, T., Lametti, S.: Binding of Aflatoxin M1 to different protein fractions in Ovine and Caprine Milk. J. Dairy Sci. **90**, 532–540 (2007)
6. Hussain, I., Anwar, J., Asi, M.R., Munawar, M.A., Kashif, M.: Aflatoxin M1 contamination in milk from five dairy species in Pakistan. Food Control **21**, 122–124 (2010)
7. Kos, J., Janić Hajnal, E., Jajić, I., Krstović, S., Mastilović, J., Šarić, B., Jovanov, P.: Comparison of ELISA, HPLC-FLD and HPLC-MS/MS methods for determination of Aflatoxin M1 in natural contaminated milk samples (2016)
8. Kim, E.K., Shon, D.H., Ryu, D., Park, J.W., Hwang, H.J., Kim, Y.B.: Occurrence of aflatoxin M1 in Korean dairy products determined by ELISA and HPLC. Food Addit. Contam. **17**, 59–64 (2000)
9. Lopez, C., Ramos, L., Ramadan, S., Bulacio, L., Perez, J.: Distribution of aflatoxin M1 in cheese obtained from milk artificially contaminated. Int. J. Food Microbiol. **64**, 211–215 (2001)
10. Pei, S.C., Zhang, Y.Y., Eremin, S.A., Lee, W.J.: Detection of aflatoxin M1 in milk products from China by ELISA using monoclonal antibodies. Food Control **20**, 1080–1085 (2009)
11. Kos, J.: Aflatoksini: analiza pojave, procena rizika i optimizacija metodologije određivanja u kukuruzu i mleku. Doktorskadisertacija (2015). http://nardus.mpn.gov.rs/bitstream/handle/

12. Anklam, E., Stroka, J., Boenke, A.: Acceptance of analytical methods for implementation of EU legislation with a focus on mycotoxins. Food Control **13**, 173–183 (2002)
13. FDA (2005) se. 527.400 whole milk, low fat milk, skim milk – Aflatoxin M1 (CPG 7106.10) FDA/ORA Compliance Policy Guides
14. Dashti, B., Al-Hamli, S., Alomirah, H., Al-Zenki, S., Bu, Abbas A., Sawaya, W.: Levels of aflatoxin M1 in milk, cheese consumed in Kuwait and occurrence of total aflatoxin in local and imported animal feed. Food Control **20**, 686–690 (2009)
15. Panahi, P., Kasaee, S., Mokhtari, A., Sharifi, A., Jangjou, A.: Assessment of Aflatoxin M1 contamination in raw milk by ELISA in Urmia. Iran. Am. Eurasian J. Toxicol. Sci. **3**, 231–233 (2011)
16. Hussain, I., Anwar, J.A.: Study on contamination of aflatoxin M1 in raw milk in the Punjab province of Pakistan. Food Control **19**, 393–395 (2008)
17. Ghanem, I., Orfi, M.: Aflatoxin M1 in raw, pasteurized and powdered milk available in the Syrian market. Food Control **20**, 603–605 (2009)
18. Kamkar, A., Jahed Khaniki, G.h.R., Alavi, S.A.: Occurrence of aflatoxin M1 in raw milk produced in Ardabil of Iran. Iran. J. Environ. Health. Sci. Eng. **8**, 123–128 (2011)
19. Sani, A.M., Khezri, M., Moradnia, H.: Determination of Aflatoxin M1 in Milk by ELISA Technique in Mashad (Northeast of Iran). ISRN Toxicology Volume 2012 (2012). Article ID 121926, 4 p
20. Iqbal, S.Z., Asi, M.R.: Assessment of aflatoxin M1 in milk and milk products from Punjab, Pakistan. Food Control **30**, 235–239 (2013)
21. Bellio, A., Bianchi, D.M., Gramaglia, M., Loria, A., Nucera, D., Gallina, S., Gili, M., Decastelli, L.: Aflatoxin M1 in cow's milk method validation for milk sampled in Northern Italy. Toxins **8**(3), 57 (2016). Academic Editors: Michelangelo Pascale and Maria C. DeRosa
22. Officila Gazzette of Bosnias and Herzegovina, maximum levels for certain contaminants in food, No 37/09 and 39/12

Phenol Content and Antioxidant Activity of Different Blueberry Species from Prozor Region

Maja Kazazic[1(✉)], Jasmina Aliman[2], Suada Djogic[2], and Maida Djapo-Lavic[1]

[1] Faculty of Education, Dzemal Bijedic University of Mostar, Mostar, Bosnia and Herzegovina
maja.kazazic@unmo.ba
[2] Agro Mediterranean Faculty, Dzemal Bijedic University of Mostar, 88000 Mostar, Bosnia and Herzegovina

Abstract. Blueberries are considered as functional foods due to their high content of bioactive compounds. This study was undertaken to investigate total phenol content, total anthocyanins and antioxidant activities of wild bilberry and two highbush blueberry cultivars Bluecrop and Goldtraube from Bosnia and Hercegovina. Extracts were investigated for their antioxidant properties using ABTS radical scavenging capacity assay and ferric reducing antioxidant potential (FRAP) assay. Total phenolic content was determined by spectrophotometric method using Folin-Ciocalteu reagent. Results indicated highest phenolic content (280,55 ± 7,22 mg GA/100 g FW) in wild bilberry. Wild bilberry showed the highest antioxidant activity regardless of the method used. Content of total anthocyanins detected in wild bilberries (178,7 ± 0,46 mg cyaniding-3-glycoside)/100 g FW) was higher than in highbush blueberry cultivars investigated. Significant relationship between antioxidant activity and total phenolic content indicates that the phenolic compounds are the major contributors to the antioxidant properties of these fruits.

Keywords: Total phenols · Antioxidant activity · Anthocyanins

1 Introduction

Last decades global interest in food with a high content of bioactive compounds is rapidly growing. Fruit berries are more studied, especially highly colored berries, because of their high content of bioactive compounds. Numerous studies have showed that blueberry has a very favourable impact on human health [1–10]. Blueberries are very rich sources of bioactive compounds such as flavonoids, phenolic acids, tannins and anthocyanins. Phenolic compounds as powerful antioxidants have a positive role by preventing cardiovascular disease, diabetes and cancer [11–13]. Different studies showed that antioxidant activity is significantly correlated with content of bioactive compounds.

Antioxidant activity and phenolic content in wild billberry and cultivar blueberries were determined with the aim of identifying the effect of regional variations [14–16].

M. Brka et al. (Eds.): AgriConf 2019, IFMBE Proceedings 78, pp. 268–274, 2020.
https://doi.org/10.1007/978-3-030-40049-1_34

The highbush blueberry (*Vaccinium corymbosum* L.) is grown in the United States, Canada, Chile, England, the Netherlands, Belgium, Germany, Switzerland, Poland, New Zealand and Australia. The world's blueberry production is constantly growing from 266802 tonnes in 2006 to 596813 tonnes in 2017 [17]. Recently cultivation of highbush blueberry started in Bosnia and Hercegovina at numerous plantations covering an area of 0.1 to 30 hectares. Interestingly so far, very few studies have been conducted on the blueberry cultivar samples in Bosnia and Hercegovina [18, 19].

The aim of this study was to investigate total phenolic content, anthocyanins and antioxidant activity of two cultivated highbush blueberries and wild bilberry grown in central region of Bosnia and Hercegovina.

2 Materials and Methods

2.1 Chemicals

All chemicals used were purchased from Sigma Aldrich Ldt. (Germany).

2.2 Plant Material and Extraction

Two species of blueberries were investigated: Bluecrop, Goldtraube (*Vaccinium corymbosum* L.), and wild bilberry (*Vaccinium myrtillus* L.). All samples were collected during July and August 2018 in Prozor region at 1270 m of elevation and frozen at −20 °C until analysis. Berries were harvested at the fully ripe stage. Maturity stage was decided based on full blue colouration.

20 mL of acidified methanol (0.1% HCl) was added to 4 g of sample. Subsequently, the samples were subjected to extraction for 60 min in a shaker at a rate of 200 rpm and then centrifuged.

2.3 Total Phenolic Content

The total phenolic content was determined with Fiolin-Ciocalteu reagent according to the method of Singelton [20]. Standard curve was generated using gallic acid as a standard. Different concentrations of gallic acid were prepared in distilled water, and their absorbance values were measured at 765 nm. For sample measurement, 5 ml (1/10 dilution) of Folin Ciocalteu reagent and 900 µL of distilled water were added to 100 µL of plant extract. After 5 min, samples were combined with 4 ml of 15% sodium carbonate (Na_2CO_3) and incubated for 120 min. Absorbance at 765 nm was measured. Data are presented as average values of three measurements for each sample.

2.4 ABTS Method for Determination of Antioxidant Activity

The free radical scavenging capacity of blueberry extracts was investigated using the ABTS radical cation decolorization assay [21]. This method is based on the reduction of ABTS+• radicals by antioxidants of the plant extracts tested. ABTS cation radical (ABTS•) was prepared by dissolving 19.5 mg of ABTS and 3.3 mg of potassium persulfate in 7 ml of distilled water and allowing the mixture to stand in the dark at room temperature for 12–16 h before use. In this study, a solution of ABTS$\bullet^{+}$was diluted in ethanol to obtain 0.70 ± 0.02 absorbance at 734 nm. An aliquot of 20 μL sample was mixed with 2 ml of the ABTS radicals in the cuvette and the absorbance was measured at 734 nm after 6 min. All solutions were prepared on the day of the experiment, and all measurements were performed in triplicate.

2.5 FRAP Method for Determination of Antioxidant Activity

The ferric reducing power of plant extract was determined using method of Benzie and Strain [22], which is based on reduction of a colorless ferric complex (Fe^{3+} tripyridyl-triazine) to a blue colored ferrous complex (Fe^{2+} -tripyridyltriazine) by the action of electron-donating antioxidants. Working FRAP reagent was prepared by mixing 20 ml of 300 mM acetate buffer (pH 3.6) with 20 ml of 10 mM TPTZ (2, 4, 6-tri-(2-pyridyl)-5-triazine) in 40 mM HCl acid with 20 ml of 20 mM ferric chloride and 24 ml of distilled water. The standard curve is made using different concentrations of $FeSO_4 \times 7H_2O$ (0.1, 0.2, 0.4, 0.6, 0.8 and 1.0 mM). 50 μL of the sample solution is added to 950 μL of freshly prepared FRAP reagent. The reaction mixture was incubated for 4 min at room temperature. The reduction is monitored by measuring the change in absorbance at 593 nm and the results were expressed as mmol of Fe (II)/100 g of fresh matter. All measurements were performed in triplicate.

2.6 Statistical Analysis

All the experiments for determination of total phenolics, total anthocyanins and antioxidant properties using ABTS and FRAP assay were conducted in triplicates. The values are expressed as the mean ± standard deviation (SD). The statistical differences between mean values were stablished by one-way analysis of variance (ANOVA) and the Tukey's test. Correlations between antioxidant activity, total phenolics and total anthocyanins, were determined using Pearson's Correlation Coefficient Test.

3 Results and Discussion

The content of total phenols in cultivars Goldtraube, Bluecrop and wild bilberry are presented in Table 1 and are in the range from 166.12 ± 6.18 to 280.55 ± 7.22 mg GA/100 g FW respectively. At the moment, there are very few published data on blueberry in Bosnia and Hercegovina but there are many studies from other regions [23–25].

Table 1. Results of total phenolic content, total anthocyanins content, ABTS radical scavenging activity and ferric-reducing antioxidant power as a function of the species

Cultivar	Total phenolic Content (mg GA/100 g FW)	Total Anthocyanins (mg cyanidin-3-glycoside)/100 g FW	ABTS Radical Scavenging Activity (mmol TE/100 g FW)	Ferric-Reducing Antioxidant Power (mM Fe^{2+}/ 100 g FW)
Goldtraube	166,12 ± 6,18[a]	83,15 ± 0,98[a]	0,14 ± 0,015[a]	0,72 ± 0,038[a]
Bluecrop	224,29 ± 1,16[b]	125,81 ± 0,66[b]	0,18 ± 0,06[b]	1,55 ± 0,012[b]
Wild blueberry	280,55 ± 7,22[c]	178,7 ± 0,46[c]	0,23 ± 0,015[c]	3,48 ± 0,36[c]

*All columns show the means ± standard deviation (n = 3); Means with different letter in the columns are significantly different according to the ANOVA with Tukey test ($p < 0.05$)

Total phenol content in Bluecrop cultivar was 224,29 ± 1,16 which is in accordance with results by [26] in Serbia (199 mg GAE/100 g FW) [26]. These results are significantly higher than those presented by [23] for the same cultivar in different regions of Turkey [23]. In their research the amount of total phenols in Bluecrop cultivar varied from 83.23 to 123.51 mg GAE/100 g, depending on the region. In the northwestern Croatia, the content of phenols in the Bluecrop cultivar was reported to be 291–368 mg/100 g FW [24].

Wild bilberry had the highest content of total phenols, 280.55 ± 7.22 mg GA/100 g FW, which is higher than detected by [23] in Turkey (200–215 mg GAE/100 g FW) [23]. Significantly higher total phenol content in wild bilberry was detected in several studies [27–30]. Different amounts of total phenols presented in these studies can be attributed to different factors such as climate, temperature, testing methods, standards used, maturation and genetics [31].

When it comes to the content of total anthocyanins, wild bilberry showed highest content of this parameter compared to other cultivars. In our study, Goldtraube had 83.15 mg cyanide-3-glucoside/100 g FW, followed by Bluecrop 125.81 mg cyanide-3-glucoside/100 g FW and wild bilberries with 178.7 mg cyanide-3-glucoside/100 g FW. Similar results were reported in the article by [25] for the content of anthocyanins in Bluecrop cultivar from North Croatia (120 mg/100 g FW) [25]. Stajčić et al. (2015) reported a higher content of total anthocyanins in wild bilberries (447.83–4.16 mg CGE/100 g FW) in Serbia as well as Jovanevi et al. for wild bilberries in Montenegro (average 360 mg CGE/100 g FW) [29, 32].

Reductive capacity determined by FRAP method was in the range from 0.72 for Goldtraube to 3.48 mM Fe^{2+}/100 g FW for wild berries. In the study by [24] reductive capacity was in the range of 37.13–48.33 mmol Fe^{2+}/100 mg FW for Bluecrop cultivar [24]. Bluecrop reductive capacity determined by Okan et al. was in the range of 454.83 to 771.03 μmol Trolox/100 g for blueberries from different regions of Turkey [23].

Table 2. Correlation coefficients (r) of TP, TA and antioxidant capacity (FRAP and ABTS)

Source	TP	FRAP	ABTS	TA
TP	1			
FRAP	0.948527	1		
ABTS	0.938451	0.957609	1	
TA	0.993339	0.975608	0.962888	1

The ability of blueberry extracts to capture ABTS radicals ranged from 0.14 to 0.23 mM TE/100 g FW. [24] showed that in the Bluecrop cultivar an average content of 15.85–16.13 mmol TE/100 mg FW was measured [24].

The correlation coefficient between the phenolic content and anthocyanins content and antioxidant activities of two cultivars of blueberries Bluecrop, Goldtraube (*Vaccinium corymbosum* L.), and wild bilberry (*Vaccinium myrtillus* L.) were determined (Table 2).

Phenolic compounds, including anthocyanins, exhibit strong antioxidant activity and they have been shown in many studies to contribute significantly to the total antioxidant capacity of fruit samples [1, 33, 34]. From the results, the strongest positive correlation was found to be between total anthocyanins and FRAP activity ($r = 0.975$). A strong positive correlation also exists between phenolics and antioxidant potential ($r = 0.94$). The differences in the total phenol content between different species are relevant for the obtained changes of the antioxidant activity. When comparing results from different studies we need to have in mind that solvents, temperatures, irradiation power or pressure, and extraction time are determining extraction yields and the activity of antioxidant extracts [35].

4 Conclusion

In this study, the contents of total phenols, anthocyanins and antioxidant activity were determined in bilberries and blueberries to provide nutrition information on sources of phenolic antioxidant compounds in fresh fruit. Due to their high content of phenolics, bilberries and blueberries are already considered functional foods. Which phenolic compound contributes most to positive health effects of berries is still unknown. On the one hand, anthocyanins are considered as the most effective antioxidants among all phenolic compounds in the berries. A high positive correlation between the antioxidant activities and the content of total phenols and anthocyanins ($R^2 > 0.90$) indicates that some other phenolic compounds in addition to anthocyanins were also responsible for antioxidant activity of berry fruits.

In the future, extensive work on bioavailability, metabolism, molecular targets, and bioactivity of phenolics in the human body is required.

Conflict of Interest. The authors have no conflict of interest to this paper.

References

1. Prior, R.L., Cao, G., Martin, A., Sofic, E., McEven, J., O'Brien, C., Lischner, N., Ehlenfeldt, M., Kalt, W., Krewer, G., Mainland, C.M.: Antioxidant capacity as influenced by total phenolic and anthocyanin content, maturity and cultivar of Vaccinium species. J. Agric. Food Chem. **46**(7), 2686–2693 (1998)
2. Wu, X., Cao, G., Prior, R.L.: Absorption and metabolism of anthocyanins in elderly women after consumption of elderberry or blueberry. J. Nutr. **132**(7), 1865–1871 (2002)
3. Scalbert, A., Manach, C., Morand, C., Remesy, C., Jimenez, L.: Dietary polyphenols and the prevention of diseases. Crit. Rev. Food Sci. Nutr. **45**, 287–306 (2005)
4. Scalzo, J., Currie, A., Stephens, J., McGhie, T., Alspach, P.: The anthocyanin composition of different Vaccinium, Ribes and Rubus genotypes. BioFactors **34**, 13–21 (2008)
5. Crawford, K., Mellentin, J.: Successful Superfruit Strategy: How to Build a Super-Fruit Business. Woodhead Publishing, Cambridge (2008)
6. Krikorian, R., Shidler, M.D., Nash, T.A., Kalt, W., Vinqvist-Tymchuk, M.R., Shukitt-Hale, B., Joseph, J.A.: Blueberry supplementation improves memory in older adults. J. Agric. Food Chem. **58**, 3996–4000 (2010)
7. Basu, A., Du, M., Leyva, M.J., Sanchez, K., Betts, N.M., Wu, M., Aston, C.E., Lyons, T.J.: Blueberry decrease cardiovascular risk factors in obese men and women with metabolic syndrome. J. Nutr. **140**, 1582–1587 (2010)
8. Carey, A.N., Gomes, S.M., Shukitt-Hale, B.: Blueberry supplementation improves memory in middle-aged mice fed a high-fat diet. J. Agric. Food Chem. **62**(18), 3972–3978 (2014)
9. Nil, S.H., Park, S.W.: Edible berries: bioactive components and their effect on human health. Nutrition **30**(2), 134–144 (2014)
10. Singh, R.: Current Alzheimer's management with berries fruit therapy. J. Public Health Nutr. **1**(2), 17–24 (2018)
11. Stevenson, D., Lowe, T.: Plant-derived compounds as antioxidants for health–are they all really antioxidants? Funct. Plant Sci. Biotechnol. **3**(1), 1–12 (2009)
12. Adams, L.S., Phung, S., Yee, N., Seeram, N.P., Li, L., Chen, S.: Blueberry phytochemicals inhibit growth and metastatic potential of MDA-MB-231 breast cancer cells through modulation of the phosphatidylinositol 3-kinase pathway. Cancer Res. **70**, 3594–3605 (2010)
13. Samad, N.B., Debnath, T., Ye, M., Hasnat, M.A., Lim, B.O.: In vitro antioxidant and anti-inflammatory activities of Korean blueberry (*Vaccinium corymbosum* L.) extracts. Asian Pac. J. Trop. Biomed. **4**, 807–815 (2014)
14. Taruscio, T.G., Barney, D.L., Exon, J.: Content and profile of flavonoid and phenolic acid compounds in conjunction with the antioxidant capacity for a cultivar of northwest Vaccinium berries. J. Agric. Food Chem. **52**(10), 3169–3176 (2004)
15. Brambilla, A., Scalzo, R.L., Bertolo, G., Torreggiani, D.: Steam-blanched highbush blueberry (*Vaccinium corymbosum* L.) juice: phenolic profile and antioxidant capacity in relation to cultivar selection. J. Agric. Food Chem. **56**(8), 2643–2648 (2008)
16. Yang, G., Yue, J., Gong, X., Qian, B., Wang, H., Deng, Y., Zhao, Y.: Blueberry leaf extracts incorporated chitosan coatings for preserving postharvest quality of fresh blueberries. Postharvest Biol. Technol. **92**, 46–53 (2014)
17. Food and Agriculture Organization of the United Nations (2019): FAOSTAT Database. FAO, Rome. Accessed 30 Jan 2019 from FAOSTAT

18. Stanojčić, S., Akagić, A., Oručević Žuljević, S., Spaho, N., Gaši, F., Vranac, A., Bijedić, A.: Effects of storage on the bioactive components and sensory properties of processed and wild blueberry products. In: 28th International Scientific-Expert Conference of Agriculture and Food Industry. Book of Abstracts, Sarajevo, September 27–29 (2017)
19. Keskin-Šašić, I., Tahirović, I., Topčagić, A., Klepo, L., Salihović, M., Ibragić, S., Toromanović, J., Ajanović, A., Velispahić, E.: Total phenolic content and antioxidant capacity of fruit juices. Glasnik hemičara I tehnologa Bosne I Hercegovine **39**, 25–28 (2012)
20. Singleton, V.L.: Colorimetry of total phenolics with phosphomolybdic phosphotungstic acid reagents. Am. J. Enol. Vitic. **16**, 144–158 (1965)
21. Re, R., Pellegrini, N., Proteggente, A., Pannala, A., Yang, M., Rice-Evans, C.: Antioxidant activity applying an improved ABTS radical cation decolorisation assay. Free Radical Biol. Med. **26**, 1231–1237 (1999)
22. Benzie, I.F., Strain, J.J.: The ferric reducing ability of plasma (FRAP) as a measure of 'antioxidant power': the FRAP assay. Anal. Biochem. **239**, 70–76 (1996)
23. Okan, O.T., Deniz, I., Yayli, N., Şat, G.I., Öz, M., Serdar, G.T.: Antioxidant activity, sugar content and phenolic profiling of blueberry cultivars: a comprehensive comparison. Notulae Botanicae Horti Agrobotanici Cluj-Napoca **46**(2), 639–652 (2018)
24. Dragovic-Uzelac, V.D., Savic, Z., Brala, A., Levaj, B., Kavacevic, D.B., Bisko, A.: Evaluation of phenolic content and antioxidant capacity of blueberry cultivars (*Vaccinium corymbosum* L.) grown in the Northwest Croatia. Food Technol. Biotechnol. **48**(2), 214–221 (2010)
25. Stevenson, D., Scalzo, J.: Anthocyanin composition and content of blueberry from around the world. J. Berry Res. **2**(4), 179–189 (2012)
26. Milivojević, J., Maksimović, V., Dragišić Maksimović, J., Radivojević, D., Poledica, M., Ercisli, S.: A comparison of major taste-and health-related compounds of Vaccinium berries. Turk. J. Biol. **36**, 738–745 (2012)
27. Rimpapa, Z., Toromanović, J., Tahirović, I., Šapčanin, A., Sofić, E.: Total content of phenols and anthocyanins in edible fruits from Bosnia. Bosnian J. Basic Med. Sci. **7**(2), 119–122 (2007)
28. Ochmian, I., Oszmiański, J., Skupień, K.: Chemical composition, phenolics, and firmness of small black fruit. J. Appl. Bot. Food Qual. **83**, 64–69 (2009)
29. Stajčić, M.S., Tepić, N.A., Djilas, M.S., Šumić, M.Z., Čanadanović-Brunet, M.J., Ćetković, S.G., Vulić, J.J., Tumbas, T.V.: Chemical composition and antioxidant activity of berry fruit. Acta Periodica Technol. **43**, 93–105 (2012)
30. Rohloff, J., Uleberg, E., Nes, A., Krogstad, T., Nestby, R., Martinussen, I.: Nutritional composition of bilberries (Vaccinium myrtillus L.) from forest fields in Norway – Effects of geographic origin, climate, fertilization and soil properties. J. Appl. Bot. Food Qual. **88**, 274–287 (2015)
31. Skrovankova, S., Sumczynski, D., Mlcek, J., Jurikova, T., Sochor, J.: Bioactive compounds and antioxidant activity in different types of berries. Int. J. Mol. Sci. **16**(10), 24673–24706 (2015)
32. Jovanevi, M., Balijagi, J., Menkovi, N., Šavikin, K., Zduni, G., Jankovi, T., Deki-Ivankovi, M.: Analysis of phenolic compounds in wild populations of bilberry (*Vaccinium myrtillus* L) from Montenegro. J. Med. Plants Res. **5**, 910–914 (2011)
33. Özgen, M., Tulio, A.Z., Chanon, A.M., Janakiraman, N., Reese, R.N., Miller, A.R.: Phytonutrient accumulation and antioxidant capacity at eight developmental stages of black raspberry fruit. HortScience **41**, 1082 (2006)
34. Wada, L., Ou, B.: Antioxidant activity and phenolic content of Oregon cranberries. J. Agric. Food Chem. **50**, 3495–3500 (2002)
35. Hidalgo, G.I., Almajano, M.P.: Red fruits: extraction of antioxidants, phenolic content, and radical scavenging determination: a review. Antioxidants (Basel) **6**(1), 7 (2017)

Dynamics of Microbial Populations Activities After the Application of Nicosulfuron

Saud Hamidović[1](✉), Blažo Lalević[2], Berina Borovac[1], Anera Kazlagić[1], Sanel Haseljić[1], Vera Raičević[2], and Mirha Đikić[1]

[1] Faculty of Agriculture and Food Sciences, University of Sarajevo, 71000 Sarajevo, Bosnia and Herzegovina
s.hamidovic@ppf.unsa.ba

[2] Faculty of Agriculture, University of Belgrade, 11000 Belgrade, Republic of Serbia

Abstract. When a pesticide gets into soil, it takes part in a number of physical, chemical and biological processes that depend not only on the compound itself, but a number of other factors such as physical, chemical, and biological characteristics of soil and climatic factors. Microorganisms play an important role in pesticide degradation as they are able to utilize the biogenic elements from those compounds, as well as energy for their physiological processes. On the other hand, pesticides are more or less toxic substances that can have adverse effect on microbial populations and prevent their development, reduce their abundance, deplete their taxonomic complexity and create communities with a lower level of diversity and reduced physiological activity. The aim of this study was to analyze post nicosulfuron application dynamics of microbial populations' activity in the soil. By standard methodology, soil pH value was determined as slightly acid to slightly alkaline, humus and available K content as moderate to high, and available P content as very low to moderate. Microbial abundance (total number of bacteria, number of ammonification bacteria, fungi and actinomycetes) was determined by agar plate method. The application of nicosulfuron in all samples and at both locations has resulted in increased number of bacteria, fungi and ammonification bacteria and reduction of the number of actinomycetes. The highest positive effect on the number of microorganisms was 15 days after treatment with nicosulfuron.

Keywords: Soil · Pesticides · Microorganisms · Nicosulfuron

1 Introduction

The soil is an ecosystem with high diversity of microorganisms, whose enzymes play the central role in soil metabolic activity. They support plant supplying with the basic biogenic elements (N, P, K), degrade pesticides, produce bioactive substances and much more. Microorganism are also indicators of contamination with heavy metals, water-logging, and soil compaction. Pesticides are an indispensable segment of agricultural production. The main aims in their development is efficiency and selectivity, sufficient retention time and favorable toxicological and ecotoxicological characteristics in line with minimum unwanted impact on the environment. Frequent and long-term usage

M. Brka et al. (Eds.): AgriConf 2019, IFMBE Proceedings 78, pp. 275–280, 2020.
https://doi.org/10.1007/978-3-030-40049-1_35

of pesticides has negative effects on soil characteristics (Sethi et al. 2013) and global environment (Castro 2005). However, many research reported stimulatory or inhibitory impact of pesticides on microbial populations (Devare et al. 2007). Pesticides can affect microorganisms, altering their biodiversity, number, activity and rhythm of reproduction (Johansen et al. 2001; Rhine et al. 2003). Pesticides from the sulphonylureas group, due to their high degree of efficacy, low amount of usage and toxicity to mammals, have quickly become one of the most commonly used pesticides in plant production. The results of many studies have shown that depending on the compound properties, the amount and frequency of application, as well as the environmental conditions, pesticides can reduce or increase the number and activity of microorganisms in the soil. The objective of our study was to monitor the number of different groups of microorganisms in soil after use of nicosulfuron.

2 Materials and Methods

The experiment was set up at locations Divjak and Turbe. At the Divjak location wheat was sown in 2017 and the precrop was corn. In corn crop, in 2017, nicosulfuron was used in combination with herbicide Dikamba. On the Turbe location, corn was sown and treated with nicosulfuron in the spring of 2018. Soil samples for chemical and microbiological analysis were taken on three occasions during the research. Soil control samples at the Turbe location were taken before the herbicide application in the early spring of 2018. The second and third sampling was done 15 and 45 days after herbicide treatment. Samples were taken from two depths: 0–10 cm and 10–30 cm.

The activity of microbial populations was monitored by examining the total number of bacteria, ammonification bacteria, fungi, and actinomycetes on the appropriate nutrient mediums. Nutrient agar (Torlak, Serbia) was used for ammonification bacteria isolation, Tryptic soy agar (Torlak, Serbia - TSA) for bacterial isolation, Sabouraud maltose agar (Thermo Scientific, UK)) for fungi isolation and Starch ammonia agar (SA) for the isolation of actinomycetes. Number of microorganisms was expressed as colony forming units (CFU) per gram of absolutely dry soil.

Before setting up the experiment, soil samples were tested for the pH value of H2O and KCl, the humus content and easily accessible forms of potassium and phosphorus. The actual pH value was determined by the free ions present in the liquid phase of the soil. The exchangeable pH value was determined by the presence of H+ ions, but also of the Al3+ and Fe3+ ions that are included in the soil solution by replacing neutral salts from the soil adsorption complex. The soil humus content was determined by a bichromatic method based on wet burning of soil organic matter using K-bichromate.

Easy accessible phosphorus form in soil was determined by AL method (Egner et al. 1960) by extraction of the soil with ammonium lactate. The phosphorus determined by the AL method refers to a water and weak acids soluble fractions, which are most important for plant nutrition. After the extraction of the soil samples, the phosphorus concentration in the extract was determined by the blue method. The potassium content was determined by a flame photometer and an atomic absorption spectrophotometer (Agilent, USA).

3 Results and Discussion

Soil as a natural complex encompasses physical factors (content and ratio of mineral particles, content of organic matter), chemical factors (pH, ion exchange capacity and mineral content) and microbiological factors (species and number of microorganisms). Soils which have a high content of organic matter and clay require usage of higher doses of herbicides to achieve a satisfactory effect. Components of such soils have a high herbicide adsorption capacity. In these types of soil, the residual effect of the herbicide is more pronounced and the possibility of damage to crops is higher (Ostojić 2004).

pH value as a chemical parameter is an important factor that influences the alterations of ionizing organic compounds. It affects the biotic and abiotic degradation of ionic and nonionic compounds (Weber et al. 2000). In acid soils, the herbicide molecules bind firmly to the soil adsorption complex and their persistence is more pronounced (Loux and Reese 1993). The optimal pH for most microorganisms is close to neutral (7.0). Fungi tolerate a pH of 2.0 to 8.0, although their growth is generally better in acidic pH. Bacterial growth is usually stimulated by almost neutral pH values. Actinomycetes prefer an alkaline pH range. Results of the chemical analysis of the soil samples showed the neutral pH for both research sample sites.

Mineralization (dehumification) is a significant microbiological process, and dehumifiers are mostly fungi, actinomycetes and bacteria (Govedarica and Jarak 1995). The average humus content in the examined samples at Turbe's site was 3.6%, which can be characterized as a humous soil. At the Divjak site the average humus content in the tested samples was 3%, which can be characterized as medium humid soil (Table 1). Higher available P and K content in Turbe compared to Divjak was noticed.

Table 1. Basic chemical properties of soil

Location	pH in H_2O	pH in KCl	Humus (%)	Available P_2O_5	Available K_2O
Turbe control	7.3	6.4	3.39	20.13	23.1
Turbe 0–10 cm	7.0	6.4	4.27	36.88	60.7
Turbe 10–30 cm	7.3	6.5	3.14	29.56	43.7
Divjak control	6.7	5.7	2.24	0.69	7.6
Divjak 0–10 cm	8.0	7.1	3.82	0.75	18.4
Divjak 10–30 cm	6.7	5.8	2.87	0.50	10.2

Different herbicides may exhibit different impacts on individual groups of microorganisms in the soil. They may lead to changes in their number and composition, and thus to changes in soil fertility. Greaves et al. (1987) state that herbicides applied in the recommended amounts can stimulate the number and development of one group of microorganisms and inhibit others. Numerous studies have shown different effects of various herbicides on the number and biochemical activity of microorganisms (Govedarica and Živanović 1994; Milošević et al. 1995, 2000; Govedarica et al. 1995, 2000, 2001). Studies have shown that the total number of microorganisms, especially bacteria, can be used as an indicator of the general biological activity of the soil (Milošević et al. 2001; Radivojević et al. 2003).

The number of bacteria recorder prior to the application of herbicide was approximately equal in both sites (Table 2). The highest number of bacteria was recorded in the surface layers of soil from Turbe location at a depth of 0–10 cm. The lowest number of bacteria 20×10^6/g of soil was recorded in the deep layers of soil (10–30 cm) at Divjak site. Increased bacterial activity was partially transient because 45 days after herbicide application, their number was reduced, but still higher than the control level. These results indicate that the nicosulfuron application had stimulating effect on bacterial activity at both sites. It is known that some bacteria have the ability to use organic pollutants (in this case herbicides) as unique carbon and energy source, which certainly has an impact on the microbiological properties of the soil. These bacteria are able to metabolize herbicides to simple components that can serve other microorganisms as a source of nutrients and energy. This results in stimulation of the number and activity of microorganisms in the soil.

Table 2. Total number of bacteria ($\times 10^6$ CFU/g) in soil samples

Locality	Divjak			Turbe		
Samples	10–30 cm	0–10 cm	Control	10–30 cm	0–10 cm	Control
Before treatment	20	36	31	25	44	35
15 days after	90	133	109	134	142	90
45 days after	62	42	71	45	59	79

The effect of herbicide on fungi is variable and depends on the application dosage (Milošević et al. 2000; Milošević and Govedarica 2001). As regards to the presence of fungi in the samples, no high number of these microorganisms was recorded. This may be related to the neutral pH value of the soil at both sites, because fungi prefer acidic environment. The highest number of these microorganisms was recorded 15 days after the herbicide application. This value decreased again after one month but was substantially higher than the pre-treatment period. This indicates that the application of nicosulfuron had a stimulating effect on the number of fungi in the examined soil (Table 3).

Table 3. Total number of fungi ($\times 10^4$ CFU/g) in soil samples

Locality	Divjak			Turbe		
Samples	10–30 cm	0–10 cm	Control	10–30 cm	0–10 cm	Control
Before treatment	1	0	4	4	7	6
15 days after	5	8	8	14	17	4
45 days after	7	5	8	6	5	8

Actinomycetes are heterotrophic microorganisms with complex enzymatic systems that enable them to utilize different simple molecules (organic acids, sugars) as well as complex ones (proteins, polysaccharides). They are numerous in neutral and alkaline soils. After bacteria, they are the second most prevalent group of microorganisms in soil. They are aerobic and grow the best in the surface layers of soil. Based on the results, it can be concluded that the application of nicosulfuron had an inhibitory effect on the number of actinomycetes in the soil samples. Prior to the herbicide application, the number of actinomycetes was significantly higher in relation to the next two sampling periods. Because of the higher content of humus at the site of Turbe, a greater number of actinomycetes were recorded as they actively participate in the processes of humification and mineralization of organic residues (Table 4).

Table 4. Total number of actinomycetes ($\times 10^4$ CFU/g) in soil samples

Locality	Divjak			Turbe		
Samples	10–30 cm	0–10 cm	Control	10–30 cm	0–10 cm	Control
Before treatment	46	37	27	66	33	36
15 days after	9	4	4	15	7	9
45 days after	6	2	0	3	11	3

Ammonification bacteria are numerous in all types of soil. Their number ranges up to several million in grams of soil (Table 5). The use of nicosulfuron herbicide had a stimulating effect on the number of these bacteria 15 days after treatment. However, a month later their number was significantly reduced and was lower than the initial state, i.e. before the application of the herbicide. As with previous studies, the number of these microorganisms was higher in Turbe's location. As an ultimate outcome it can be concluded that the application of nicosulfuron had an inhibitory effect on the number of ammonification bacteria in all examined samples at both sites. Ammonification bacteria are a significant physiological group of microorganisms whose number depends on the application of agrotechnical measures. Numerous examples of agrotechnical measures negative effects include the presence and activity of microorganisms involved in protein digestion and their translation into affordable forms of plant nutrition (Bhuyan et al. 2014).

Table 5. Total number of ammonification bacteria ($\times 10^6$ CFU/g) in soil samples

Locality	Divjak			Turbe		
Samples	10–30 cm	0–10 cm	Control	10–30 cm	0–10 cm	Control
Before treatment	36	43	45	65	59	61
15 days after	78	88	43	99	78	96
45 days after	27	51	37	30	14	27

4 Conclusion

Our results suggest the various influence of nicosulfuron pesticide on microbial diversity in soil from Turbe and Divjak location. The highest positive effect on the number of microorganisms was 15 days after treatment with nicosulfuron. This results can be useful for further research of herbicide application considering the beneficial soil microflora importance in crop production.

References

Bhuyan, S.I., Tripathi, O.P., Khan, M.L.: Effect of season, soil and land use pattern on soil N mineralization, ammonification and nitrification: a study in Arunachal Pradesh, eastern Himalaya. Int. J. Environ. Sci. **5**, 88–97 (2014)

Castro, J.I.R.: Stable isotope probing analysis of the influence of liming on root exudate utilization by soil microorganisms. Environ. Microbiol. **7**(6), 828–838 (2005)

Devare, M., Londono, R.L., Thies, J.E.: Neither transgenic Bt maize (MON863) nor tefluthrin insecticide adversely affect soil microbial activity or biomass: a 3-year field analysis. Soil Biol. Biochem. **39**, 2038–2047 (2007)

Egner, H., Riehm, H., Domingo, W.R.: Untersuchungen über die chemische Bodenanalyse als Grundlage für die Beurteilung des Nahrstoffzustandes der Boden, II: Chemische Extractionsmetoden zu Phosphorund Kaliumbestimmung. Kungliga Lantbrukshügskolans Annaler **26**, 199–215 (1960)

Govedarica, M., Jarak, M.: Microbiology. Faculty of Agriculture, Novi Sad. (2003)

Govedarica, M., Jarak, M.: Soil microbiology. Faculty of Agriculture, Novi Sad. (1995)

Johansen, K.A., Iwaki, D.D., Lengyel, J.A.: Patterning and morphogenesis of the Drosophila embryonic hindgut. A. Dros. Res. Conf. **42**, 696C (2001)

Loux, M.M., Reese, K.D.: Effect of soil type and pH on persistence and carryover of imidazolinone herbicides. Weed Technol. **7**, 452–458 (1993)

Milošević, N., Govedarica, M., Jarak, M., Đorđević, S.: Pesticides and Microorganisms. Stylos, Novi Sad. (2001)

Ostojić, Z.: Biotest (bio assay) for proving residual effect of persistent herbicides on crops. **1**, 13–19 (2004)

Radivojević, Lj., Stanković-Kalezić, R.: Effect of trifularin and pendimethalin on soil microorganisms. Acta herbol. **9**, 77–84 (2000)

Rhine, E.D., Fuhrmann, J.J., Radosevich, M.: Microbial community responses to atrazine exposure and nutrient availability: linking degradation capacity to community structure. Mocrob. Ecol. **46**, 145–160 (2003)

Sethi, S., Datta, A., Gupta, B., Gupta, S.: Optimization of cellulase production from bacteria isolated from soil. ISRN Biotechnol., 1–7 (2013). Article ID 985685

Weber, J.B., Wilkerson, G.G., Linker, H.M., Wilcut, J.W., Leidy, R.B., Senseman, S., Witt, W.W., Barrett, M., Vencill, W.K., Shaw, D.R., Mueller, T.C., Miller, D.K., Brecke, B.J., Talbert, R.E., Peeper, T.F.: A proposal to standardize soil/solution herbicide distribution coefficients. Weed Sci. **48**, 75–88 (2000)

Evaluation of Microbiological Parameters of Drinking Water from the Dobrenica Water Supply System in Ostrožac na Uni

Samira Dedić[1(✉)], Huska Jukić[2], Miloš Rodić[3], Zlatko Jusufhodžić[3], and Aida Džaferović[1]

[1] Faculty of Biotechnical Sciences, University of Bihać, Kulina Bana 2, 77000 Bihać, Bosnia and Herzegovina
samira.dedic@yahoo.com
[2] Faculty of Health Studies, University of Bihać, Nositelja hrvatskog trolista 4, Bihać, Bosnia and Herzegovina
[3] Public Institution "Veterinary Institute" Bihać, Omera Novljanina 6, Bihać, Bosnia and Herzegovina

Abstract. The Dobrenica water supply system in Ostrožac on Uni was originally intended only for drinking water supply for the Sedra Hotel. Later on, the left and the right side of the village Ostrožac on Uni, i.e. Donji Srbljani, were connected to the Dobrenica water supply. Since the Sedra Hotel has been closed for three years, the maintenance of the water supply system and the chlorination of water has ceased. At the moment, the water is chlorinated at the entrance to the hotel facility due to special circumstances (holding facilities for migrants), while the villagers still receive unsafe and nonchlorinated water. Water samples were analysed for basic microbiological parameters: *Echcerichia coli* and coliform bacteria, total aerobic colony count at 22 °C and 37 °C, intestinal enterococci, *Pseudomonas aeruginosa*, and *Clostridium perfringens*. Microbiological analyses of water samples were performed in the laboratory of the Public Institution Veterinary Institute in Bihać, which has accredited methods required by the Standard BAS EN ISO/IEC 17025: 2006 issued by the Institute for Accreditation of BiH, and MACs are taken from the Guidelines for drinking-water quality ("Official Gazette of BiH", No. 40/2010, 43/2010, 30/2012 and 62/2017). The results of the analyses indicate the presence of coliform bacteria, *Echcerichia coli, Pseudomonas aeruginosa, Clostridium perfringens*, and intestinal enterococci. A high percentage of unsafe water samples indicate the need to monitor the micro- biological quality of water for safer drinking water supply to the population. In order to make the water safe for further use, it is necessary to first rinse the pipes and then proceed with regular chlorination.

Keywords: Water · Microbiological parameters · Bacteria · Chlorination

1 Introduction

The entire life on Earth depends on the available quantities of water of a certain quality. However, water used for human consumption must be hygienically safe. The quality of natural water is sensitive and it varies, becoming one of the limiting factors for their

M. Brka et al. (Eds.): AgriConf 2019, IFMBE Proceedings 78, pp. 281–287, 2020.
https://doi.org/10.1007/978-3-030-40049-1_36

use. Drinking water can be a carrier of toxic substances, pathogenic microorganisms and a cause of poisoning or epidemics. We perform microbiological analysis of water and determine the possible presence of bacteria that could endanger human health.

The bacteriological quality of drinking water is of paramount importance and monitoring must be a priority [1]. Drinking water quality has a direct connection to human health and the provision of safe drinking water to consumers is one of the most important public health priorities [2].

Microbiological drinking-water testing is generally based on the detection of coliform bacteria and the total number of bacteria. Coliform bacteria belong to the primarily non-pathogenic bacteria that are mostly found in the lower intestinal tract or the colon of warm-blooded animals and humans. Feces, excreted from the organism, enter wastewater, which in turn can enter natural waters. If large numbers of coliforms are found in water, it is a possible faecal contamination indicator, i.e. water is contaminated by pathogens. Examples of such pathogens include *Salmonella* spp., *Shigella* spp., *Vibrio cholera* and *Escherichia coli* [3]. The number of faecal coliform bacteria per gram ranges between 10^6 and 10^9. The concentration of coliform bacteria in domestic wastewater ranges from 10^{11} to 10^{13} in 100 ml of wastewater [4]. The ones in the environment can survive for a long time and thus provide a reliable indication of faecal contamination long after being excreted from the digestive system. The most important in this group is *Escherichia coli* bacteria, since it is rarely present in the digestive system of humans [5].

Total coliform bacteria are one of the microbiological parameters of water testing and faecal coliforms are drinking-water quality indicators. Sometimes water-induced epidemics can occur due to lack of treatment or irregular treatment of raw water. Dysentery, abdominal typhus, paratyphus, cholera, etc. can be transmitted by water. The causative agents of these infections: *Shigella dysenteriae*, *Salmonella typhosa*, *Vibrio cholera* and *Echcerichia coli* come mostly from human faeces and urine in the water, which come from diseased people [5].

Due to legislative acts (Water Act BiH, *Guidelines for drinking-water quality* "Off. Gazette of BiH", No. 40/10, 43/10, 30/12 and 62/17), a downward trend of unsafe drinking water samples is noted. Drinking water quality monitoring is carried out by sanitary inspection and public health institutions in accordance with the guidelines for drinking water quality, which include sample number and locations [6, 7]. Access to clean and safe drinking water continues to be a problem in developing countries, especially in rural areas. Due to inadequate drinking-water supply, villagers often seek the solution for themselves to meet their basic water need.

According to data from the Water Management Strategy in FBiH, about 60% of the population of the FBiH is connected to the public water supply system. The rest of the population (around 40%) satisfies their needs for water through individual, group or local water supply, which is not monitored by municipal water management. In rural areas, only 20% of the population is connected to the public water supply system. Regular water quality testing is performed only for public (municipal or cantonal) waterworks [8]. There are 16 registered local rural water supply systems in Bihać municipality, which don't have a built-in water disinfection system, and generally don't control the quality of drinking water; their regular maintenance is almost non-existent. Local water supply systems in the municipality provide the population of the following settlements or local communities: Dobrenica, Pritoka, Ripač, Lohovo, Ću-kovi, Mali and Veliki Skočaj, Međudražje, Zavlje, Vučijak, Donji Srbljani, Orašac, Kulen Vakuf and Martin Brod. None of these systems is under the control of the relevant institutions [9]. The aim of this study is to monitor the microbiological parameters of the quality of water samples taken at the customers' consumption tapes: Ostrožac on Uni, Donji Srbljani, and the source of Dobrenica, because the locals receive unsafe and unchlorinated water.

2 Material and Method of Work

Microbiological analysis of drinking water samples was carried for these parameters: *Echcerichia coli* and coliform bacteria, a colon collection at 22 °C and 37 °C, intestinal enterococci, *Pseudomonas aeruginosa* and *Clostridium perfringens* (including spores). Membrane filtration was performed first, followed by anaerobic incubation at m-CP agar at 44 °C for 21 ± 3 h. The numbers of opaque yellow colonies pass into pink or red after exposure to the action of ammonium hydroxide in the period of 20 to 30 s. Microbiological analyses of water samples were performed in the laboratory of the Public Institution Veterinary Institute in Bihać, which has accredited methods required by the Standard BAS EN ISO/IEC 17025: 2006 issued by the Institute for Accreditation of BiH, and MACs are taken from the Guidelines for drinking-water quality ("Official Gazette of BiH", No. 40/2010, 43/2010, 30/2012 and 62/2017) [6, 7], which define the scope of the study, the frequency and number of samples per water supply system as well as the maximum permissible concentration for the examined parameters (MDK) etc. Forty samples of household tap water were collected in Donji Srbljani from September to the end of December 2018.

Microbiological parameters
Escherichia coli and coliform bacteria - Part 2 Most probable number method (EN ISO 9308-2 :2014, IDT;ISO 9308-2:2012,IDT)
Intestinal enterococci (*Enterococcus faecalis*) - Part 2: Membrane filtration method (EN ISO 7899-2:2000, IDT; ISO 7899-2:2000, IDT)
Pseudomonas aeruginosa - Membrane filtration method (EN ISO 16266:2008, IDT ISO 16266:2006, IDT)
Enumeration of culturable micro-organisms– Colony count, incubation at 36ºC and 22ºC (EN ISO 6222:1999 IDT; ISO 6222:1999; IDT)
Enumeration of *Clostridium perfringens* colonies – Membrane filtration method - (EN ISO 14189:2016, IDT; ISO 14189:2013, IDT)

Water supply "Dobrenica" in Ostrožac on Uni was built in 1984. The length of the water supply is 2700 m from the source to the hotel Sedra and the total length of the water supply network is 4000 m. The source of the Dobrenica is located at the foot of the Prošići village and its capacity is 8 l/s. The water supply source is made of a sold 7 × 7 m construction, with an average depth of 1.5 m. It consists of a pool accumulating water, measuring 5 × 7 m and an overflow system, so that excess water from the spring flows into the river Una.

The necessary tests (source, microbiological and chemical analysis) were carried out, prior to the construction of water supply, by a licensed house from Zagreb. In the documents submitted by the authorized house to the investor "Dubrava" Cazin, on the basis of the microbiological results of the test of water, it was concluded that water from the source Dobrenica is not recommended for drinking without prior chlorination. In the report, the reason for such a finding that surface- and groundwater interact, where *Escherichia coli* bacteria occur.

Based on the findings during the construction of the water supply system, the technical documentation had envisaged and installed the equipment for liquid chlorination at the source, using liquid chlorine, i.e. sodium chloride hypochlorite (NaClO). During the exploitation of the water supply since its inception, until 2015, this system was regularly maintained according to the instructions of the competent authorities, which carried out the chemical and microbiological analysis of the Dobrenica water supply. The water supply was originally intended for supplying drinking water exclusively to the Sedra hotel. The Hotel Sedra could not be opened until it received findings on the chemical and biological safety of drinking water. During the war, the left and the right side of the village Ostrožac on Uni, i.e. Donji Srbljani, were connected to the Dobrenica water supply.

The authorized house which made the project documentation has defined the exact criteria under which water from the Dobrenica source can be used for drinking. In order to solve the problem of pollution and water purification, five sludge discharge pipes in diameter of φ 52 mm were made for regular annual rinsing of the deposit from the pipeline. A pipe φ 150 mm in diameter was used for the construction of the pipeline. The height difference between angles the water source and the structure is 27 m, and the water pressure is 2.7 bars. Since the Sedra Hotel has been closed for three years, the maintenance of the water supply system and the chlorination of water has ceased.

The Sedra Hotel became the holding facility for migrants in 2018. The International Organization for Migration (IOM), which takes care of migrants, has worked on the problem of microbiological quality of water, trying to meet the drinking-water parameters. However, the IOM has solved the problem of water quality by chlorinating water at the entrance to the hotel facility instead of taking into account the opinion of the project house, which strictly stated that the most effective way to deal with the water quality problem is to chlorinate at the source. At the moment, the water is chlorinated at the entrance to the hotel facility due to special circumstances while the villagers still receive unsafe and nonchlorinated water.

In order to obtain reliable results of quantitative bacteriological testing, it is very important to ensure that water is not contaminated during taking and transporting the sample as well as preventing reproduction of the existing bacteria but ensuring they remain alive and capable of it.

Water samples for this analysis were taken in 1000 ml clean glass bottles, previously sterilized for 1 h in a drying oven at 160–180 °C. The bottles were sealed with glass stoppers and aluminium caps beforehand. The samples were transported in refrigeration units within one hour after taking and kept in a refrigerator at a temperature of up to 4 °C and sown within 12 h.

3 Results and Discussion

Based on the analysis of water samples of household tap water collected in the period from the beginning of September to the end of December 2018 (Table 1), we can see that several samples are bacteriologically unsafe. Thus, *Escherichia coli* and total coliforms were present in the samples throughout the four months. Their presence in water is usually the result of problems with the water treatment system or with water supply pipes. The presence of *Enterococcus* and *Pseudomonas occurs*. The presence of *Pseudomonas aeruginosa* indicates to old faecal contamination and inadequate maintenance of the water system in facilities that pose a health hazard to the sensitive population. While all of the samples were positive throughout the 4 months, it occurs due to the poor purification and disinfection of water, i.e. Non-chlorinated water. The summer period also carries the risk of contamination due to the favourable conditions for the reproduction of pathogens and increased anthropogenic activity in the source power supply zone. Since there is no continuous monitoring of the mentioned source, we can only speculate on what kind of data would be obtained by systematic maintenance and water analysis throughout the year.

Table 1. Results of species and number of microbiological isolated bacteria of water samples taken on taps of consumers

Analysis parameters	September		October		November		December	
	TNS	PS	TNS	PS	TNS	PS	TNS	PS
E. coli and coliform bacteria	10	6	10	4	10	6	10	6
Intestinal enterococci	10	6	10	6	10	2	10	4
Pseudomonas aeruginosa	10	8	10	6	10	8	10	6
Clostridium perfringens	10	8	10	4	10	2	10	4
Total aerobic colony count at 22 °C	10	10	10	10	10	10	10	10
Total aerobic colony count at 36 °C	10	10	10	10	10	10	10	10

TNS - total number of samples, PS - positive samples

One of the general measures to protect the population from contagious diseases is to provide safe drinking water. This measure is organized and implemented by the local government bodies and other competent authorities in accordance with the Law - which means that the local community is obliged to provide safe drinking water to all citizens on its territory [10]. In cases where the parameters do not meet the *Guidelines for drinking-water quality* ("Official Gazette of BiH", No. 40/2010, 43/2010, 30/2012 and 62/2017), water supply should be limited. Moreover, the competent authorities and sanitary inspection should be notified within 24 h. The consumers should be informed and given appropriate recommendations, unless the competent authorities assess that the exceeded values cannot endanger human health. It is also necessary to investigate the cause of the parameter deviation and to take emergency measures to remove the cause of the unsafe water. After the measures have been taken, it is necessary to perform the analysis again in the laboratory of the public health institute.

Wilhelm and Malik investigated the frequency of fecal-bacteria in surface water. They reported a positive correlation between ammonia concentration and fecal contamination indicator bacteria in drinking water samples [11]. Ujević and Gajšak in their reports on the health safety of drinking water point out that local water supply systems, from the public health point of view, pose the greatest risk since water is distributed to consumers without any treatment, and often without disinfection [12]. They state that analyses show that out of 617 local water supply system samples, 348 were unsafe (56.4%), and the most common cause of this was microbiological contamination. 309 samples (50.1%) were microbiologically unsafe. In our study, all 40 samples (100%) that were analyzed for specific microbial water quality parameters were microbiologically unsafe. These analyses should be taken throughout the year, with more samples taken at more locations, especially directly from the reservoir, and with other physicochemical parameters included. UK data show that 0.03 to 0.04% of (chlorinated) drinking water samples from the supply system were positive for *Echcerichia coli.* In France, 0.4 to 1.0% of samples contained coliforms (no data on thermotolerant coliforms or *Echcerichia coli*) [13]. Bearing in mind the bacteriologically unsafe water of the Dobrenica water supply, the Canton should invest some financial resources in monitoring the water quality of the local water supply systems. Ensuring the safety of drinking water is one of the basic measures for protecting human health, because unsafe water can cause various diseases. It is necessary to protect drinking water sources, to purify water and to regularly check the water quality.

4 Conclusion

Most samples of household tap water collected throughout the four months were polluted and potentially dangerous for public health. In order for the water to be used for further use it is necessary to first rinse the pipes and then proceed with regular chlorination. Local water supplies that don't have proper water treatment are considered unsafe for use and must be under constant surveillance as they are a serious threat to human health and the cause of the hydrological epidemics. This study is limited to a smaller number of water samples and the assessment of certain microbiological parameters of household water samples. It is therefore necessary to regularly assess the quality of drinking water sources, main distribution tanks, distribution systems and pipes to ensure that water is safe for human consumption. Moreover, further studies should take into account the additional water quality parameters and the analysis of water throughout the year.

Acknowledgements. The authors have no conflict of interest to state regarding the results presented within this manuscript.

References

1. Bharti, N., Katyal, D.: Water quality indices used for surface water vulnerability assessment. Int. J. Environ. Sci. **2**(1), 154–173 (2011)
2. Boe-Hansen, R.: Microbial growth in drinking water distribution systems. Ph.D. thesis, Lyngby, Denmark (2001)
3. Smruti, S., Sanjeeda, I.: Microbiological analysis of surface water in Indore, India. Res. J. Recent Sci. **1**, 323–325 (2012)
4. Henze, M., van Loosdrecht, C.M., Ekama, G.A., Brđanović, D.: Biological Wastewater Treatment. Principles, Modeling and Design. IWA Publishing, London (2008)
5. Duraković, S.: Primjenjena mikrobiologija, Prehrambeno tehnološki inženjering. Durieux, d. o.o., Zagreb (1996)
6. Pravilnik o zdravstvenoj ispravnosti vode za piće ("Sl. Glasnik BiH", br. 40/2010, 43/2010, 30/2012)
7. Pravilnik o izmjenama pravilnika o zdravstvenoj ispravnosti vode za piće ("Sl. Glasnik BiH", br 62/2017)
8. Strategija upravljanja vodama FBiH (2010–2022). http://www.fzofbih.org.ba/userfiles/file/strategija_vode.pdf
9. Strateški plan za vode i okolišnu sanitaciju (2007). http://www.mojemjesto.ba/files/documents/Strateski%20plan%20za%20vode%20i%20okolisnu%20sanitaciju.pdf
10. Mehinović, N., Popović, M., Azabagić, S., Aljukić, A.: Fizičko-hemijski i mikrobiološki kvalitet vode s javnih česmi i lokalnih izvorišta na području Tuzlanskog kantona (2011). file:///C:/Users/User/Downloads/42-203-1-PB%20(4).pdf
11. Wilhelm, L.J., Malik, T.L.: Fecal-indicator bacteria in surface waters of Santee River Basin and coastal drainages, North and South Carolina, 1995–98: U.S. Geological Survey Fact Sheet FS-085-98 (1998). 6 p.
12. Ujević, B.M., Gajšak, F.: Izvještaj o zdravstvenoj ispravnosti vode za ljudsku potrošnju u republici Hrvatskoj za 2017 (2018). godinu. https://www.hzjz.hr/wp-content/uploads/2018/06/Izvjestaj-RH-voda-za-ljudsku-potrosnju2017_v1.pd
13. van Lieverloo, J.H.M., Blokker, E.J.M., Medema, G.: Quantitative microbial risk assessment of distributed drinking water using faecal indicator incidence and concentrations. J. Water Health **5**, 131–149 (2007)

Seasonal Variations in Deposition of Fat in the Body of Rainbow Trout Oncorhynchus Mykiss (Walbaum 1792) and Its Effect on Yield and Quality of Meat in Cages at Commercial Breeding Farm

Enver Karahmet(✉), Senita Isaković, Samir Muhamedagić, Almir Toroman, and Nermina Đulančić

Faculty of Agriculture and Food Sciences, University of Sarajevo, Zmaja od Bosne 8, 71000 Sarajevo, Bosnia and Herzegovina
enverkarahmet@yahoo.com

Abstract. The aim of the experiment was to determine the effects of seasonal changes in the degree of fat deposition in various parts of the body and its impact on yield VSI (Viscero Somatic Index) of rainbow trout Oncorhynchus mykiss (Walbaum 1792), under the same conditions of feeding in cage farming systems of commercial fish farms. The task was to identify what is and what impact seasonal variation and environment on yield and quality of the meat of rainbow trout, which is fed with same nutrients to achieve a certain weight. As a basic experimental material has been used rainbow trout produced for the Bosnian and foreign markets of varying body weight: First group of fish between 150 ± 5 g, group II 200 ± 5 g and III group 250 ± 5 g. The same is provided from the local fish farm Salakovac on the Neretva River. A total of 180 fish samples were analyzed.

This test showed there were highly significant statistical differences in the content of: fat (0.000), moisture (0.000), and protein (0.000) relative to the weight of the samples. Gathering data confirmed there were statistically significant differences between the seasons March-November in fat content (0.019), and November-December (0.033). At the same time proteins showed statistically significant differences only in relation to season of March-August (0,001). The results of analysis of the value of Viscero Somatic Index (VSI) showed very highly significant statistical difference in relation to season (0,000).

Keywords: Fat deposition · Rainbow trout · Yield (VSI) · Seasonality · Cage farming

1 Introduction

Fish meat possesses high nutritional quality and is therefore a particularly recommended human dietary component. Information concerning the chemical and fatty acid composition of freshwater fish is valuable to nutritionists who are interested in finding sources of low-fat, high protein foods, with desirable fatty acid compositions and acceptable amount of total cholesterol (Ljubojević et al. 2013).

M. Brka et al. (Eds.): AgriConf 2019, IFMBE Proceedings 78, pp. 288–294, 2020.
https://doi.org/10.1007/978-3-030-40049-1_37

Seasonal trend of growth and distribution of energy have been described for many fish populations. However, detailed information between growth and quantitative changes in fat content are missing. In particular, very little is known about the dynamics of the relative importance of some fat tissue for its storage, while the goals of this research and they were to describe and quantify seasonal growth models, disposal and utilization of fat for trout, during its breeding in cages system (Forsberg 2006; Jorgensen et al. 1997). The chemical and fatty acid compositions of fish vary greatly between different species and within the same species.

Farmed fish contain high n-3 PUFA (12.3–36.3% vs. 5.48–37.2% in the wild) and have higher muscle fat and n-6 PUFA contents (mainly 18: 2n-6) than their wild counterparts. The aquaculture management, diet, and season can affect fillet composition and fatty acids, while season (i.e. food availability and maturation) largely affects lipid quality in wild fish (Grigorakis 2015).

Fatty acid composition also varies largely depending on the fish species, on their being wild fish or farm-raised, on the age of fish and on the origin of diets and its composition (Ćirković et al. 2011; Yeganeh et al. 2012; Ljubojević et al. 2013).

The conditions of the natural environment vary significantly during the year, water temperature and the environment, day light shows significant changes during the breeding period. It is well known and documented that fish growth and its quality are strongly influenced by water temperature and day light. Further factors affecting the production parameters in cage fish farming are still partial pressure of oxygen, NH4, turbidity, etc., the social factor and the season. Fish production is directly related to the quantity of "pure" meat (Seather et al. 1996; Branas 2000; Morkere and Rorvik 2001).

It is well known that the growth and chemical composition of meat can be influenced by feeding regime, the influence of the environment and of seasonal variation. Farming of rainbow trout has been developed during the past few hundred years, showing high efficiency in many aspects of its culture using a well-established system.

Research and development of aquaculture are continually focused on increasing production efficiency and sales by increasing the density of flocks, improving technology of water circulation, development of genetically superior line of fish, in the sense of the growth, control ripening, improvement fish food diets and market development. Understanding seasonal changes and their impact on growth, survival and level of movement is a crucial problem for salmon producers according (Kiesslig 2005).

2 Material and Methods

As a basic experimental material that has been used rainbow trout intended for the Bosnian and foreign markets of varying body weight: first group of fish between 150 ± 5 g, group II 200 ± 5 g and III group 250 ± 5 g. The same is provided from the local fish farm Salakovac on the Neretva River. A total of 180 fish samples. During the two-year experiment has been analyzed and described the impact of climate changes on the deposition of fat in the body of rainbow trout Oncorhynchus mykiss (Walbaum 1792) in cage farming on commercial fish farm. Fish are caught from the net cages 24 h after feeding to make the intestine tract as empty as possible. This was very important because the fish food diet contained between 18% and 22% of the crude fat and 38% and 44% of the crude protein, which directly influenced the evaluation of the chemical composition of the visceral part of the fish. Then, they were anesthetized in water with a solution (p-aminobenzoic ethyl ester, 50 mg^{-1}) for easier manipulation and avoiding the emergence of stress in fish samples.

Periodic fish sampling was closely connected with their natural cycles. The first sampling was conducted in late March, and this season has included the period from November to March, the second was at the end of July, covering the period from April to July, a third sampling was done at the end of November and the season included the period from August to November.

3 Results

The obtained results were closely associated with age categories fish on which they performed measurements. The obtained test results of VSI for first year showed that there were highly significant statistical differences among the studied seasons. Likewise, the results showed a statistically significant difference between the VSI seasons. In addition, the results of statistical analysis showed that there were highly significant differences of VSI tested an average weight of samples in relation to the season. It turned out that the chemical composition of muscle tissue samples of fish was affected by natural factors, the direct impact on the regime of eating, which in fact affects mostly the chemical composition of muscle tissue.

Survey results showed that the differences in average body weight, average weight without the head, gills, Viscero Somatic Index (VSI) and gonads between fish populations surveyed in different seasons was statistically significant (Table 1).

Table 1. Testing between season and weight of samples

Source	Dependent variable	Type III Sum of squares	df	Mean square	F	Sig.
Samples weight	Moisture	166,930 (a)	1	166,930	42,865	**,000**
	Fat	12,342 (b)	1	12,342	23,865	**,000**
	Proteins	5,992 (c)	1	5,992	20,514	**,000**
	Ash	,018 (d)	1	,018	,487	,487
Season	Moisture	7,576	2	3,788	,973	,381
	Fat	3,576	2	1,788	3,457	**,035**
	Proteins	3,236	2	1,618	5,540	**,005**
	Ash	,033	2	,017	,441	,645
Samples weight * season	Moisture	5,672	2	2,836	,728	,485
	Fat	1,493	2	,747	1,444	,240
	Proteins	,694	2	,347	1,189	,308
	Ash	,066	2	,033	,878	,418

a R Squared = ,278 (Adjusted R Squared = ,260)
b R Squared = ,167 (Adjusted R Squared = ,145)
c R Squared = ,154 (Adjusted R Squared = ,133)
d R Squared = ,032 (Adjusted R Squared = ,007)

This test showed there were highly significant statistical differences in the content of fat (0.000), moisture (0.000) and protein (0.000) relative to the weight of the samples. LSD test confirmed there were statistically significant differences between the seasons March-November in fat content (0.019) and November-December (0.033). At the same time proteins showed statistically significant differences only in relation to season of March-August (0,001).

LSD test showed that significant differences existed between the season March-November in fat content (0.019) and November-December (0,033). At the same time, proteins showed statistically significant differences only in relation to the season of March to August (0.001).

The obtained data show that there is a highly significant statistical difference between the average moisture, fat and protein in different weight categories of fish samples per year at the level (0,000). Weight of the samples did not affect the average ash content (Fig. 1).

The research results VSI between the first and second year did not show statistically significant differences in relation to the weight of the samples. The graph gives an overview showing the correlation between year, weight of the samples and season was highly significant at $R^2 = 0.5677$ (Table 2).

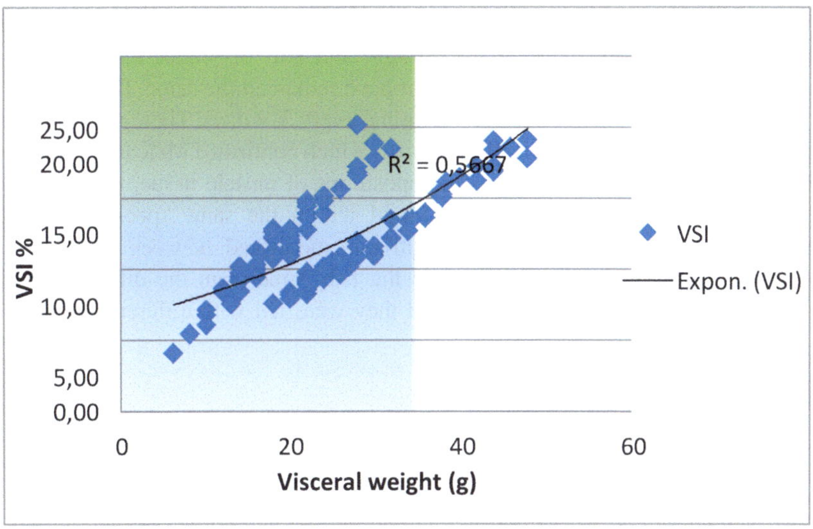

Fig. 1. VSI correlation between years

Table 2. ANOVA between groups

	Sum of square	df	Mean square	F	Sig.
Between groups	938,238	11	85,294	22,305	**,000**
Within the group	412,996	108	3,824		
Total	1351,233	120			

Analysis of variance showed a very high significance of difference between groups of the VSI values (0.000) as it can be clearly seen in Table 3.

4 Discussion

Many studies have been associated with programs for monitoring natural effects on growth, reproduction and enhanced efficiency of feed conversion with salmonids. Distribution of energy is a response to seasonal changes with regard to differences in the conditions of feeding, indicating that they are independent (Rabah 2005; Johnson et al. 2004). Of course, during the summer and especially autumn, larger fish have a higher level of deposited fat in the body. In the autumn, the ratio of deposited fats in flesh of fish with fats in other seasons was higher (Oppedal et al. 2011).

These changes are associated with changes in appetite, activity, changes of chemical status and certainly changes in sexual activity. Changes in seasonal behavior and tissue composition of the fish indicate that acclimatization is very important and affects fish with many physiological factors, reducing or increasing the enzymatic activity of the organism. Results of these studies have confirmed that the level of fat was under the influence of the season which is consistent with the research: Thakur et al. (2003), Aursand et al. (1994), Johnson et al. (2006).

The tested fish showed a response to seasonal changes especially in food intake during the studied period. During the autumn and part of winter fat reserves have increased, but their growth and development have remained the same. During the summer, the fish took food only early in the morning, every 3–4 days. Their growth in length was more intense and faster during the summer, which confirmed Metcalfe et al. (2002).

Regarding chemical and fatty acid compositions of muscle tissue, it was expected that some major differences would be found within the same species and among different species in percentages of the nutrients monitored between the tested fish species in the present experiment because the fish came from the different environments, different species were included, and they were fed with different diets.

5 Conclusions

Based on the presented data, it is noticeable that the season's impact on the content of certain ingredients of table fish is expressed. This is particularly reflected in the dynamic depending fish during the studied period. The experiment showed that this is set physiologically energy model and a performance test adequate response of physiological and other parameters, and the regulation of growth and the production of energy reserves in the fish body. Standard feeding confirmed that the deposition of fat in the trout was more intense during the late summer or early autumn, and then fell again during the winter period until early spring.

The results of the statistical analysis showed that there are highly significant statistical differences in the VSI in the examined average weights of the samples compared to the season for both years. The chemical composition of the muscle tissue of the examined fish samples was shown to be influenced by natural factors, which is a direct influence on the mode of taking food. This test showed there are highly significant statistical differences in the content of: fats (0,000), moisture content (0.000) and protein (0,000) relative to the weight of samples. The results of this study are, however, insufficient to show how seasonal variation has an impact on trout.

Conflict of Interest. The authors declare that there are no conflicts of interest regarding the publication of this paper.

References

Ljubojević, D., Trbović, D., Lujić, J., Bjelić-Cabrilo, O., Kostić, D., Novakov, N., Ćirković, M.: Fatty acid composition of fishes from inland waters. Bul. J. Agric. Sci. **19**(Supplement 1), 62–71 (2013). Agricultural Academy

Forsberg, O.I., Guttormsen, A.G.: Modeling optimal dietary pigmentation strategies in farmed Atlantic salmon: application of mixed-integer non-linear mathematical programming techniques. Aquaculture **261**(1), 118–124 (2006)

Jorgensen, E.H., Johansen, S.J.S., Jobling, M.: Seasonal patterns of growth, lipid deposition and lipid depletion in anadromous Arctic charr. J. Fish Biol. **51**, 312–326 (1997)

Grigorakis, K.: Fillet proximate composition, lipid quality, yields and organoleptic quality of mediterranean farmed marine fish: a review with emphasis on new species. Crit. Rev. Food Sci. Nutr. **57**(14), 2956–2969 (2015)

Ćirković, M., Trbović, D., Ljubojević, D.: Meat quality of fish farmed in polyculture in carp ponds in Republic of Serbia. Meat Technol. **52**, 106–121 (2011)

Yeganeh, S., Shabanpour, B., Hosseini, H., Imanpour, M.R., Shabani, A.: Comparison of farmed and wild common carp (Cyprinus carpio): seasonal variations in chemical composition and fatty acids profile. Czech J. Food Sci. **30**, 503–511 (2012)

Seather, B.S., Johnsen, H.K., Jobling, M.: Seasonal changes in food consumption and growth of Arctic charr exposed to either simulated natural or a 12:12 LD photoperiod at constant water temperature. J. Fish Biol. **48**, 1113–1122 (1996)

Branas, E., Linner, J.: Growth effects in Artic charr reared in cold water: feed frequency, access to bottom and stocking density. Aquac. Int. **8**, 381–389 (2000)

Mørkøre, T., Rorvik, K.A.: Seasonal variations in growth, feed utilisation and product quality of farmed Atlantic salmon *(Salmo salar)* transferred to seawater as 0+ smolts or 1+ smolts. Aquaculture **199**, 145–157 (2001)

Kiessling Anders, J., Pickova, J.G., Eales, B., Dosanjh, D.Higgs: Age, ration level, and exercise affect the fatty acid profile of chinook salmon (Oncorhynchus tshawytscha) muscle differently. Aquaculture **243**(1–4), 345–356 (2005)

Rabah, S.: Light microscope study of Oncorhynchus kisutch muscle development. Egypt. J. Aquat. Res. **30**(1), 3030–3313 (2005)

Johnston, I.A., Manthri, S., Bickerdike, R., Dingwall, A., Luijkx, R., Campbell, P., Nickell, D., Alderson, R.: Growth performance, muscle structure and flesh quality in out-of-season Atlantic salmon *(Salmo salar)* smolts reared under two different photoperiod regimes. Aquaculture **237**, 281–300 (2004)

Oppedal, F., Tim, D., Lars, S.H.: Environmental drivers of Atlantic salmon behaviour in sea-cages: a-review. Aquaculture **311**, 1–18 (2011)

Thakur, D.P., Morioka, K., Itoh, Y., Obatake, A.: Lipid composition and deposition of cultured yelowtail *Seriola quiquradiata* muscle at different anatomical locations in relation to meat texture. Fish. Sci. **69**, 487–494 (2003)

Aursand, M., Beivik, B., Rainuzzo, J.R., Jorgensen, K., Mohr, V.: Lipid distribution and composition of commercially farmed Atlantic salmon (Salmo salar). J. Sci. Food Agric. **64**, 239–248 (1994)

Johnston, I.A., Li, X., Vieira, V.L.A., Nickell, D., Dingwall, A., Alderson, R., Campbell, P., Bickerdike, R.: Muscle and flesh quality traits in wild and farmed Atlantic salmon. Aquaculture **256**, 323–336 (2006)

Metcalfe, B.N., Bull, C.D., Mangel, M.: Seasonal variation in catch-up growth reveals state-dependent somatic allocation in salmon. Evol. Ecol. Res. **4**, 871–881 (2002)

Determination of Antioxidant and Heavy Metals in Cold-Pressed Edible Oils

Enisa Nezirević-Nizić[1(✉)], Selma Čorbo[2], Sedina Podrug[3], and Munevera Begić[2]

[1] Federal Institute of Agropedology, Dolina 6, 71 000 Sarajevo, Bosnia and Herzegovina
enisa1978@hotmail.com
[2] Faculty of Agriculture and Food Science, University of Sarajevo, Zmaja od Bosne 8, 71 000 Sarajevo, Bosnia and Herzegovina
[3] Argeta d.o.o. Sarajevo, Donji Hadžići 138, 71 000 Sarajevo, Bosnia and Herzegovina

Abstract. The aim of this research was to determine the antioxidant and heavy metals content in various types of cold-pressed oil produced in Bosnia and Herzegovina. Samples of cold-pressed pumpkin, sunflower and linseed oils were collected during two seasons and analyzed. During the first season, cold-pressed pumpkin oil had a significantly higher content of total phenols, chlorophylls and carotenoids. Cold-pressed linseed oil, during the first season, had a significantly higher content of chlorophylls and carotenoids in comparison with the same oil in the second season. Cold-pressed sunflower oil from the first season had a significantly higher content of total phenols and chlorophylls in comparison with the sunflower oil from the second season. In the first season, cold-pressed sunflower oil compared to other oils had a significantly higher content of Pb, Cu and Cd. In the second season, cold-pressed pumpkin and linseed oil contained a significantly higher amount of Cu compared to cold-pressed sunflower oil. The content of Cd is significantly higher in cold-pressed linseed and sunflower oil compared with cold-pressed pumpkin oil. In the second season cold-pressed linseed oil had significantly higher amount of heavy metals compared to linseed oil from the first season.

Keywords: Cold-pressed oils · Total phenolic compound · Chlorophylls · Carotenoids · Heavy metals

1 Introduction

Cold-pressed edible oils are obtained from various oilseeds and fruit by mechanical extraction. Oils designated as cold-pressed according to Codex Alimentarius, should be obtained "without altering the oil, by mechanical procedures only, e.g. expelling or pressing, without the application of heat". Cold-pressed edible oils are produced in small plants using small or medium-sized equipment, and processing oilseed from the local environment, applying very simple production technology [1]. In the recent years the consumers are paying more and more attention to those aspects of their life which improves its quality. Therefore, the diet, in addition to the way and conditions of life, is

M. Brka et al. (Eds.): AgriConf 2019, IFMBE Proceedings 78, pp. 295–302, 2020.
https://doi.org/10.1007/978-3-030-40049-1_38

one of the key factors influencing human health and well-being [2]. Cold-pressed oil are important part of diet, primarily due to high-quality compounds such as phenolic compounds, carotenoids, essential fatty acids. Since the increased interest in the production of cold-pressed oils in Bosnia and Herzegovina, the aim of this research was to determine the antioxidant and heavy metals content from different cold-pressed oil produced in Bosnia and Herzegovina. In addition, it is necessary to stimulate the increase of oilseed production and the production of cold-pressed oils in Bosnia and Herzegovina.

2 Materials and Methods

2.1 Materials

Cold-pressed oils from hull-less pumpkin (Golica) and sunflower were obtained from seeds grown in Novi Grad, and linseed oil from linseed grown in Bosanski Petrovac. Basic agro technical measures were applied. The oilseeds were harvested at stage of full technological maturity. The pumpkin seeds were manually harvested in August when the oil content was 40–50%. Sunflower seeds were mechanically harvested in September, oil content was 40–58%, and linseed in early August when the oil content was 35–47%. Upon receipt of raw materials, the seeds were cleaned, weighed (max.1% of impurity), dried (6–8% of moisture in seeds), ground, heated at 20–25 °C, and cold pressed in press type of Kern Kraft, capacity of 40–50 kg/hour). The obtained cold-pressed oils were stored in dark glass bottles at room temperature, in dark green glass bottles of 0.25 to 1000 ml. All samples of oil in the original packaging, are kept shortly (up to several days) in the dark and cold room at a temperature of 5 to 8 °C, until the time for analysis. Samples of cold-pressed oils were taken during two harvest seasons.

2.2 Methods

Total phenol compound content was measured by using spectrophotometer applying method that is based on color reaction of phenols with Folin-Ciocalteau reagent. 1 ml of the sample was mixed with 15 ml of distilled water, 5 ml of Folin-Ciocalteau reagent and with 15 ml of 20% Na_2CO_3 solution. After 2 h of incubation, absorbance of all samples was measured at 765 nm, using SHIMADZU UV-visible spectrophotometer UVmini-1240. Total phenol content was quantified by gallic acid calibration curve [3]. Chlorophyll content was determined, in undiluted sample of oil by measuring absorbance at 670, 630 and 710 nm, by using spectrophotometer. The results were as the content of pheophytin *a* [4]. Carotenoid content was determined, in 10% diluted sample of oil, by measuring absorbance at 445 nm, by using spectrophotometer. The results were as the content of β-carotene [5]. Heavy metals (Pb, Cu, Cd) are determined using atomic absorber spectrophotometer SHIMADZU AA-6650. Preparation and incineration of samples using the Milestone Ethos D microwave oven, 2005 [6]. Statistical analysis and comparison of the results were done by t-test ($p < 0.05$, significance level), by using Excel 2007.

3 Results and Discussions

3.1 Antioxidant Content in Cold-Pressed Oils

The mean values for the content of phenol, chlorophyll, carotenoid and the t-test in the tested samples of cold-pressed oil are shown in Table 1.

Table 1. Average content of antioxidants in cold-pressed oils

Compound (mg/100 g)	I season		
	Pumpkin oil	**Linseed oil**	**Sunflower oil**
Total phenolic compound content	4.48 ± 0.38^{aA}	1.35 ± 0.09^{bA}	3.68 ± 0.90^{aA}
Chlorophyll (feofitin *a*)	26.97 ± 0.18^{aA}	1.42 ± 0.55^{cA}	3.24 ± 0.75^{bA}
Carotenoid (ß-carotenoid)	92.38 ± 0.69^{aA}	12.25 ± 0.28^{bA}	1.26 ± 0.13^{cA}
Compound (mg/ 100 g)	**II season**		
	Pumpkin oil	**Linseed oil**	**Sunflower oil**
Total phenolic compound content	3.70 ± 0.26^{aB}	1.07 ± 0.06^{bA}	0.69 ± 0.24^{bB}
Chlorophyll (feofitin *a*)	18.65 ± 0.49^{aB}	0.68 ± 0.50^{bB}	0.45 ± 0.15^{bB}
Carotenoid (ß-carotenoid)	10.15 ± 0.30^{aB}	2.14 ± 0.08^{bB}	1.30 ± 0.06^{cA}

*means with different lowercase letters indicate significant statistical difference ($p \leq 0.05$) between different oils in one season

*means with different capital letters indicate significant statistical difference ($p \leq 0.05$) between the same oil in two seasons

Total phenolic compound content, chlorophyll and carotenoid content are significantly higher in the first season in comparison with oil obtained in the second season (Table 1). Cold-pressed linseed oil obtained in the first season had significantly higher content of chlorophyll (1.42 mg/100 g) and carotenoid (0.68 mg/100 g) in comparison with linseed oil obtained in the second season (0.68 mg/100 g chlorophyll and 2.14 mg/100 g carotenoid). Likewise, cold-pressed sunflower oil from the first season had significantly higher content of phenolic compounds and chlorophyll as compared to sunflower from the second season. Cold-pressed pumpkin oil, during both season had significantly higher amount of total phenolic compound content, chlorophyll and carotenoid as compared to cold-pressed sunflower and linseed oil. In the first season, linseed had the lowest chlorophyll content (1.42 mg/100 g). This value is significantly lower in comparison to cold-pressed sunflower oil (3.24 mg/100 g). However, in the second season no significant differences were found in chlorophyll content between these cold-pressed oil. Cold-pressed sunflower oil had significantly lower carotenoid content both seasons (1.26 and 1.30 mg/100 g), as compared to cold-pressed linseed oil (12.25 and 2.14 mg/100 g).

Statistical analysis (t-test) showed that the total phenolic compounds content was significantly higher in cold-pressed pumpkin oil as compared to cold-pressed linseed or sunflower oil. Also, cold-pressed pumpkin oil from the second season had a significantly lower phenolic content as compared to the same oil from the first season. Significant differences were found in cold-pressed sunflower oil (3.68 and 0.69 mg/100 g). However, no significant differences were found in cold-pressed linseed oil (1.35 and 1.07 mg/100 g). Siger et al. [7] reported high amount of phenolic compounds in cold-pressed pumpkin oil (2.46 mg/100 g oil), linseed oil (1.14 mg/100 g) and sunflower oil (1.20 mg/100 g). Lower content of phenolic compound in cold-pressed pumpkin oil (0.98 mg/g) reported Parry et al. [8]. The values from our research are in accordance with the results by other authors. Vujasinović [9] reported that total phenolic compound content in extra virgin cold-pressed pumpkin oil ranged between 0.79 and 2.36 mg/100 g, and in cold-pressed pumpkin oil between 0.32 and 0.79 mg/100 g. Higher amount found Ardabili et al. [10] 6.62 mg/100 g oil. The total phenolic compound content in cold-pressed linseed oil was 1.35 mg/100 g and 1.07 mg/100 g (Table 1), which is in accordance with Mińkowski et al. [11] who state phenolic content in linseed oil of 1.17 mg/100 g. Depending on the extraction method, there is a difference in the content of total phenol in linseed oil. Total phenolic compound in linseed oil obtained by double pressing 26 mg/100 g, while oil obtained by single pressing contains 10 mg/100 g of oil and about 2 mg/100 g in triple pressed oil. Losses of total phenols content of linseed oil during triple press is due to application of high pressure and extensive heat [12]. Authors Hamed and Abo-Elwafa [13] cited a total phenols content of 38.46 mg/100 g. Herchi et al. [14] have determined significant seasonal influences on the content of total phenolic acids in linseed (whole seeds) and other bioactive compounds (flavonoids, sterols, tocopherols). The phenolic acid content ranged from 86 to 94 mg/100 g of oil. The difference in chlorophyll content is significant. Cold-pressed oils obtained during the second season have significantly lower chlorophyll content. These results (Table 1) are consistent with similar studies [15], according to which the chlorophyll content in linseed oil ranges from 0.234 to 6.571 mg/100 g of oil, depending on the variety. Also, Choo et al. [16] found 0.08 to 0.576 mg/100 g chlorophyll in linseed oil. Although differences in this study were found to be significant, authors Herchi et al. [14] concluded that there is no significant difference in chlorophyll content in linseed oil by season. During five seasons, the chlorophyll content ranged from 0.22 to 0.28 mg/100 g of oil. Mińkowski et al. [17] reported chlorophyll content in linseed oil of 0.04 mg/100 g of oil. According to Vujasinović [9] the temperature and the time of roasting can effect on the degradation of carotenoids. The content of total carotenoids is the highest (55.91 mg/kg) in oil obtained from pumpkin seed roasted at 90 °C, 50 min, while the lower (38.60 mg/kg) is in oil from seeds roasted at 130 °C, 50 min. The authors explain these changes by the fact that increased temperature and longer roasting time enhance oxidative processes, leading to the oxidation of carotenoids, thereby reducing the nutritional value of the obtained oil. Herchi et al. [14] have determined that there is no significant seasonal effect on carotenoid content in linseed oil. During the five seasons the carotenoid content varied from 0.72 to 0.82 mg/100 g of oil. In another study by Herchi et al. [15] in linseed oil, the carotenoid content varies from 0.752 to 4.715 mg/100 g.

3.2 Heavy Metals Content in Cold-Pressed Oils

Some metals are naturally present in the seeds, and their content in the obtained oil may vary [18]. The content of heavy metals in edible oils is an important for oil quality, in terms of storage and their impact on human nutrition and health. Heavy metals accelerate the oxidation of edible oils [19]. Heavy metals have toxic effects on human health, and their content in oils needs to be controlled in order to prevent excessive accumulation in the food chain [20]. Table 2 shows the mean value of heavy metal content in cold-pressed oils. In the first season, cold-pressed sunflower oil compared to other oils had a significantly higher content of Pb, Cu and Cd. In cold-pressed linseed Pb and Cd were not found. The content of Pb is significantly higher in cold-pressed sunflower oil compared to cold-pressed linseed oil or cold-pressed pumpkin oil. In the second season, cold-pressed pumpkin and linseed oil contained a significantly higher amount of Cu compared to cold-pressed sunflower oil. The content of Cd is significantly higher in cold-pressed linseed and sunflower oil compared with cold-pressed pumpkin oil. Heavy metals, Pb and Cd, were found in linseed oil obtained in the second.

Table 2. Average content of heavy metals in cold-pressed oils

Element (mg/kg)	I season		
	Pumpkin oil	Linseed oil	Sunflower oil
Pb	0.0085 ± 0.00^{bB}	-	0.095 ± 0.00^{aA}
Cu	0.0050 ± 0.00^{bB}	0.0032 ± 0.00^{b}	0.089 ± 0.00^{aA}
Cd	0.0022 ± 0.00^{bA}	-	0.052 ± 0.00^{aA}
Element (mg/kg)	I season		
	Pumpkin oil	Linseed oil	Sunflower oil
Pb	0.0122 ± 0.00^{cA}	0.022 ± 0.00^{bC}	0.049 ± 0.00^{aB}
Cu	0.0169 ± 0.00^{aA}	0.011 ± 0.00^{aB}	0.0083 ± 0.00^{bC}
Cd	0.0043 ± 0.00^{bA}	0.037 ± 0.00^{aB}	0.060 ± 0.00^{aA}

*means with different lowercase letters indicate significant statistical difference ($p \leq 0.05$) between different oils in one season

*means with different capital letters indicate significant statistical difference ($p \leq 0.05$) between the same oil in two seasons

Wroniak et al. [21] in cold-pressed pumpkin oil found 0.10 mg/kg Pb, 1.10 mg/kg Cu and 0.005 mg/kg Cd. Authors state that the content of Cu in the cold-pressed pumpkin oil found has exceeded the limits. These values are higher as compared to results from this study (Table 2). Wroniak et al. [21] stated for cold-pressed sunflower oil, the following values: 0.05 to 0.07 mg/kg Pb, then 0.03 to 0.05 mg/kg Cu, and 0.005 to 0.015 mg/kg Cd. Those values are similar with analyzed cold-pressed sunflower oil. However, Cu content is lower in both seasons, but the content of Cd is higher as compared to the above mentioned research. Pehlivan et al. [20] stated that the content of Pb in cold-pressed sunflower oil ranged from 0.0012 to 0.0026 mg/kg, Cu from 0.0105 to 0.0231 mg/kg, and Cd from 0.0007 to 0.0045 mg/kg. In sunflower oil Farzin and Moassesi [19] found Pb content is 0.01582 mg/kg, Cu 0.00025 mg/kg and Cd 0.00395 mg/kg. Similar results for heavy metal content in sunflower oil have been published by Dugo et al. [22], 0.005 to 0.0155 mg/kg Pb, 0.055 to 0.153 mg/kg Cu, and from 0.00051 to 0.00415 mg/kg Cd. Authors Llorent-Martínez et al. [23] reported in sunflower oil content of Cu from 0.003 to 0.014 mg/kg, while Pb and Cd were not found. The higher content of Cu, from 0.018 to 0.114 mg/kg. For the sunflower cold pressed oil, Krygier et al. [24] indicate 0.030 mg/kg Pb, 0.016 mg/kg Cu and 0.009 mg/kg Cd. These results are lower than the results obtained for sunflower oil (Table 2). The content of individual metals differs depending on the type of oil and the oilseed from which it is obtained. Ashraf [25] based on the analysis of several types of edible oils, lists the following values of heavy metal content. In sunflower oil, the content of Cd is 2.88 mg/kg, followed by Cu of 0.062 mg/kg and Pb in an amount of 0.011 mg/kg. The highest content of Pb was found in sesame oil (0.017 mg/kg), Cu in olive (0.286 mg/kg) and Cd (6.34 mg/kg) in corn. Saastamoinen et al. [26] found that the content of Pb ranged from 0.040 to 0.054 mg/kg and Cd was 0.39 to 0.92 mg/kg. Depending on the variety, the difference in the content of heavy metals can be in the same type of oil. Compared with analyzed linseed oil, the content of heavy metals is lower (0.022 mg/kg Pb) in the second season, and 0.037 mg/kg Cd while Pb and Cd in this oil were not detected in the first season. Ansari et al. [18] state that sunflower varieties accumulate a larger amount of Cd and Pb. The amount of Cd in sunflower oil, depending on the variety, ranges from 1.70 to 6.18 mg/kg, while the content of Pb in oil ranges from 0.79 to 4.4 mg/kg. The presence of Cd can be derived from soil and fertilizer. In addition, metallic contamination may be due to proximity to the industrial zone or highway. Since both Cd and Pb accumulate have toxic effects, the danger lies in the constant consumption of low-content foods Cd [19]. Yankov and Tahsin [27] found that heavy metals accumulate in the upper layer of soil, and a limited amount is distributed in the lower layers. The content of Cu and Zn in sunflower seed is reduced with the distance from the pollutant. The amount of heavy metals that can be transferred from the soil to the seed and then into the oil differs from certain metals. In the example of rapeseed, Darracq et al. [28] state that certain heavy metals (Cr, Hg, Pb) are less transferred from soil to rapeseed, while Cd, Cu, Zn, Ni are much more transferable. In addition to heavy metal transfer from the ground, contamination from the air is possible. Palizban et al. [29] reported that Pb contamination is higher in unwashed oilseeds compared with washed oilseed, because oilseed can contaminated by heavy metals from air.

4 Conclusion

The quality of cold-pressed oil depends on the quality of raw material, purity, maturity, and other operations that precede pressing (drying, storing, etc.). The antioxidant properties of cold-pressed oils depend on the presence of natural antioxidant (phenolic compounds, carotenoids, tocopherols, etc.). During the first season, the highest content of total phenols (4.48 mg/100 g), carotenoids (92.38 mg/100 g) and chlorophyll (26.97 mg/100 g) was found in cold-pressed pumpkin oil. The lowest carotenoid content was reported in cold-pressed sunflower oil (1.26 mg/100 g) during the first season as well chlorophyll (0.45 mg/100 g) during the second season. The highest content of heavy metals is found in cold-pressed sunflower oil compared to other cold-pressed oils. In the second season, the cold-pressed pumpkin oil contained more Pb

(0.0122 mg/kg) and Cu (0.0169 mg/kg) compared to other cold-pressed oils. The content of heavy metals increased in cold-pressed linseed oil in the second season compared to the first season.

References

1. Dimić, E.B., Premović, T.Đ., Takači, A.A., Vujasinović, V.B., Radoćaj, O.F., Dimić, S.B.: Effect of seed quality on oxidative stability of cold-pressed sunflower oil. Chem. Ind. **69**(2), 175–184 (2015)
2. Makała, H.: Cold-pressed oils as functional food. Plant Lipids Science, Technology, Nutritional Value and Benefits to Human Health, pp. 185–200 (2015)
3. Ough, C.S., Amerine, M.A.: Methods for Analysis of Musts and Wines, 2nd edn. Wiley, New York (1998)
4. Pokorny, J., Kalinova, L., Dysseler, P.: Determination of chlorophyll pigments in crude vegetable oils. Pure Appl. Chem. **67**, 1781–1787 (1995)
5. British Standard Methods (BSM) of analysis. Determination of carotene in vegetable oils, B.S 684, section 2.20 (1977)
6. Welz, B., Sperling, M.: Atomic Absorption Spectrometry. Wiley-VCH, Weinheim (1999). ISBN 3-527-28571-7
7. Siger, A., Nogala-Kalucka, M., Lampart-Szczapa, E.: The content and antioxidant activity of phenolic compounds in cold-pressed plant oils. J. Food Lipids **15**, 137–149 (2008)
8. Parry, J., Hao, Z., Luther, M., Su, L., Zhou, K., Yu, L.: Characterization of cold-pressed onion, parsley, cardamom, mullein, roasted pumpkin, and milk thistle seed oils. J. Am. Oil. Chem. Soc. **83**(10), 847–854 (2006)
9. Vujasinović, V.B.: Influence of thermal treatment of pumpkin *seed Cucurbita pepo L.* on nutritive value and oxidative stability of oil. Ph. D. thesis. University of Novi Sad, Faculty of technology Novi Sad, pp. 28–29 (2011)
10. Ardabili, A.G., Farhoosh, R., Khodaparast, M.H.H.: Chemical composition and physico-chemical properties of pumpkin seeds (*Cucurbita pepo subsp. pepo var. Styriaka*) grown in Iran. J. Agr. Sci. Tech. **13**, 1053–1063 (2011)
11. Mińkowski, K., Zawada, K., Ptasznik, S., Kalinowski, A.: Effect of phenolic compounds in seeds on oxidative stability and antiradical activity of n-3-PUFA-rich oils pressed from them. Żywność **4**(89), 118–132 (2013)
12. Kasote, D.M., Badhe, Y.S., Hegde, M.V.: Effect of mechanical press oil extraction processing on quality of linseed oil. Ind. Crops Prod. **42**, 10–13 (2013)
13. Hamed, S.F., Abo-Elwafa, G.A.: Enhancement of oxidation stability of flax seed oil by blending with stable vegetable oils. J. Appl. Sci. Res. **8**(10), 5039–5048 (2012)
14. Herchi, W., Bahashwan, S., Sakouhi, F., Boukhchina, S.: Influence of harvest year in the physicochemical properties and antioxidant activity of flaxseed hull oils from Tunisia. Food Sci. Technol. **35**(1), 175–182 (2015)
15. Herchi, W., Al Hujaili, A.D., Sakouhi, F., Sebei, K., Trabelsi, H., Kallel, H., Boukhchina, S.: Flaxseed hull: chemical composition and antioxidant activity during development. J. Oleo Sci. **63**(7), 1–9 (2014)
16. Choo, W.S., Birch, J., Dufour, J.P.: Physicochemical and quality characteristics of cold-pressed flaxseed oils. J. Food Compos. Anal. **20**, 202–211 (2007)
17. Mińkowski, K., Grześkiewicz, S., Jerzewska, M., Ropelewska, M.: Chemical composition profile of plant oils with high content of linolenic acids. Żywność **6**(73), 146–157 (2010)

18. Ansari, R., Gul Kazi, T., Jamali, M.K., Arain, M.B., Wagan, M.D., Jalbani, N., Afridi, H.I., Shah, A.Q.: Variation in accumulation of heavy metals in different verities of sunflower seed oil with the aid of multivariate technique. Food Chem. **115**, 318–323 (2009)
19. Farzin, L., Moassesi, M.E.: Determination of metal contents in edible vegetable oils produced in Iran using microwave-assisted acid digestion. J. Appl. Chem. Res. **8**(3), 35–43 (2014)
20. Pehlivan, E., Arslan, G., Gode, F., Altun, T., Özcan, M.M.: Determination of some inorganic metals in edible vegetable oils by inductively coupled plasma atomic emission spectroscopy (ICP-AES). Grasas Aceites **59**(3), 239–244 (2008)
21. Wroniak, M., Kwiatkowska, M., Krygier, K.: Characteristic of selected cold pressed oils. Żywność **2**(47), 46–58 (2006)
22. Dugo, G., La Pera, L., La Torre, G.L., Giuffrida, D.: Determination of Cd(II), Cu(II), Pb(II), and Zn(II) content in commercial vegetable oils using derivative potentiometric stripping analysis. Food Chem. **87**, 639–645 (2004)
23. Llorent-Martínez, E.J., Ortega-Barrales, P., Fernández-de Córdova, M.L., Domínguez-Vidal, A., Ruiz-Medina, A.: Investigation by ICP-MS of trace element levels in vegetable edible oils produced in Spain. Food Chem. **127**, 1257–1262 (2011)
24. Krygier, K., Wroniak, M., Dobczyński, K., Kiełt, I., Grześkiewicz, S., Obiedziński, M.: Characteristic of commercial cold pressed vegetable oils. Rośliny Oleiste **19**, 573–582 (1998)
25. Ashraf, M.W.: Levels of selected heavy metals in varieties of vegetable oils consumed in Kingdom of Saudi Arabia and health risk assessment of local. Nutr. Food Sci. **1**(3) (2014)
26. Saastamoinen, M., Pihlava, J.M., Eurola, M.: Content of lignan, cadmium and some other chemical components in linseed and oil hemp varieties in south-western part of Finland. NJF Rep. **6**(2), 112–115 (2010)
27. Yankov, B., Tahsin, N.: Accumulation and distribution of Pb, Cu, Zn and Cd in sunflower (*Helianthus annuus L.*) grown in an industrially polluted region. Helia **24**(34), 131–136 (2001)
28. Darracq, S., Bernhard-Bitaud, C., Bourrie, B., Evrard, J., Burghart, P., Pages, X., Lacoste, F.: Heavy metals transfer from soil to rapeseed oil. In: Waste Contaminants: Lifecycle and Entry into Food Chain, pp. 61–64 (2015)
29. Palizban, A., Badii, A., Asghari, G., Mardani-Nafchi, H.: Lead and cadmium contamination in seeds and oils of *Brassica napus L.* and Carthamus tinctorius grown in Isfahan province/Iran. Iran. J. Toxicol. **8**(27), 1196–1202 (2015)

Fatty Acid Composition and Stability of Cold-Pressed Vegetable Oils

Munevera Begić[1(✉)], Enisa Nezirević-Nizić[2], Selma Čorbo[1], Sedina Podrug[3], Zilha Ašimović[1], and Ševal Muminović[1]

[1] Faculty of Agriculture and Food Science, University of Sarajevo, Zmaja od Bosne 8, 71 000 Sarajevo, Bosnia and Herzegovina
m.begic@ppf.unsa.ba
[2] Federal Institute of Agropedology, Dolina 6, 71 000 Sarajevo, Bosnia and Herzegovina
[3] Argeta d.o.o. Sarajevo, Donji Hadžići 138, 71 000 Sarajevo, Bosnia and Herzegovina

Abstract. The aim of this study was to determine the composition of fatty acids and stability of different cold-pressed vegetable oils manufactured in Bosnia and Herzegovina. Three kinds of cold-pressed oils were used during the experiment (pumpkin oil, linseed oil and sunflower oil). The composition of fatty acids (both saturated and unsaturated), peroxide value, and content of free fatty acids has been determined on tested samples. The largest amount of saturated fatty acids during the two-year research was found in pumpkin oil with the largest share of palmitic acid. The content of unsaturated fatty acids in all the tested samples was larger in the first year of research. The highest content of unsaturated fatty acids in two years of research was found in linseed oil with largest share of α-linolenic acid. The highest peroxide value and content of free fatty acids were found in linseed oil. Both parameters that were tested had lower values in the second year of testing in all the samples.

Keywords: Cold-pressed oils · Peroxide value · Fatty acid composition · Free fatty acid contents

1 Introduction

Cold-pressed edible oils are products extracted from certain raw materials through cold pressing at the temperature of up to 50 °C. Fatty acids are essential part of human nutrition and are not only a source of energy but also important components of cell material. Fatty acids are part of phospholipids that form cell membranes. Most vegetable oils contain two essential fatty acids – linoleic and α-linolenic acid. Besides that, oils contain antioxidants, vitamin E, carotenoids and phenolic acids [1]. These oils have been known as a functional nutrition because of the presence of bioactive substances. Certain components present in oils in nutrition, like essential fatty acids, sterols, and tocopherols, have very positive health benefit effect against cardiovascular disease, cancer, obesity and other [2].

M. Brka et al. (Eds.): AgriConf 2019, IFMBE Proceedings 78, pp. 303–312, 2020.
https://doi.org/10.1007/978-3-030-40049-1_39

Pumpkin oil is among the oldest edible oils [3]. Pumpkin seed which is an old name for oil pumpkin seed, was actually the first oil seed in our region, i.e. the first raw material of the oil industry. Pumpkin seed is rich in oil, protein, fiber, vitamins and minerals. The content of the protein in pumpkin seed is, for the most part, around 35%, with albumin and globulin as dominant proteins that make up for 59% of the total protein [4]. Limiting amino acids of the protein of pumpkin seed bread are lysine, valine, threonine and isoleucine [5, 6]. Cold-pressed pumpkin oil has the following beneficial properties: unsaturated fatty acids lower high cholesterol levels; slow down vascular congestion processes, while antioxidants (vitamin A, E and selenium) slow down aging processes in humans [7].

Linseed oil can be a complete and balanced source of fatty acids. The oil contains ω-3 and ω-9 fatty acids and apart from the traditional use, new applications are found in other areas as well, like: food industry, medicine, cosmetics, and so on. Also, it is widely used in the production of paint and varnish that dry fast creating a hard layer. It is a typical representative of fast-dry oils for its high content of unsaturated fatty acids. From medical point of view, linseed oil cures skin disease like eczema, psoriasis, acne or sun burns [8].

High quality of sunflower oil is mostly derived from unsaturated fatty acids, tocopherols, sterols, carotenoids and other compounds. Sunflower oil plays an important role in human nutrition for its high energy and biological value, and its unsaturated fatty acids (linoleic and oleic) contribute to the reduction of cardiovascular diseases. Sunflower seed is also appreciated for its protein content rich in essential amino acids [9]. Linoleic acid, that sunflower oil is rich in, is very important for human organism because it affects the cholesterol metabolism and stimulates its excretion [10].

2 Materials and Methods

2.1 Materials

Cold-pressed edible non-refined vegetable oils obtained from hull-less pumpkin (seeds without hulls), standard linoleic type of sunflower (seeds with hulls) and linseed (seeds without hulls) seeds have been used for the investigation of this research. Pumpkin and sunflower oil have been obtained from independent growers and manufacturers of Agricultural Cooperative "Agrojapra" from Donji Agići, Novi Grad, and linseed oil from Bosanski Petrovac. Adequate agrotechnical measures for the cultivation of different oil seeds have been carried out at the specific locations. Picking of the pumpkin was done in the full maturity phase (in August) when the content of oil in the seeds is between 40 and 50%. The fruits were hand-picked in order to avoid the damage. Harvesting of sunflower (the beginning of September) and linseeds (July and August) was done mechanically at the moment of complete technological maturity, when the sunflower seed oil content is at 40 to 58%, and linseed oil content at 35 to 47%. Six samples of pumpkin seed, sunflower seed, and linseed oil have been taken for analysis, obtained by cold pressing methods. All the cold-pressed oil samples were kept in the original packaging and adequately stored (dark and cold rooms at the temperature of 5 to 8 °C) up until the moment of testing.

2.2 Methods

Methods for quality testing of pumpkin, linseed, and sunflower cold-pressed oils were determining composition of fatty acids, peroxide value and free fatty acids content. To determine the composition of fatty acids gas chromatograph GC/MS-6890 II, Hewlett-Packard with mass selective detector (MSD) 6890 II: FKKT-UL was used [11]. Capillary columns: SP 2560, 100 m × 0,25 mmlD, 0,20 μm; Detector: flame ionization detector FID, 260 °C, separation 100.1; Carrier gas: helium, 20 cm^3/s. Temperature of the owen was 140 °C/min (DGF, C-VI 10a, 1985). Determination of peroxide value was performed by Wheeler method [12], and the results are expressed in mmol/kg of oils. Free fatty acids (FFA) in the cold-pressed oil samples are determined by standard titrimetric method [13], and the results are expressed in percentage of the amount of oleic acid present. For all the tested parameters mean value and standard deviation have been taken into account. Statistical analysis and comparison of the results were done by t-test ($p < 0.05$, significance level), by using Excel 2007.

3 Results and Discussion

The content of saturated fatty acids (SAFA) in the tested samples of pumpkin, linseed and sunflower oil are presented in Table 1.

Table 1. Saturated fatty acid composition of cold-pressed oils

Fatty acid (%m/m)	I year		
	Pumpkin oil	Linseed oil	Sunflower oil
Lauric acid C12:0	0.88 ± 0.01^{aA}	–	0.60 ± 0.05^{bA}
Myristic acid C14:0	0.38 ± 0.08^{aA}	–	0.28 ± 0.01^{aA}
Palmitic acid C16:0	12.27 ± 0.25^{aA}	7.30 ± 0.10^{bA}	7.51 ± 0.10^{bA}
Stearic acid C18:0	5.11 ± 0.10^{aB}	5.68 ± 0.05^{aA}	3.91 ± 0.09^{bB}
$\sum$SAFA	$\mathbf{18.64 \pm 0.10^{aA}}$	$\mathbf{12.98 \pm 0.23^{bA}}$	$\mathbf{12.30 \pm 0.25^{bA}}$
Fatty acid (%m/m)	II year		
	Pumpkin oil	Linseed oil	Sunflower oil
Lauric acid C12:0	0.78 ± 0.01^{aA}	–	0.09 ± 0.10^{bB}
Myristic acid C14:0	0.23 ± 0.20^{aA}	–	0.17 ± 0.03^{bB}
Palmitic acid C16:0	12.00 ± 0.11^{aA}	6.17 ± 0.04^{bA}	6.65 ± 005^{bB}
Stearic acid C18:0	6.89 ± 0.10^{aA}	4.83 ± 0.35^{bB}	4.32 ± 0.10^{bA}
$\sum$SAFA	$\mathbf{19.90 \pm 0.12^{aA}}$	$\mathbf{11.00 \pm 0.42^{bB}}$	$\mathbf{11.23 \pm 0.12^{bB}}$

***means with different lowercase letters indicate significant statistical difference ($p \leq 0.05$) between different oils in one season

***means with different capital letters indicate significant statistical difference ($p \leq 0.05$) between the same oil in two seasons

From the previous table it can be noted that the total content of saturated fatty acids in pumpkin oil in the first year of investigations amounts 18.64%, and slightly higher in the second year when it is 19.90%, and the difference is not statistically significant.

Pumpkin oil, in comparison to the linseed (7.30%) and sunflower (7.51%) oil has significantly higher content of palmitic acid (12.27%). Total amount of saturated fatty acids is significantly higher in pumpkin oil (18.64%) in comparison to linseed (12.98%), i.e. sunflower (12.30%) oil. The results of research in terms of content of saturated fatty acids in pumpkin oil are in accordance with the research conducted by Tsaknis et al. [14], where they established the content of palmitic acid is 12.70% and stearic 6.00%. Similar results for pumpkin oil were discovered by Radočaj and Dimić [15] who obtained the following results: 0.04% lauric, 0.12% myristic, 13.35% palmitic and 6.40% stearic acid. The research conducted by Neđeral Nakić et al. [16] shows that the content of palmitic acid is rather similar in the oil extracted from naked and husk pumpkin seeds. The content of palmitic acid equals 12.01% (naked pumpkin seeds) and 12.00% (husk pumpkin seeds), and stearic 5.25% (naked pumpkin seeds) and 4.97% (husk pumpkin seeds).

The total amount of saturated fatty acids of the tested linseed oil is 12.98% in the first year of research and 11.00% in the second. The ratio of certain fatty acids in linseed oil is dependent on the time of harvest, i.e. temperature. The content of saturated fatty acids (palmitic and stearic) changed during the two seasons due to changes in temperature during the maturing season. Linseed oil contains approximately 11.01% of saturated fatty acids, i.e. palmitic and stearic acids [17]. In the linseed oil, the content of saturated fatty acids varies from 7.97 to 12.30%, and unsaturated from 84.90 to 92.03% [18]. El-Beltagi et al. [19] notes the content of palmitic acid of 7.1% and stearic 3.7% in linseed oil. Radočaj and Dimić [15] found the following values in the linseed oil:

5.04% of palmitic and 3.45% of stearic acid.

The total amount of saturated fatty acids in the tested sunflower oil is significantly higher in the first (12.30%) than in the second year of research (11.23%). The tested sample of the sunflower oil contains 7.51% of palmitic and 3.91% of stearic acid in the first, i.e. 6.65% of palmitic and 4.32% of stearic acid in the second year of research. Chowdhury et al. [20] note in their research the content of palmitic acid at 6.52% and stearic at 1.98% in sunflower oil. Similar is claimed by Krygier et al. [21] for cold pressed sunflower oil, where they found 6.2% of palmitic and 4.2% of stearic acid.

The results of research of the content of unsaturated fatty acids in cold-pressed pumpkin, linseed and sunflower oil are presented in Table 2.

Total amount of unsaturated fatty acids in pumpkin oil composition is 81.45% in the first and 80.13% in the second year of research. Based on the analysis of the three kinds of cold-pressed oils, it has been established that pumpkin oil has the highest amount of oleic acid (Table 2), in comparison to other oils. The content of linoleic acid is the highest in comparison to the results of other conducted research, while the content of oleic acid is lower (Table 2). Lower content of linoleic acid in the pumpkin oil was noted by Rafalowski et al. [22], at the amount from 39 to 42%, and higher content of oleic acid (36 to 37%). Similar is noted by Tsaknis et al. [14], according to whom pumpkin oil contains 42.10% of linoleic and 38.10% of oleic acid. The results of research conducted by Radočaj and Dimić [15] show 47.52% of linoleic and 31.02% of oleic acid in pumpkin oil. However, the highest content of linoleic acid in pumpkin oil was found by Gulay Kirbaslar et al. [23] at the amount of 70.17%, but they also found

Table 2. Unsaturated fatty acid composition of cold-pressed oils

Fatty acid (%m/m)	I year		
	Pumpkin oil	Linseed oil	Sunflower oil
Oleic acid C18:1	33.72 ± 0.07^{aA}	26.19 ± 0.06^{bA}	27.04 ± 0.03^{bA}
Linoleic acid C18:2	47.73 ± 0.06^{bB}	16.48 ± 0.16^{cA}	60.60 ± 1.56^{aA}
α-linolenic acid C18:3	Trace	55.23 ± 0.20^{A}	Trace
∑MUFA + PUFA	**81.45 ± 0.17^{cA}**	**97.90 ± 0,70^{aA}**	**87.64 ± 0.84^{bA}**
Fatty acid (%m/m)	II year		
	Pumpkin oil	Linseed oil	Sunflower oil
Oleic acid C18:1	28.93 ± 1.00^{aB}	22.64 ± 0.45^{cB}	25.59 ± 0.40^{bB}
Linoleic acid C18:2	51.20 ± 0.26^{bA}	15.74 ± 0.05^{cB}	56.39 ± 0.40^{aB}
α-linolenic acid C18:3	Trace	50.60 ± 0.25^{B}	Trace
∑MUFA + PUFA	**80.13 ± 0.77^{cB}**	**88.98 ± 0.75^{aB}**	**81.98 ± 0.80^{bB}**

***means with different lowercase letters indicate significant statistical difference ($p \leq 0.05$) between different oils in one season
***means with different capital letters indicate significant statistical difference ($p \leq 0.05$) between the same oil in two seasons
*Sum of monounsaturated - MUFA and polyunsaturated – PUFA fatty acids

low content of oleic acid (17.62%). Unlike the above named authors, Rezig et al. [24] found 34.77% of linoleic and 44.11% of oleic acid in pumpkin oil. The content of fatty acids differs in pumpkin from species to species. Stevenson et al. [25] found that the content of linoleic acid varies from 36.2 to 62.8%, and oleic from 17.0 to 39.5%, based on the tests performed on 12 different pumpkin species. Neđeral Nakić et al. [16] for pumpkin oil, notes the following values of unsaturated fatty acids: 46.58% (naked pumpkin seeds) and 51.51% (husk pumpkin seeds) of linoleic acid; 35.12% (naked pumpkin seeds) and 30.46% (husk pumpkin seeds) of oleic acid.

The total content of unsaturated fatty acids in linseed oil equals 97.90% in the first and 88,98% in the second year of research. Linseed oil is distinguished by a high amount of α-linolenic acid whose content equals 55.23% in the first and 50.60% in the second year. In the first year of research linseed oil contained a total of 26.19% of monounsaturated and 71.71% of polyunsaturated acids. In the second year of research the content of monounsaturated acids was 22.64% and 66.34% of polyunsaturated. The noted results are in accordance with the results of Pop et al. [17] which show the total content of unsaturated fatty acids at 88.97% in linseed oil, i.e. 53.21% of α-linolenic, 18.51% of oleic and 17.25% of linoleic acid. The authors of similar research [18] on the linseed, that contains dominantly unsaturated fatty acids, note values of α-linolenic acid from 48.08 to 57.58%, and linoleic from 11.18 to 16.30%. The specified values are in accordance with the obtained results (Table 2). Radočaj and Dimić [15] established the following amounts of acids in linseed oil: 53.49% of α-linolenic, 18.07% of oleic and 17.07% of linoleic acid. The highest content of α-linolenic acid for linseed oil was found by Choo et al. [16], which ranged from 51.80 to 60.42% of α-linolenic acid, from 13.44 to 22.21% of oleic and from 15.17 to 17.44% of linoleic acid. Mińkowski et al. [27] found the same results when conducting the research on linseed oil and the values are as follows: 52.7% of α-linolenic, 19.7% of oleic and 16.2% of linoleic acid. Lower content of α-linolenic acid in the linseed oil is noted by El-Beltagi et al. [19], according to whom the content of α-linolenic acid equals 30.48%, of oleic 22.0%, and linoleic 18.3%. Similar results were found by Rusníková et al. [28] where they noted smaller content of α-linolenic acid at 38.0%, but higher content of linoleic acid (27.06%) and oleic as well (15.06%). The effect of the location on the fatty acid composition was tested by Zając et al. [29], where they found that the location has a small impact on the quality of the linseed oil. In another study Zając et al. [30] confirmed once again that the species of linseed have a larger impact on the composition of fatty acids than the season of the year has.

Sunflower oil have a high content of polysaturated fatty acids (linoleic). In the first year this oil showed a content of 60.60%, which is statistically significant in comparison to the second year (56.39%). The content of monounsaturated (oleic) fatty acids in this oil equals 27.04%. Chowdhury et al. [20] note the content of linoleic at 46.02%, and oleic at 45.39% in sunflower oil. In addition, the results of research conducted by Rusníková et al. [28] show that the content of linoleic acid in sunflower oil equals 50.46% and oleic 28.62%. Similar results have been found by Gulay Kirbaslar et al. [23] for sunflower oil, according to whom the content of linoleic acid equals 48.58%, and oleic 32.69%. Lower values of linoleic acid for sunflower oil have been found by Khoufi et al. [31] where they vary from 9.34 to 13.52%. Higher content of linoleic acid for cold pressed sunflower oil is noted by Krygier et al. [21], at the amount of 66.1%, and 18.6% of oleic acid. The results of research conducted by Onemli [32] show that higher daily temperatures, and lower oscillations in temperature during the seed development contribute to the higher content of oleic acid in sunflower oil, i.e. higher temperature has a positive effect on the fatty acid composition. Turhan et al. [33] found that the location, i.e. temperature has a significant effect on the composition of fatty acids (linoleic, oleic, palmitic and stearic) in sunflower oil. Differences in content of linoleic acid vary from 52.94 to 64.54%, and oleic from 21.87 to 36.25%, stearic from 3.30 to 4.87% and palmitic from 5.71 to 6.85%.

The results of peroxide value and content of free fatty acids in the tested cold-pressed oils are presented in Table 3.

Table 3. Peroxide value and free fatty acids of cold-pressed oils

Parameter	I year		
	Pumpkin oil	Linseed oil	Sunflower oil
Peroxide value (mmol/kg)	6.53 ± 0.14^{aA}	6.75 ± 0.12^{aA}	5.60 ± 0.30^{aA}
Free fatty acids (% oleic acid)	0.22 ± 0.02^{aA}	0.23 ± 0.02^{aA}	0.12 ± 0.03^{aA}
Parameter	II year		
	Pumpkin oil	Linseed oil	Sunflower oil
Peroxide value (mmol/kg)	2.22 ± 0.06^{bC}	5.54 ± 0.55^{bA}	4.79 ± 0.16^{bB}
Free fatty acids (% oleic acid)	0.17 ± 0.02^{aA}	0.19 ± 0.01^{aA}	0.11 ± 0.01^{aA}

***means with different lowercase letters indicate significant statistical difference ($p \leq 0.05$) between different oils in one season

***means with different capital letters indicate significant statistical difference ($p \leq 0.05$) between the same oil in two seasons

High content of unsaturated fatty acids makes cold pressed oils susceptible to oxidative changes, especially the linseed oil containing α-linolenic acid. The results of the research show that the peroxide value in oils equals 6.53 mmol/kg for pumpkin oil, 5.60 mmol/kg for sunflower oil and 6.75 mmol/kg for linseed oil (Table 3). Peroxide values are significantly higher in the first year, and linseed oil had a higher peroxide value in comparison to other oils during the whole period. Grujić et al. [34], show the peroxide value of 0.23 mmol/kg of oil for sunflower oil before bottling, and a value of 7.80 mmol/kg of oil after 6-month storage in dark glass bottle, i.e. 8.60 mmol/kg of oil kept in PET bottles. Peroxide value of pumpkin seed oil was 0.74 mmol/kg of oil, while sunflower oil has a value of 1.08 in regular sunflower and up to 2.00 mmol/kg of oil in high oleic hybrids [35]. Peroxide values for cold-pressed pumpkin oil equals 2.17 mmol/kg in average, according to Vujasinović [36], while the value varies from 0.99 to 3.15 mmol/kg. For virgin pumpkin oil average value equals 2.54 mmol/kg, and it varies from 1.00 to 7.80 mmol/kg. Sunflower seed of oleic type, after storage for 12 to 24 months before the processing, has a lower peroxide values of 1.2, i.e. 1.3 mmol/kg of oil, while the α-linolenic type has a peroxide value significantly higher and it varies from 8.3 to 12.8 mmol/kg of oil. Besides that, processing of pumpkin oil, i.e. toasting before pressing influences the peroxide value. Vujasinović et al. [37] state that the lower peroxide value was noted in untreated pumpkin seed (0.68 mmol/kg). In the seeds toasted at 130 °C for 60 min, peroxide value was slightly higher (2.55 mmol/kg), and the highest in the seeds toasted at 130 °C for 30 min (2.92 mmol/kg). Storing oil at high temperatures causes increase of peroxide value. In linseeds oil, stored at 18 °C peroxide value increases in six months and reaches a value of 14 mmol/kg, while at 6 °C after the same period of time peroxide value reaches a value of 10 mmol/kg of oil [38]. The content of free fatty acids in cold pressed oils extracted from sunflower seeds was equal to 0.12% in the first year, and for pumpkin and linseed oil it varied from 0.22 to 2%. The content of free fatty acids, in all tested samples, was lower in the second year of research.

Similar results to ours can be found at Poaina et al. [39]. According to their allegations the content of free fatty acids in pumpkin seed oil was between 0.68% and 0.72% in sunflower oil. These high values are a consequence of hydrolytic activities of lipolytic enzymes during the preparation of the seeds for processing. Free fatty acids are responsible for development of unpleasant taste (hydrolytic rancidity), and they are more sensitive to oxidation. They are removed from the oil by refining. Therefore, any oil acidity increase must be avoided. Vidrih et al. [35] established the content of free fatty acids (% of oleic acid) in sunflower oil, which, according to them, ranges from 0.05% (regular) to 0.032% (high oleic hybrid), and 0.44% for pumpkin oil.

The content of free fatty acids in pumpkin oil varies from 0.38% to 0.39% [15, 40]. However, slightly higher contents are noted by Tsaknis et al. [14], where raw pumpkin oil contains 0.97% of free fatty acids, and 0.04% after refining. The research results are lower for pumpkin oil in comparison to the stated results (0.22% in the first and 0.17% in the second year). The content of free fatty acids in linseed oil, according to Radočaj and Dimić [15], equals 0.10%, and in pumpkin oil 0.38%. Variations in content of free fatty acids in linseed oil varies from 0.25% to 0.98% [16].

4 Conclusions

The climate where oil samples were taken from has certain agro ecological and climatic conditions for cultivation of pumpkin, sunflower and linseed. Processing of oil seeds was done by cold pressing and satisfying quality results were found, especially with pumpkin oil (18.64% of SAFA in the first and 19.90% in the second year). Linseed and sunflower oil have the largest content of saturated fatty acids in the first year (12.98% and 12.30%) in comparison to the second year of research (11.00% and 11.23%). With all the tested samples, during the two years of research, the largest content of saturated fatty acids was the palmitic acid. The highest content of unsaturated fatty acids in both years of research was found in linseed oil (97.90% in the first and 88,98% in the second year of research), and the content of α-linolenic acid was dominant. In the first year of testing, pumpkin and sunflower oil had the highest content of unsaturated fatty acids (linoleic acid for the most part) in comparison to the second year. Based on the stated results, it can be concluded that all the tested oil samples showed a better stability in the second year of testing.

References

1. Perry, J., Hao, Z., Luther, M., Su, L., Zhou, K., Yu, L.: Characterization of cold-pressed onion, parsley, cardamom, mullein, roasted pumpkin, and milk thistle seed oils. J. Am. Oil Chem. Soc. **83**(10), 847–854 (2006)
2. Makała, H.: Cold-pressed oils as functional food. Plant Lipids Science, Technology, Nutritional Value and Benefits to Human Health, pp. 185–200 (2015)
3. Marković, P.: The development of oil production in Yugoslavia. Business community of vegetable oil and fat producers of Yugoslavia, Belgrade, Serbia (1988)
4. Idouraine, A., Kohlhepp, E.A., Weber, C.W., Warid, W.A., Martinez-Tellez, J.J.: Nutrient constituents from eight lines of naked squash (*Cucurbita pepo L.*). J. Agric. Food Chem. **44**, 721–724 (1996)
5. Štrucelj, D.: Contribution to the knowledge of lipid and protein components of pumpkin seeds and changes in the processing. Ph.D. thesis. Faculty of Food Technology and Biotechnology University of Zagreb, Zagreb, Croatia (1981)
6. Mansour, E.H., Dworshak, E., Gasi, A., Barna, E., Gergely, A.: Nutritive value of pumpkin (*Cucurbita pepo Kakai 35*) seed products. J. Agric. Food Chem. **61**, 73–78 (1993)
7. Berenji, J.: Production of pumpkin seeds and pumpkin oil. Development Foundation of Vojvodina "Halo" Subotica, pp. 4–22 (2014)
8. Šimetić, S.: Flax in production and use. Seed Prod. **25**(3–4), 217–221 (2008)
9. Vratarić, M.: Importance of sunflower as a culture and production of sunflower in the world and in the Republic of Croatia. In: Sunflower *Helianthus annuus L.*, pp. 1–13 (2004)
10. Krizmanić, M., Mijić, A., Liović, I., Sudarić, A., Sudar, R., Duvnjak, T., Krizmanić, G., Bilandžić, M.: Environmental impact on oil content and fatty acid composition of new OS-hybrid combinations of sunflower. Agriculture **19**(1), 41–47 (2013)
11. Härtig, C.: Rapid identification of fatty acid methyl esters using a multidimensional gas chromatography-mass spectrometry database. J. Chromatogr. A **1**(1177), 159–169 (2008)
12. ISO-International standard of animal and vegetable oils and fats. Determination of peroxide value. ISO, Genève, Switzerland 3960 (1977)

13. ISO-International standard of animal and vegetable oils and fats. Determination of acid value and of acidity. ISO Genève, Switzerland 660 (1983)
14. Tsaknis, J., Lalas, S., Lazos, E.S.: Characterization of crude and purified pumpkin seed oil. Grasas Aceites **48**(5), 267–272 (1997)
15. Radočaj, O., Dimić, E.: Physico-chemical and nutritive characteristics of selected cold-pressed oils found in the European market. La Rivista Italiana Delle Sostanze Grasse **XC**, 219–228 (2013)
16. Neđeral Nakić, S., Rade, D., Škevin, D., Štrucelj, D., Mokrovčak, Ž., Bartolić, M.: Chemical characteristics of oils from naked and husk seeds of *Cucurbita pepo L.* Eur. J. Lipid Sci. Technol. **108**, 936–943 (2006)
17. Pop, G., Alexa, E., Laza, A., Mihoc, M., Militaru, A.: Nutritional quality of linseed and oil hemp varieties. In: 6th Central European Congress on Food, Serbia, pp. 262–267 (2012)
18. Bayrak, A., Kiralan, M., Arslan, N., Ipek, A., Cosge, B., Khawar, K.M.: Fatty acid compositions of linseed (*Linum Usitatissimum L.*) genotypes of different origin cultivated in Turkey. Biotechnol. Biotechnol. Equip. **24**(2), 1836–1842 (2010)
19. El-Beltagi, H.S., Salama, Z.A., El-Hariri, D.M.: Evaluation of fatty acids profile and the content of some secondary metabolites in seeds of different flax cultivars (*Linum usitatissimum L.*). Gen. Appl. Plant Physiol. **33**(3–4), 187–202 (2007)
20. Chowdhury, K., Banu, L.A., Khan, S., Latif, A.: Studies on the fatty acid composition of edible oil. Bangladesh J. Sci. Ind. Res. **42**(3), 311–316 (2007)
21. Krygier, K., Wroniak, M., Dobczyński, K., Kiełt, I., Grześkiewicz, S., Obiedziński, M.: Characteristic of commercial cold pressed vegetable oils. Rośliny Oleiste **19**, 573–582 (1998)
22. Rafalowski, R., Żegarska, Z., Kuncewicz, A., Borejszo, Z.: Fatty acid composition, tocopherols and β-carotene content in Polish commercial vegetable oils. Pak. J. Nutr. **7**(2), 278–282 (2008)
23. Gülay Kırbaslar, F., Türker, G., Özsoy-Günes, Z., Ünal, M., Dülger, B., Ertas, E., Kızılkaya, B.: Evaluation of fatty acid composition, antioxidant and antimicrobial activity, mineral composition and calorie values of some nuts and seeds from Turkey. Rec. Nat. Prod. **6**(4), 339–349 (2012)
24. Rezig, L., Chouaibi, M., Msaada, K., Hamdi, S.: Chemical composition and profile characterization of pumpkin (*Cucurbita maxima*) seed oil. Ind. Crop. Prod. **37**, 82–87 (2012)
25. Stevenson, D.G., Eller, F.J., Wang, L., Jane, J.L., Wang, T., Inglett, G.E.: Oil and tocopherol content and composition of pumpkin seed oil in 12 cultivars. J. Agric. Food Chem. **55**, 4005–4013 (2007)
26. Choo, W.S., Birch, J., Dufour, J.P.: Physicochemical and quality characteristics of cold-pressed flaxseed oils. J. Food Compos. Anal. **20**, 202–211 (2007)
27. Mińkowski, K., Zawada, K., Ptasznik, S., Kalinowski, A.: Effect of phenolic compounds in seeds on oxidative stability and antiradical activity of n-3-PUFA-rich oils pressed from them. Żywność **4**(89), 118–132 (2013)
28. Rusníková, L., Straková, E., Suchý, P.: Assessment of oils of selected legumes and oil plants with regard to animal nutrition. Acta Vet. Brno **82**, 283–288 (2013)
29. Zając, T., Borowiec, F., Micek, P.: Comparison of productivity, chemical composition and fatty acid profile of yellow and brown linseeds. Rośliny Oleiste **22**, 442–453 (2001)
30. Zając, T., Klima, K., Borowiec, F., Witkowicz, R., Barteczko, J.: Yielding of linseed varieties in various site conditions. Rośliny Oleiste **23**, 275–286 (2002)
31. Khoufi, S., Khamassi, K., Da Silva, J.A.T., Rezgui, S., Ben Jeddi, F.: Watering regime affects oil content and fatty acid composition of six sunflower lines. J. New Sci. **7**(1), 1–9 (2014)

32. Onemli, F.: Impact of climate changes and correlations on oil fatty acids in sunflower. Pak. J. Agri. Sci. **49**(4), 455–458 (2012)
33. Turhan, H., Citak, N., Pehlivanoglu, H., Mengul, Z.: Effects of ecological and topographic conditions on oil content and fatty acid composition in sunflower. Bulgarian J. Agri. Sci. **16**(5), 553–558 (2010)
34. Grujić, S., Aleksić, V., Vukić, M., Petrović, Z.: The effect of packing material on storage stability of sunflower oil. Qual. Life **2**(3–4), 75–83 (2011)
35. Vidrih, R., Vidakovič, S., Abramovič, H.: Biochemical parameters and oxidative resistance to thermal treatment of refined and unrefined vegetable edible oils. Czech J. Food Sci. **28**(5), 376–384 (2010)
36. Vujasinović, V.B.: Influence of thermal treatment of pumpkin seed *Cucurbita pepo L.* on nutritive value and oxidative stability of oil. Ph.D. thesis. University of Novi Sad, Faculty of technology Novi Sad, pp. 28–29 (2011)
37. Vujasinović, V., Djilas, S., Dimić, E., Basic, Z., Radočaj, O.: The effect of roasting on the chemical composition and oxidative stability of pumpkin oil. Eur. J. Lipid Sci. Technol. **114**, 568–574 (2012)
38. Masłowski, A., Andrejko, D., Slaska-Grzywna, B., Sagan, A., Szmigielski, M., Mazur, J., Rydzak, L., Sobczak, P.: Impact of the storage temperature and time on the selected quality properties of rapeseed, linseed and cameline seed oil. Agric. Eng. 115–124 (2013)
39. Poiana, M.-A., Alexa, E., Moigradean, D., Popa, M.: The influence of the storage conditions on the oxidative stability and antioxidant properties of sunflower and pumpkin oil. In: Proceedings 44th Croatian and 4th International Symposium on Agriculture, University of Zagreb, Faculty of Agriculture, Zagreb, pp. 449–453 (2009)
40. Ardabili, A.G., Farhoosh, R., Khodaparast, M.H.H.: Chemical composition and physicochemical properties of pumpkin seeds (*Cucurbita pepo subsp. pepo var. Styriaka*) grown in Iran. J. Agric. Sci. Technol. **13**, 1053–1063 (2011)

Innovative Technologies in Agriculture and Human Health

Impact of Parental BMI on Adolescents' BMI: Moderating Effects of Gender

Irzada Taljic(✉) and Haris Memisevic

Faculty of Educational Sciences, University of Sarajevo,
Skenderija 72, 71000 Sarajevo, Bosnia and Herzegovina
irzadata@gmail.com

Abstract. Nutritional status is influenced by many factors. One of the most important factors is parental BMI not only because of genetics but also because of food choices parents make for children. The aim of this study was to analyze the relationship between parents' BMI and children's BMI. The sample for this study consisted of 280 adolescents, 131 boys (46.8%) and 149 girls (53.2%) living in Canton Sarajevo, Bosnia and Herzegovina. We measured adolescents' BMI and we used parents' self-reported measures to calculate parents' and childrens' BMI. The correlation was significant between mothers' BMI and childrens' BMI (r = .24, $p < .01$) but not between fathers' BMI and childrens' BMI (r = .08, $p = .16$). In relation to the child's gender the correlation between mothers' BMI is stronger for the boys (r = .28, $p < .01$) than for the girls (r = .20, $p = .013$) and fathers' BMI was correlated with boys' BMI (r = .19, $p = .032$), but not with girls' BMI (r = .012, $p = .88$). The results of this study point to the complex relationship between parents' and childrens' BMI.

Because of the complexity of this issue and the need for more precise results, it is recommended to collect more data, with larger sample size and to include additional environmental factors into the prediction model.

Keywords: Adolescents · Parents · BMI · Predictor

1 Introduction

Proper nutrition and healthy diet should be adopted and encouraged from a young age. Parents and caretakers of children are the most responsible for early food choices. Parents who prepare meals are making food choices for their children and have a responsibility for creating good, proper food habits which, in turn, influence nutritional status and BMI (Body Mass Index) category [1, 2]. Studies show that parents are a behavioural model regarding eating habits of their children. Thus, a positive parental attitude towards healthy food choices represents a safe way to instil healthy eating habits in their children [3].

The link between genetics and BMI as well as genetics and nutritional behaviour is a subject of numerous studies. There are numerous health risks related to obesity. Besides risks for individuals who are overweight, excessive body weight also carries a risk for the offspring. Studies have shown that mother's excessive body weight during pregnancy can result in prematurity [4]. Numerous studies have also shown that the

M. Brka et al. (Eds.): AgriConf 2019, IFMBE Proceedings 78, pp. 315–320, 2020.
https://doi.org/10.1007/978-3-030-40049-1_40

parental BMI is a significant predictor of obesity in children and adolescents [5–7]. There are some contradicting reports on the influence of mothers' BMI on the BMI of their children. Some studies have found the influence of mothers' BMI on the BMI of daughters and sons [8] while the others have found a significant link between BMI of mothers and BMI of daughters [7, 9, 10]. There are many factors contributing to the BMI. In the study among adolescents in Iran [7], BMI of the children was significantly related to parental BMI ($p < .001$), gender ($p = .02$), parents' education level ($p < .001$), father's employment status ($p < .001$), and family income ($p < .001$). On the other hand, some studies revealed no significant associations between marital status, occupation, and education of parents and BMI-for-age of the students [8]. However, that study showed that both parents' BMI were significantly associated with students' BMI-for-age.

Adolescents are specific group to study food behaviour because of different factors influencing nutritional behaviour in that period of life. To the best of authors' knowledge there are studies conducted in Bosnia and Herzegovina about BMI of adolescents and influence of some factors [11, 12] but there are no studies on relationship between parents' BMI and childrens' BMI.

2 Subjects and Methods

This study was conducted in Canton Sarajevo, Bosnia and Herzegovina. The sample for this study consisted of 280 adolescents aged 14 to 15 years (mean age = 14.7, SD = 0.5). In relation to the gender of participants, there were 131 boys (46.8%) and 149 girls (53.2%).

Study protocol and the schools (12 schools in 9 municipalities) were approved by Ministry of Education and Sciences of Canton Sarajevo. In the schools, children were selected randomly. Adolescents were measured for BMI percentiles (body weight and body height) while wearing light sports clothes in the school gyms and categorized according to the World Health Organization cut-off points [13]. Formula used for calculating BMI was the same as for the adults and compared with z-scores or percentiles. During childhood and adolescence the ratio between weight and height varies with sex and age, so the cut-off values that determine the nutritional status of those aged 0–19 years are gender- and age-specific. Parents' weight and height measurements were self-reported and classified according to the WHO classification for BMI cut-off points. This is the most common method used for assessing parental overweight and obesity [13, 14]. Additionally, socioeconomic status (SES) was determined through self-reported level of education and an income level.

Statistical analysis were performed using Statistical Software SPSS 21.

3 Results and Discussion

The mean BMI score of adolescents in this sample was 21.5 ± 3.45. There were no statistically significant differences in the BMI of adolescents in relation to the gender (Table 1).

Table 1. The mean BMI score of girls and boys

BMI score	M	SD	t (278)*
Boys	21.3	3.9	−0.72
Girls	21.6	3.3	

N = 280, *p = .48

Table 2 shows that most of the adolescents in this sample were in the category of normal BMI (191 or 68.2%). There were 49 underweight adolescents in the sample (17.5%), 33 adolescents (11.8%) in the category of overweight and 7 adolescents (2.5%) in obese category. Again, the distribution of boys and girls in these categories did not differ statistically according to the results of Chi square test (X^2 = 1.1, p = .57).

Table 2. BMI categories of children

BMI category			
Underweight	Normal weight	Overweight	Obesity
49 (17.5%)	191 (68.2%)	33 (11.8%)	7 (2.5%)

In relation to mothers' BMI (Table 3), they were mostly in normal weight category (46.4%) followed by overweight (37.9%) and fathers were mostly in overweight category (47.1%) followed by normal weight (30.7%). There were more obese fathers (20.0%) than mothers (10.0%) and the mothers were more underweight (5.7%) than fathers (2.1%).

Table 3. BMI categories of parents

BMI category				
	Underweight	Normal weight	Overweight	Obesity
Mother	16 (5.7%)	130 (46.4%)	106 (37.9%)	28 (10.0%)
Father	6 (2.1%)	86 (30.7%)	132 (47.1%)	56 (20.0%)

Correlations between children's BMI and BMI of their mothers and fathers were also calculated. The correlation was significant between mothers' BMI and childrens' BMI (r = .24, $p < .01$) but not between fathers' BMI and childrens' BMI (r = .08, p = .16).

We further wanted to examine whether this pattern of correlations is the same for girls and boys. The results show (Table 4.) that the correlation between mothers' BMI is stronger for the boys (r = .28, $p < .01$) than for the girls (r = .20, p = .013). However, the fathers' BMI was now correlated with boys' BMI (r = .19, p = .032), but not with girls' BMI (r = .012, p = .88). It is important to note that most of these statistically significant correlations were rather small except for the correlation between mothers' BMI and boys' BMI, which was moderate.

Table 4. Correlations between childrens' BMI and BMI of their mothers and fathers

Variable	Girls	Boys
BMI mother	.20**	.28
BMI Father	.012	.19*

** Correlation is significant at the 0.01 level (2-tailed)
* Correlation is significant at the 0.05 level (2-tailed)

Significance of the influence of genetics is shown in the study of the relationship between BMI of parents, BMI of grandparents and BMI of children [14]. That national study involving 2591 children aged 5–19 years, proved for the first time, the link between BMI of children and the obesity of grandparents. In cases where parents and grandparents were in normal weight category, 7.9% of children were overweight. In cases where parents were overweight and grandparents in normal weight category, 17.9% of children were overweight, and where parents were obese and grandparents in normal weight category, 31.9% of children were overweight. It is significant that in the cases where parents were in normal weight category and grandparents were obese prevalence of overweight children was (17.4%). In the study among adolescents [15] Multivariable Logistic Regression showed that the father's BMI was significantly associated with male BMI (OR: 2.02) and female BMI (OR: 1.59), whereas the mother's BMI was significantly associated with female BMI only (OR: .514).

We then performed a regression analysis predicting children's BMI by using following predictor variables: mothers' BMI, fathers' BMI, income, mother's educational level and father's educational level. These results are shown in Table 5.

Table 5. Regression analisys with mothers' BMI, fathers' BMI income, mothers' educational and fathers' educational level as predictors of childrens' BMI

Variable	B	SEB	β
Mothers' BMI	.23	.05	.26*
Fathers' BMI	.03	.05	.04
Income	.32	.29	.07
Mother's education	1.05	.62	.11
Father's education	−.37	.56	−.04

Note. R^2 = .08, (N = 280, p < .01), *p < .01.

This prediction model presented in Table 5, although statistically significant, explained only 8% of variance in childrens' BMI and the only significant factor contributing to children's BMI was mothers' BMI.

In study conducted in Luxembourg [16] children of foreign nationality had 2.9 times more risk to be overweight or obese than other children of Luxembourg nationality (OR = 2.90, 95% CI: 1.38–6.10). Children living in household with at least one parent who was obese were 6.51 times more likely to be overweight or obese compared to those in household with both parents normal (OR = 6.51, 95% CI: 2.48–17.08). Overall, nationality and weight status of parents were the main determinants of children's weight status.

Association between socio-economic status and BMI of adolescents in Japan [17] shown among girls, low economic status was significantly positively associated with BMI z-scores in the crude model (coefficient: .35; $p = .001$). In the final model adjusted for other possible covariates, low economic status remained significantly positively associated with BMI z-score (coefficient: .22; $p = .044$). The same association was not found for boys.

4 Conclusions

The results in this study have shown that adolescents and their parents are mostly in the normal weight category. The correlation was significant between mothers' BMI and childrens' BMI. When examined by gender, results show higher correlation between mothers' BMI and the boys' BMI than for the girls. However, fathers' BMI was correlated with boys' BMI but not with girls' BMI. Correlations between parents' education level and an income have no significant differences on childrens' BMI status. There is a need for including more environmental factors in order to obtain a better prediction model.

Promoting healthy diets and educating parents on the consequences of overweight and as well as of underweight on children's health in adulthood are effective strategies to control their childrens normal weight.

References

1. Affinita, A., Catalani, L., Cecchetto, G., De Lorenzo, G., Dilillo, D., Donegani, G., Fransos, L., Lucidi, F., Mameli, C., Manna, E., Marconi, P., Mele, G., Minestroni, L., Montanari, M., Morcellini, M., Rovera, G., Rotilio, G., Sachet, M., Zuccotti, G.V.: Breakfast: a multidisciplinary approach. Ital. J. Pediatr. **39**(44), 1–10 (2013)
2. Pearson, N., Biddle, S.J., Gorely, T.: Family correlates of breakfast consumption among children and adolescents. A systematic review. Appetite **52**(1), 1–7 (2008)
3. Silvia Scaglioni, S., Salvioni, M., Galimberti, C.: Influence of parental attitudes in the development of children eating behaviour. Br. J. Nutr. **99**(Suppl. 1), S22–S25 (2008)
4. Weng, S.F., Redsell, S.A., Swift, J.A., Yang, M., Glazebrook, C.P.: Systematic review and meta-analyses of risk factors for childhood overweight identifiable during infancy. Arch. Dis. Child. **97**, 1019–1026 (2012)
5. Stettler, N., Tershakovec, A.M., Zemel, B.S., Leonard, M.B., Boston, R.C., Katz, S.H., Stallings, V.A.: Early risk factors for increased adiposity: a cohort study of African American subjects followed from birth to young adulthood. Am. J. Clin. Nutr. **72**(2), 378–383 (2000)
6. Lobstein, T., Baur, L., Uauy, R.: IASO International Obesity TaskForce. Obesity in children and young people: a crisis in public health. Obes. Rev. **5**(Suppl. 1), 4–104 (2004)

7. Shafaghi, K., Shariff, Z.M., Mohd Taib, M.N., Abdul Rahman, H., Mobarhan, M.G., Jabarri, H.: Parental body mass index is associated with adolescent overweight and obesity in Mashhad, Iran. Asia Pac. J. Clin. Nutr. **23**(2), 225–231 (2014)
8. Johannsen, D.L., Johannsen, N.M., Specker, B.L.: Influence of parents' eating behaviors and child feeding practices on children's weight status. Obesity (Silver Spring) **14**(3), 431–439 (2006)
9. Whitaker, R.C., Deeks, C.M., Baughcum, A.E., Specker, B.L.: The relationship of childhood adiposity to parent body mass index and eating behavior. Obes. Res. **8**(3), 234–240 (2000)
10. Bralić, I., Vrdoljak, J., Kovačić, V.: Associations between parental and child overweight and obesity. Coll. Antropol. **29**(1), 481–486 (2005)
11. Taljić, I., Hadžić, A., Djukic Ratkovic, D.: Perception and satisfaction with body image according to the nutritional status among adolescents boys from urban and rural areas of the Sarajevo Canton. In: 10th International Scientific and Professional Conference Proceedings: With Food to Health, pp. 9–14. Josip Juraj Strossmayer University of Osijek, Faculty of Food Technology Osijek and Faculty of Pharmacy, University of Tuzla, Osijek, Croatia (2017)
12. Taljić, I., Toroman, A.: Category of nutritional status of adolescents according to gender and type of settlement: results of Canton Sarajevo. Hrana i ishrana. **57**(1), 13–18 (2016)
13. WHO. http://apps.who.int/bmi/index.jsp?introPage=intro_3.html. Accessed 18 May 2019
14. Davis, M.M., McGonagle, K., Shoeni, R.F., Stafford, F.: Grandparental and parental obesity influences on childhood overweight: implications for primary care practice. JABFM **21**(6), 549–554 (2008)
15. Wurbach, A., Zellner, K., Kromeyer Hauschild, K.: Meal patterns among children and adolescents and their associations with weight status and parental characteristics. Public Health Nutr. **12**, 1115–1121 (2009)
16. Tchicaya, A., Lorentz, N.: Relationship between children's body mass index and parents' obesity and socioeconomic status: a multilevel analysis applied with Luxembourg data. Health. **6**, 2322–2332 (2014)
17. Mizuta, A., Fijuwara, T., Ojima, T.: Association between economic status and body mass index among adolescents: a community-based cross-sectional study in Japan. BMC Obes. **3**(47), 2–9 (2016)

Information and Communication Technologies for Smart and Sustainable Agriculture

Hamid El Bilali(✉), Francesco Bottalico, Giovanni Ottomano Palmisano, and Roberto Capone

International Centre for Advanced Mediterranean Agronomic Studies (CIHEAM-Bari), via Ceglie 9, 70010 Valenzano, (Bari), Italy
elbilali@iamb.it

Abstract. Agriculture faces different environmental, economic and social challenges that make the case for transition towards sustainability. Digitization is one of the most significant ongoing transformation processes in global agriculture. Information and Communication Technologies (ICTs) range from traditional communication aids (e.g. telephones, televisions), internet and mobile applications, to Big Data analytics and information systems, Cloud computing, Internet of Things, remote sensing and drones, blockchain and artificial intelligence. Different terms have been used to refer to the application of ICTs in agriculture e.g. digital agriculture, e-agriculture, smart agriculture, precision agriculture. This review paper analyses the potential of ICT in agricultural sector. In particular, it explores the benefits of ICT in terms of agriculture smartness and sustainability, provides an overview on the main technologies used, delineates the contours of the smart farming market and landscape, and analyses the potential drawbacks of such a 'digital revolution' in agriculture. Smart agriculture is presented by some scholars and practitioners as the 'Third Green Revolution', after the plant breeding and genetics revolutions. It is widely recognised that ICTs can help reducing inefficiencies, increasing resource productivity, decreasing management costs, and improving traceability and transparency. However, some caution is needed to make the ongoing 'digital revolution' in agriculture inclusive for small-scale farmers and women, especially in developing countries. Moreover, some threats and challenges relating to digital and smart technologies need to be appropriately addressed through policies to support the development of the necessary market and legal architecture for ICT and smart farming, with due consideration to ethical questions.

Keywords: Smart farming · E-agriculture · Digital agriculture · Precision farming · Sustainable agriculture · ICT

1 Introduction

Information and communication technology (ICT) includes services, devices, applications and networks. ICT is defined by the World Bank (2011) as "… *any device, tool, or application that permits the exchange or collection of data through interaction or transmission*" (p. 3) and it adds that ICT "*includes anything ranging from radio to satellite imagery to mobile phones or electronic money transfers*" (p. 3). Indeed, ICTs

M. Brka et al. (Eds.): AgriConf 2019, IFMBE Proceedings 78, pp. 321–334, 2020.
https://doi.org/10.1007/978-3-030-40049-1_41

range from traditional communication aids (e.g. telephones, televisions, radios), internet and mobile applications, to Cloud Computing, Big Data analytics and information systems, Internet of Things (IoT), remote sensing and drones using geographic information systems (GIS), blockchain and artificial intelligence (FAO and ITU 2019; World Bank 2011). The importance of ICT for addressing sustainability challenges was recognized in the 8th Millennium Development Goal "Global partnership for development", with the target to "*...make available the benefits of new technologies, especially information and communications technologies (ICTs)*" (United Nations 2015a) in the fight against poverty. It is also highlighted in the 2030 Agenda for Sustainable Development (United Nations 2015b) and the Sustainable Development Goals (SDGs) where ICT is considered in relation to SDG 4 (Quality education), 5 (Gender equality), 9 (Industry, innovation and infrastructure) and 17 (Partnerships for the goals).

Agriculture is strongly related to many sustainability challenges such as biodiversity loss, climate change, water scarcity, rural poverty, and malnutrition and food insecurity (Bruinsma 2011; FAO 2014, 2016; Foley et al. 2011; IAASTD 2009; Postel 2000). For that, and in order to address these pressing challenges, there have been many calls for sustainability transitions in agro-food systems (El Bilali 2018, 2019; FAO 2017; UNEP 2018). ICT is one of the instruments contemplated to make agriculture smarter and more sustainable (Bello and Aderbigbe 2014; El Bilali and Allahyari 2018; FAO et al. 2017; FAO and ITU 2019; Singh et al. 2014; Sørensen et al. 2019; Svenfelt and Zapico 2016). Thanks to their disruptive potential, ICTs hold the potential to contribute to transition towards sustainability in agriculture (Berti and Mulligan 2015; FAO et al. 2017; Nakasone et al.2014; Poppe et al. 2013; World Bank 2011). Disruptive ICT trends include Big Data (web of data, linked open data), mobile/cloud computing, location-based monitoring (remote sensing, geo information, drones, etc.), IoT (Poppe 2016; Wolfert 2015). Therefore, agriculture digitization is high on the political agenda, especially in industrialized and in transition economies (G20 2017), and ICT application in agricultural development is highlighted as a priority (Kolshus et al. 2015; UN-ESCAP 2008). ICTs are increasingly used in modern agriculture (Lehmann et al. 2012). The global smart farming market is expected to have an annual growth rate of 19.3% from 2017 to 2022 thus reaching $23.14 billion by 2022 thanks to the growing penetration of ICT in farming – in developed and developing countries alike – and the increasing attention to climate-smart agriculture (BIS Research 2018). Apart from addressing the issue of climate change, ICTs have also been put forward as a means to achieve food security and to enhance agri-food systems sustainability (Svenfelt and Zapico 2016).

This review paper analyses the potential of ICT in agricultural sector. In particular, it explores the benefits of ICT in terms of agriculture smartness and sustainability, provides an overview on the main technologies used, delineates the contours of the smart farming market and landscape, and analyses the potential drawbacks of such a 'digital revolution' in agriculture.

2 ICTs in Agriculture: E-Agriculture, Smart Agriculture and Precision Agriculture

Different terms have been used to refer to the application of ICTs in agriculture e.g. e-agriculture, digital agriculture, smart agriculture, precision agriculture.

FAO uses the term *e-agriculture* to refer to the application of ICTs in agriculture and rural development (FAO et al. 2017; Kolshus et al. 2015). E-agriculture involves the conceptualization, development and application of ICTs solutions in agriculture and rural areas (FAO 2019).

Precision agriculture (PA) is a widely cited example of the application of ICT in agriculture (Allahyari et al. 2016; Balafoutis et al. 2017; Ess and Morgan 2003; McBratney et al. 2005). It is a modern farming model that consists in the use of sensors to optimize the utilization of inputs (e.g. fertilizers, pesticides, irrigation water) (Mintert et al. 2016). It came into use in the 1980s as GPS technologies became accessible by farmers, especially in developed countries. Modern PA relies on a combination of satellite navigation and GPS technology, sensors and IoT (Schrijver et al. 2016). McKinsey & Company (Magnin 2016) define precision agriculture (i.e. precision farming) as "*a technology-enabled approach to farming management that observes, measures, and analyzes the needs of individual fields and crops*" (p. 4) and argue that it is shaped by two trends "*big-data and advanced-analytics capabilities on the one hand, and robotics—aerial imagery, sensors, sophisticated local weather forecasts—on the other*" (p. 4). The European Parliamentary Research Service (Schrijver et al. 2016) defines precision agriculture (PA) as "*a modern farming management concept using digital techniques to monitor and optimise agricultural production processes*" (p. 6) and highlights that the key point in PA is optimisation of the use of resources and inputs (e.g. fertilizers). In many countries (e.g. China, Japan), wide-scale deployment of smartphones and IoT systems led to a rapid adoption of precision farming solutions. Precision agriculture technologies (PATs) include variable rate technologies (planting/seeding, irrigation, nutrient application, pesticide application), machine guidance (auto-guidance or driver assistance), precision physical weeding technology, controlled traffic farming (a system confining all load of machinery to permanent traffic lanes) (Balafoutis et al. 2017). In variable rate technologies, data from sensors allow tailoring the amounts of inputs to current crops needs and different fields/plots, instead of applying the same amount of inputs to the whole area. PATs are nowadays present in all stages of crop production and increasingly used in livestock production (Banhazi et al. 2012; FAO et al. 2017). For instance, virtual fence technology allows managing cattle herd based on sensors and remote-sensing signals, or actuators attached to cattle (Umstatter 2011). Companies offer a variety of applications for precision crop farming such as yield monitoring and forecasting, precision irrigation, crop scouting, variable rate application, and recordkeeping. In this regard, a large market share of precision crop farming is held by precision irrigation products. As for the livestock sector, the introduction of autonomous milking robots is expected to increase the growth of smart farming solutions (BIS Research 2018).

Smart farming (cf. smart agriculture) represents the application of modern ICTs for optimising complex farming systems (AgroCares 2019; Smart AKIS 2016). This application is leading to what can be called a 'Third Green Revolution' (following the plant breeding and genetics revolutions) (Smart AKIS 2016) or 'fourth revolution' (after domestication of animals and plants, systematic use of crop rotations, and 'Green revolution') (Walter et al. 2017). Unlike with precision agriculture, smart farming (SF) focuses on access to data and their use in a smart way rather than precise measurement and determination of differences between farm plots or animals. Smart agriculture involves the use of a wide range of ICTs and mobile devices (e.g. smart phones and tablets) to access real-time data about soil, plants, climate, weather, etc. so that farmers have the information needed to make informed decisions (AgroCares 2019). SF is based upon the combined application of ICT solutions such as the IoT, precision equipment, sensors and actuators, Big Data, GPS, robotics, unmanned aerial vehicles (e.g. drones), etc. It is, hence, strongly related to three interconnected technology fields; management information systems, precision agriculture, agricultural automation and robotics (Smart AKIS 2016). The IoT support is essential for sensors and machines to interface with the farmer. Autonomous robotic tractors, machines (e.g. seeding machines), drones are self-sufficient, along with the inclusion of GPS, cameras and the IoT connectivity to enable remote operation and monitoring. Drones enable farmers to collect more information about crops (e.g. crop health) and can execute different operations such as imaging, monitoring, spraying. Sensors collect data on soil as well as air quality, weather, light conditions, irrigation. Pulling together sensors and drones is what forms the Internet of Things (IoT), a network of physical devices that are carefully outfitted with electrical connectivity and that enables data exchange, processing and aggregation. The use of the IoT enables a more immediate integration of the physical world into electronically-based systems, resulting in economic benefits, efficiency, and reduced human labour (Smart AKIS 2016).

Digital farming integrates both precision farming and smart farming concepts. Its essence lies in creating actionable intelligence and meaningful added value from data thus going beyond the mere availability of data (CEMA-Agri 2017). In a paper by the German Agricultural Society (DLG 2018), digital agriculture is understood to mean consistent application of precision agriculture and smart farming methods, use of web-based data platforms together with Big Data, and internal and external networking of the farm.

3 Benefits of ICTs: Smartness and Sustainability of Agriculture

Svenfelt and Zapico (2016) argue that ICT solutions can improve the sustainability of agriculture by increasing efficiency, networking agro-food chain actors, and enhancing traceability and transparency. Likewise, El Bilali and Allahyari (2018) put that "*ICTs can contribute to agro-food sustainability transition by increasing resource productivity, reducing inefficiencies, decreasing management costs, and improving food chain coordination*" (p. 456). According to FAO and the International Telecommunication Union (FAO and ITU 2016), benefits of e-agriculture include increasing markets efficiency, improving vertical and horizontal linkages, facilitating information sharing and networking, developing value-added services, reducing individual and institutional risk, increasing food and nutrition security and safety. Verdouw et al. (2016) and Mat et al. (2018) argue that agriculture productivity and sustainability can be dramatically improved through the revolutionary potential of IoT perspective. Schrijver et al. (2016) highlight that precision agriculture can contribute to food security and support environmental sustainability of farming. Likewise, Walter et al. (2017) put that "*ICT and data management can provide novel ways into a profitable, socially accepted agriculture that benefits the environment (e.g., soil, water, climate), species diversity, and farmers in developing and developed countries*" (p. 6150). Poppe et al. (2013) add that ICT development has a large potential impact on sustainability, resource efficiency, food safety, and waste generation and management. Meanwhile, FAO et al. (2017) enumerate as benefits of ICT improving market access, enhancing agricultural extension and advisory services, improving preparedness of farmer and reducing climate change-related risks, increasing food safety and traceability and facilitating certification, fostering financial inclusion of rural communities, improving access by smallholders to insurance and risk management products (e.g. market-based index-insurance products). Focusing on the specific benefits of PA, authors (FAO et al. 2017) put that "*While the main incentive to adopt Precision Agriculture (PA) methods is to maximise profitability, it can also tackle health and safety issues as well as reduce environmental impacts of farming practices*" (p. 9).

The role of ICTs in improving system efficiency is central in literature on ICT for sustainability (Caputo et al. 2018a; Dao et al. 2011; GeSI 2008; van Marrewijk and Hardjono 2003). ICTs have been for a long time also used to increase resource efficiency and productivity in agro-food systems (Berti and Mulligan 2015; Svenfelt and Zapico 2016; Thöni and Tjoa 2017). Indeed, ICT can help reducing the use of agricultural resources and inputs (pesticides, fertilizers, energy, water) as well as decreasing environmental externalities (Lehmann et al. 2012). Smart farming is considered to have a real potential to deliver more productivity and sustainability in agriculture by fostering a more precise and resource-efficient approach especially in the use of inputs and resources (e.g. water, energy) (Smart AKIS 2016).

The reduction of input use thanks to variable rate technologies, that are an integral component of precision farming, has positive impacts, not only environmental (Balafoutis et al. 2017; Bora et al. 2012; European Union 2017; Hedley 2015; Mutchek and Williams 2010; Saidi 2013; Schrijver et al. 2016) but also economic (Balafoutis et al. 2017; Batte and Ehsani 2006; Bergtold et al. 2009; European Union 2017; Hedley 2015; Jensen et al. 2012; Koch et al. 2004; Lambert and Lowenberg-De Boer 2000; Plant et al. 2000; Shockley et al. 2015; Swinton and Lowenberg-DeBoer 1998; Tekin 2010; Timmermann et al. 2003). ICT decision-support systems help producers to minimize production costs while maximizing production efficiency (Hedley 2015; Walter et al. 2017) and reducing environmental footprints of farm operations (Hedley 2015). Moreover, ICT-based smart irrigation systems can reduce not only water consumption (Evans and King 2010; HydroSence 2013; Mutchek and Williams 2010) but also greenhouse gas (GHG) emissions and carbon footprint (Mutchek and Williams 2010) thanks to reduced energy use. In addition, precision agriculture practices and technologies result in lower GHG emissions and enhance carbon sequestration ability of soils thus mitigating climate change (European Union 2014). This makes essential strategies to scale up climate-smart agriculture (CSA) (Westermann et al. 2018).

From the social point of view, it is argued that smart farming reduces the heavy workload of the farmers and farm workers, hence improving their wellbeing and quality of life (Smart AKIS 2016). ICTs can also help alleviating the big mismatch between demand and supply in the agricultural and rural labour market regarding skills and location (Poppe et al. 2013). Moreover, some scholars (Caputo et al. 2018b; Del Giudice et al. 2017; Walter et al. 2017; Wognum and Bremmers 2009; Wognum et al. 2011) argue that ICTs can increase food chain transparency. Indeed, ICT and information systems have been used to improve agro-food products traceability and food chains transparency (Kaloxylos et al. 2013).

A growing body of evidence shows that benefits can be increased by exploiting the synergies between different ICTs such as IoT, mobile and cloud-based Big Data analytics (Rajeswari et al. 2017), IoT and Blockchain (Lin et al. 2018; Shyamala Devi et al. 2019), IoT, artificial intelligence and Cloud computing (Bu and Wang 2019).

4 Landscape of Smart Farming: Market Perspectives and Main Actors

The market of smart farming is growing quickly. Precision farming technologies are taking up with an expected annual growth of 12% till 2020. The European Union (European Union 2017; Sarantis 2015; Schrijver et al. 2016; Zarco-Tejada et al. 2014) and the US are the most promising markets. Regional analysis of smart farming market performed by BIS Research (2018) shows that North America is at the forefront of the global market, with high market penetration in the United States. However, it is in the Asia-Pacific region that the fastest market growth is projected from 2017 to 2022 mainly due to the presence of economically advancing countries such as India and China (BIS Research 2018). Pivoto et al. (2018) show that China is one of the countries where most research on smart agriculture is being carried out, together with US, South Korea, Germany and Japan. There is still a divide in terms of the use of smart agriculture technologies and solutions not only between developing countries and the developed world, but also among the industrialised countries. In fact, Smart AKIS (2016) argues that while in the USA possibly up to 80% of farmers use some kind of smart farming technologies (SFT), in Europe it is no more than 24%, despite an increasing attention to precision agriculture (Schrijver et al. 2016). Moreover, smart farming technologies are still not affordable for the large majority of farmers (especially smallholders) in developing countries. FAO et al. (2017) put that "*The digital divide between developing and developed countries is nowhere more evident than in agriculture*" (p. 9).

In a review on smart farming, with a particular focus on Big Data, Wolfert et al. (2017) put that "*The landscape of stakeholders exhibits an interesting game between powerful tech companies, venture capitalists and often small start-ups and new entrants. At the same time there are several public institutions that publish open data...*" (p. 69). Indeed, different stakeholders (from public, civil society and private sectors) shape the smart farming arena and the market is becoming more competitive. BIS Research (2018) profiles 24 leading companies engaged in precision agriculture: AGCO Corporation, Boumatic LLC, Ag Leader Technology, Dairymaster, CNH

Industrial, DeLaval, Deere & Company, Fullwood Ltd., DICKEY-john, Philips Lighting, Raven Industries, Osram Licht AG, SST Development Group, Cree Inc., The Climate Corporation, General Hydroponics, Topcon Positioning Systems, AKVA Group, Trimble Inc., XpertSea, Afimilk Ltd., Eruvaka Technologies, Allflex USA Inc., DJI Innovations. Moreover, different international organisations are actively involved in the promotion of e-agriculture and smart agriculture through, among others, the development of resources. These include FAO (FAO 2013, 2015, 2019) and the World Bank (World Bank 2012). Different platforms and communities of practice facilitate the exchange of experiences and information on ICT use in agriculture e.g. *e-Agriculture Community of Practice* (run by FAO)[1], *ICT Observatory* and *ICTUpdate* (of the Technical Centre for Agricultural and Rural Cooperation)[2], *InfoDev* (a project of the World Bank)[3], *ICT in Agriculture Sourcebook* (provided by the World Bank)[4], *FAO-ITU e-Agriculture Strategy Guide* (FAO et al. 2017).

5 ICTs in Agriculture: Potential Negative Impacts and Challenges

ICTs come not only with opportunities to farmers and farming communities but also certain challenges. Davies (2014) highlight that ICT solutions, when developed in isolation from practices and realities of producers, run the risk of hampering transition towards sustainability in agriculture. Walter et al. (2017) put that "*Only by considering new technologies in conjunction with a diversity of crop and livestock systems, as well as the relevant markets and policies, can farming in the digital era become smart farming*" (p. 6149). Data ownership is a key concern with digital technologies use (Berti and Mulligan 2015; Walter et al. 2017) as well as cyber security (Barreto and Amaral 2018) and privacy (especially regarding Big Data and IoT) (FAO et al. 2017). The potential data misuse creates additional ethical and legal challenges for smart agriculture monitoring and regulation (Charo 2015). Moreover, as there must be accountability for mismanagement, challenges of accountability and responsibility will intensify with emerging smart agriculture technologies. Furthermore, agriculture digitalization might affect employment opportunities and job profiles in farming even further (Walter et al. 2017) with far-reaching implications in rural areas. There are also some concerns about high energy consumption and carbon footprint that imply the use of modern ICTs such as the IoT (Popli et al. 2019).

A host of challenges should be addressed to increase the uptake of ICTs in agriculture (Sylvester 2013, 2015). Profitability is a key factor hampering widespread adoption of ICTs in agriculture as it is not always easy to demonstrate that ICTs uptake will improve farm profitability (Mintert et al. 2016). High adoption costs for individual farms and limited skills and knowledge can represent important adoption hurdles

[1] http://www.e-agriculture.org.

[2] http://ictupdate.cta.int.

[3] http://www.infodev.org.

[4] http://ictinagriculture.org.

(Baumgart-Getz et al. 2012; Kutter et al. 2011; Pivoto et al. 2018), especially for small-scale farmers in developing countries. Therefore, the benefits of ICT and smart agriculture technologies might be limited to industrialized countries and widely grown crops (e.g. wheat, maize, rice) (Walter et al. 2017), which may widen the North-South technology divide.

Wolfert et al. (2014) enumerated several bottlenecks of ICT development in agriculture and, consequently, of the spread of smart agriculture, such as regional focus and cultural differences, small-scaled and isolated software development, difficult or impossible interoperability between various systems, complicated handling and integration of large data amounts. Schrijver et al. (2016) and Pivoto et al. (2018) highlight that the adoption of precision farming technologies can be a challenge for farmers as they may need to acquire new knowledge and skills. Besides capacity (cf. farmers' skills and human capital), several factors hinder the integration of ICT in agriculture in the Global South, such as connectivity, content (e.g. language, cultural appropriateness) and cost (FAO et al. 2017; IDEV 2016). Therefore, ICTs, and smart agriculture technologies, need to be tailored to use contexts and be developed in collaboration with beneficiaries and end-users to make sure that they are accessible, adequate and relevant (FAO et al. 2017; IDEV 2016; World Bank 2017). FAO (Kolshus et al. 2015) identified several challenges in making ICTs accessible for farmers such as capacity development (ability to effectively use technologies); content (adaptation of content to farmers' needs); technologies (challenge of identifying suitable technologies). In general, women, older and poor farmers, and people living in remote rural areas have difficult and limited access to ICTs (FAO et al. 2017; Kolshus et al. 2015). The E-agriculture Strategy Guide (FAO and ITU 2016) provides a framework to address the ICT opportunities and challenges for agriculture in a more efficient way. Challenges include ICT access challenges (cf. coverage, quality, costs) posing hindrances to their adoption and use, cybersecurity, e-waste. Likewise, the Overseas Development Institute (ODI), the UK Department for International Development (DFID) and FAO provided policy recommendations to foster the role of ICT in supporting agricultural development and rural livelihoods (Chapman et al. 2003): building on existing systems, while encouraging integration of different technologies; determining who should pay; ensuring equitable access to marginalised groups and rural population; promoting localised content; building capacity; using realistic, suitable technologies; building knowledge partnerships to ensure that knowledge gaps are filled.

6 Conclusions

ICTs can play a central role in transition towards smart and sustainable agriculture, which is deemed necessary to address the challenge of growing food demand in the ages of changing climate and increasing resources scarcity. New digital technologies and services help farmers to reduce inefficiencies and deliver greater efficiency in resource use. However, some precautions should be taken to make the ongoing 'digital revolution' in agriculture inclusive for disadvantaged categories such as small-scale farmers and women, especially in developing countries. In order to maximize the benefits of ICT-based smart agriculture, also in developing countries, it is necessary to develop services and applications that are relevant, locally appropriate, user-friendly, and affordable. Also, the interoperability of these technologies should be addressed to optimize their synergistic effects. Moreover, investment (public, private, and public-private) is needed to move the technologies from 'proof-of-concept' to 'proof-of-work' stage. Aside from the benefits of digital and smart technologies, there are also threats and challenges that need to be appropriately addressed. For that, there is a need to develop policies supporting the necessary market and legal architecture for ICT and smart farming, with due consideration of emerging ethical issues. Governments can enhance widespread uptake of ICTs in agriculture and rural areas by facilitating access to soft and hard infrastructure. Meanwhile, opponents and advocates of ICT – across policy, practice and science – need to engage in an open, sincere dialogue on the future of farming in the digital era.

Conflict of Interest. The authors declare no conflict of interest.

References

AgroCares: What is the difference between precision, digital and smart farming? (2019). https://www.agrocares.com/en/news/precision-digital-smart-farming. Accessed 29 June 2019

Allahyari, M.S., Mohammadzadeh, M., Nastis, S.A.: Agricultural experts' attitude towards precision agriculture: evidence from Guilan Agricultural Organization, Northern Iran. Inf. Process. Agric. **3**(3), 183–189 (2016). https://doi.org/10.1016/j.inpa.2016.07.001

Balafoutis, A., Beck, B., Fountas, S., Vangeyte, J., Wal, T., Soto, I., Gómez-Barbero, M., Barnes, A., Eory, V.: Precision agriculture technologies positively contributing to GHG emissions mitigation, farm productivity and economics. Sustainability **9**(8), 1339 (2017). https://doi.org/10.3390/su9081339

Banhazi, T.M., Babinszky, L., Halas, V., Tscharke, M.: Precision livestock farming: precision feeding technologies and sustainable livestock production. Int. J. Agric. Biol. Eng. **5**(4), 1–8 (2012). https://doi.org/10.3965/j.ijabe.20120504.006

Barreto, L., Amaral, A.: Smart farming: cyber security challenges. In: 2018 International Conference on Intelligent Systems (IS), pp. 870–876. Institute of Electrical and Electronics Engineers (IEEE), New York (2018). https://doi.org/10.1109/IS.2018.8710531

Batte, M.T., Ehsani, M.R.: The economics of precision guidance with auto-boom control for farmer-owned agricultural sprayers. Comput. Electron. Agric. **53**(1), 28–44 (2006). https://doi.org/10.1016/j.compag.2006.03.004

Baumgart-Getz, A., Prokopy, L.S., Floress, K.: Why farmers adopt best management practice in the United States: a meta-analysis of the adoption literature. J. Environ. Manage. **96**(1), 17–25 (2012). https://doi.org/10.1016/j.jenvman.2011.10.006

Bello, O.A., Aderbigbe, F.: ICT in agricultural sustainability and food security. Int. J. Emerg. Technol. Adv. Eng. **4**(3), 508–513 (2014)

Bergtold, J.S., Raper, R.L., Schwab, E.B.: The economic benefit of improving the proximity of tillage and planting operations in cotton production with automatic steering. Appl. Eng. Agric. **25**(2), 133–143 (2009). https://doi.org/10.13031/2013.26322

Berti, G., Mulligan, C.: ICT & the future of food and Agriculture, Stockholm (2015)

BIS Research: Global Smart Farming Market, Basel, Switzerland (2018)

Bora, G.C., Nowatzki, J.F., Roberts, D.C.: Energy savings by adopting precision agriculture in rural USA. Energy Sustain. Soc. **2**(1), 22 (2012). https://doi.org/10.1186/2192-0567-2-22

Bruinsma, J.: Looking ahead in worLd food and agricuLture: Perspectives to 2050. FAO, Rome (2011). https://doi.org/I2280E/1/06.11
Bu, F., Wang, X.: A smart agriculture IoT system based on deep reinforcement learning. Future Gener. Comput. Syst. **99**, 500–507 (2019). https://doi.org/10.1016/j.future.2019.04.041
Caputo, F., Buhnova, B., Walletzký, L.: Investigating the role of smartness for sustainability: insights from the Smart Grid domain. Sustain. Sci. **13**(5), 1299–1309 (2018a). https://doi.org/10.1007/s11625-018-0555-4
Caputo, F., Evangelista, F., Russo, G.: The role of information sharing and communication strategies for improving stakeholder engagement. In: Vrontis, D., Weber, Y., Tsoukatos, V., Shams, R. (eds.) Business Models for Strategic Innovation: Cross-functional Perspectives, pp. 25–43. Routledge, London (2018b)
CEMA-Agri: Digital Farming: what does it really mean? Brussels (2017). http://www.cema-agri.org/images/publications/position-papers/CEMA_Digital_Farming_-_Agriculture_4.0__13_02_2017_0.pdf
Chapman, R., Slaymaker, T., Young, J.: Livelihoods Approaches to Information and Communication in Support of Rural Poverty Elimination and Food Security. Bridging Research and Policy (2003). https://www.odi.org/resources/docs/208.pdf
Charo, R.A.: Yellow lights for emerging technologies. Science **349**(6246), 384–385 (2015). https://doi.org/10.1126/science.aab3885
Dao, V., Langella, I., Carbo, J.: From green to sustainability: information technology and an integrated sustainability framework. J. Strateg. Inf. Syst. **20**(1), 63–79 (2011). https://doi.org/10.1016/j.jsis.2011.01.002
Davies, A.R.: Co-creating sustainable eating futures: technology, ICT and citizen–consumer ambivalence. Futures **62**, 181–193 (2014). https://doi.org/10.1016/j.futures.2014.04.006
Del Giudice, M., Khan, Z., De Silva, M., Scuotto, V., Caputo, F., Carayannis, E.: The microlevel actions undertaken by owner-managers in improving the sustainability practices of cultural and creative small and medium enterprises: a United Kingdom-Italy comparison. J. Organ. Behav. **38**(9), 1396–1414 (2017). https://doi.org/10.1002/job.2237
DLG: Digital Agriculture – Opportunities, risks, acceptance. Frankfurt am Main (2018). https://www.smart-akis.com/wp-content/uploads/2018/03/Folder_Position_Digitalisierung_e_IT.pdf
El Bilali, H.: Relation between innovation and sustainability in the agro-food system. Ital. J. Food Sci. **30**, 200–225 (2018)
El Bilali, H.: Research on agro-food sustainability transitions: a systematic review of research themes and an analysis of research gaps. J. Clean. Prod. **221**, 353–364 (2019). https://doi.org/10.1016/j.jclepro.2019.02.232
El Bilali, H., Allahyari, M.S.: Transition towards sustainability in agriculture and food systems: role of information and communication technologies. Inf. Process. Agric. **5**(4), 456–464 (2018). https://doi.org/10.1016/j.inpa.2018.06.006
Ess, D.R., Morgan, M.T.: The Precision Farming Guide for Agriculturalists, vol. 3. Deere and Company, Moline (2003)
European Union: Measures at Farm Level to Reduce Greenhouse Gas Emissions from EU Agriculture (2014). http://www.europarl.europa.eu/RegData/etudes/note/join/2014/513997/IPOL-AGRI_NT(2014)513997_EN.pdf
European Union: Precision Farming: Sowing the seeds of a new agricultural revolution (2017). http://publications.europa.eu/resource/genpub/PUB_ZZ0217813ENN.1.1. https://doi.org/10.2830/601846
Evans, R.G., King, B.A.: Site-specific sprinkler irrigation in a water limited future. In: 5th National Decennial Irrigation Conference Proceedings, Phoenix Convention Center, Phoenix, Arizona (USA), 5–8 December 2010. American Society of Agricultural and Biological Engineers, St. Joseph (2010). https://doi.org/10.13031/2013.35829

FAO: ICT uses for inclusive agricultural value chains, Rome (2013)

FAO: Building a common vision for sustainable food and agriculture - Principles and Approaches, Rome (2014)

FAO: Success stories on information and communication technologies for agriculture and rural development, Bangkok (2015)

FAO: The State of Food and Agriculture - Climate Change, Agriculture and Food Security, Rome (2016). http://www.fao.org/3/a-i6030e.pdf

FAO: Strategic work of FAO for Sustainable Food and Agriculture, Rome (2017)

FAO: E-agriculture (2019). http://www.fao.org/e-agriculture/activity. Accessed 29 June 2019

FAO, IFPRI, & OECD: Information and Communication Technology (ICT) in Agriculture - A Report to the G20 Agricultural Deputies, Rome (2017). http://www.fao.org/3/a-i7961e.pdf

FAO, & ITU: E-Agriculture Strategy Guide - Piloted in Asia-Pacific countries, Bangkok (2016). http://www.fao.org/3/i5564e/I5564E.pdf

FAO, & ITU: E-agriculture in action: Blockchain for agriculture - Opportunities and challenges, Bangkok (2019)

Foley, J.A., Ramankutty, N., Brauman, K.A., Cassidy, E.S., Gerber, J.S., Johnston, M., Mueller, N.D., O'Connell, C., Ray, D.K., West, P.C., Balzer, C., Zaks, D.P.M.: Solutions for a cultivated planet. Nature **478**(7369), 337–342 (2011). https://doi.org/10.1038/nature10452

G20: Towards food and water security: Fostering sustainability, advancing innovation. G20 Agriculture Ministers' Action Plan 2017, Berlin, 22nd January 2017

GeSI: SMART 2020: Enabling the low carbon economy in the information age. A report by The Climate Group on behalf of the Global eSustainability Initiative (GeSI) (2008). https://www.theclimategroup.org/sites/default/files/archive/files/Smart2020Report.pdf

Hedley, C.: The role of precision agriculture for improved nutrient management on farms. J. Sci. Food Agric. **95**(1), 12–19 (2015). https://doi.org/10.1002/jsfa.6734

HydroSence: Innovative Precision Technologies for Optimised Irrigation and Integrated Crop Management in a Water-Limited Agrosystem, Athens (2013)

IAASTD: Agriculture at a Crossroads. International Assessment of Agricultural Knowledge, Science and Technology for Development. Global Report, Washington, DC (2009)

IDEV: ICT and Agriculture in the Global South (2016). http://worldaccord.org/wp-content/uploads/2017/01/IDEV-4500-Report-World-Accord-Part-1.pdf

Jensen, H.G., Jacobsen, L.-B., Pedersen, S.M., Tavella, E.: Socioeconomic impact of widespread adoption of precision farming and controlled traffic systems in Denmark. Precision Agric. **13**(6), 661–677 (2012). https://doi.org/10.1007/s11119-012-9276-3

Kaloxylos, A., Wolfert, J., Verwaart, T., Terol, C.M., Brewster, C., Robbemond, R., Sundmaker, H.: The use of future internet technologies in the agriculture and food sectors: integrating the supply chain. Procedia Technol. **8**, 51–60 (2013). https://doi.org/10.1016/j.protcy.2013.11.009

Koch, B., Khosla, R., Frasier, W.M., Westfall, D.G., Inman, D.: Economic feasibility of variable-rate nitrogen application utilizing site-specific management zones. Agron. J. **96**(6), 1572 (2004). https://doi.org/10.2134/agronj2004.1572

Kolshus, K., Pastore, A., Treinen, S., Van der Elstraeten, A.: E-agriculture 10 year Review Report: Implementation of the World Summit on the Information Society (WSIS) Action Line C7. ICT Applications: e-agriculture, Rome (2015). www.fao.org/3/a-i4605e.pdf

Kutter, T., Tiemann, S., Siebert, R., Fountas, S.: The role of communication and co-operation in the adoption of precision farming. Precision Agric. **12**(1), 2–17 (2011). https://doi.org/10.1007/s11119-009-9150-0

Lambert, D., Lowenberg-De Boer, J.: Precision Agriculture Profitability Review, West Lafayette, IN (USA) (2000)

Lehmann, R.J., Reiche, R., Schiefer, G.: Future internet and the agri-food sector: state-of-the-art in literature and research. Comput. Electron. Agric. **89**, 158–174 (2012). https://doi.org/10.1016/j.compag.2012.09.005

Lin, J., Shen, Z., Zhang, A., Chai, Y.: Blockchain and IoT based food traceability for smart agriculture. In: Proceedings of the 3rd International Conference on Crowd Science and Engineering – ICCSE 2018, pp. 1–6. ACM Press, New York (2018). https://doi.org/10.1145/3265689.3265692

Magnin, C.: How big data will revolutionize the global food chain (2016). https://www.mckinsey.it/file/6335/download?token=BhKeEm5a

Mat, I., Mohd Kassim, M.R., Harun, A.N., Yusoff, I.M.: Smart agriculture using Internet of Things. In: 2018 IEEE Conference on Open Systems (ICOS), pp. 54–59. Institute of Electrical and Electronics Engineers (IEEE) (2018). https://doi.org/10.1109/ICOS.2018.8632817

McBratney, A., Whelan, B., Ancev, T., Bouma, J.: Future directions of precision agriculture. Precision Agric. **6**(1), 7–23 (2005). https://doi.org/10.1007/s11119-005-0681-8

Mintert, J., Widmar, D., Langemeier, M., Boehlje, M., Erickson, B.: The challenges of precision agriculture: is big data the answer? Paper presented at the Southern Agricultural Economics Association (SAEA) Annual Meeting, San Antonio, Texas, 6–9 February 2016 (2016). https://ageconsearch.umn.edu/bitstream/230057/2/THECHALLENGESOFPRECISION AGRICULTURE_manuscript_SAEA_2016.pdf

Mutchek, M.A., Williams, E.D.: Design space characterization for meeting cost and carbon reduction goals. J. Ind. Ecol. **14**(5), 727–739 (2010). https://doi.org/10.1111/j.1530-9290.2010.00282.x

Nakasone, E., Torero, M., Minten, B.: The power of information: the ICT revolution in agricultural development. Ann. Rev. Resour. Econ. **6**(1), 533–550 (2014). https://doi.org/10.1146/annurev-resource-100913-012714

Pivoto, D., Waquil, P.D., Talamini, E., Finocchio, C.P.S., Dalla Corte, V.F., de Vargas Mores, G.: Scientific development of smart farming technologies and their application in Brazil. Inf. Process. Agric. **5**(1), 21–32 (2018). https://doi.org/10.1016/j.inpa.2017.12.002

Plant, R.E., Pettygrove, G.S., Reinert, W.R.: Precision agriculture can increase profits and limit environmental impacts. Calif. Agric. **54**, 66–71 (2000)

Popli, S., Jha, R.K., Jain, S.: A survey on energy efficient Narrowband Internet of Things (NBIoT): architecture, application and challenges. IEEE Access **7**, 16739–16776 (2019). https://doi.org/10.1109/ACCESS.2018.2881533

Poppe, K.: Big opportunities for big data in food and agriculture (2016). https://www.oecd.org/tad/events/Session2_KrijnPoppeOECDBigData.pdf. Accessed 10 Mar 2018

Poppe, K.J., Wolfert, S., Verdouw, C., Verwaart, T.: Information and communication technology as a driver for change in agri-food chains. EuroChoices **12**(1), 60–65 (2013). https://doi.org/10.1111/1746-692X.12022

Postel, S.L.: Entering an era of water scarcity: the challenges ahead. Ecol. Appl. **10**(4), 941–948 (2000). https://doi.org/10.1890/1051-0761

Rajeswari, S., Suthendran, K., Rajakumar, K.: A smart agricultural model by integrating IoT, mobile and cloud-based big data analytics. In: 2017 International Conference on Intelligent Computing and Control (I2C2), pp. 1–5. Institute of Electrical and Electronics Engineers (IEEE), New York (2017). https://doi.org/10.1109/I2C2.2017.8321902

Saidi, A.: An introduction to precision agriculture and its role in sustainable development of agriculture. J. Basic Appl. Sci. Res. **3**(1), 417–421 (2013)

Sarantis, M.: Europe Entering the Era of Precision Agriculture (2015). https://www.euractiv.com/section/science-policymaking/news/europe-entering-the-era-of-precision-agriculture. Accessed 6 Jan 2018

Schrijver, R., Poppe, K., Daheim, C.: Precision agriculture and the future of farming in Europe, Brussels (2016). https://doi.org/10.2861/020809

Shockley, J.M., Dillon, C.R., Stombaugh, T.S.: A whole farm analysis of the influence of auto-steer navigation on net returns, risk, and production practices. J. Agric. Appl. Econ. **43**, 57–75 (2015)

Shyamala Devi, M., Suguna, R., Joshi, A.S., Bagate, R.A.: Design of IoT blockchain based smart agriculture for enlightening safety and security, pp. 7–19 (2019). https://doi.org/10.1007/978-981-13-8300-7_2

Singh, M., Marchis, A., Capri, E.: Greening, new frontiers for research and employment in the agro-food sector. Sci. Total Environ. **472**, 437–443 (2014). https://doi.org/10.1016/j.scitotenv.2013.11.078

Smart AKIS: What is Smart Farming? (2016). https://www.smart-akis.com/index.php/network/what-is-smart-farming/. Accessed 13 June 2019

Sørensen, C.A.G., Kateris, D., Bochtis, D.: ICT Innovations and Smart Farming, pp. 1–19 (2019). https://doi.org/10.1007/978-3-030-12998-9_1

Svenfelt, Å., Zapico, J.L.: Sustainable food systems with ICT? In: Proceedings of the 4th International Conference on ICT for Sustainability (ICT4S 2016), Amsterdam, 29 August–1 September 2016, pp. 194–201 (2016)

Swinton, S.M., Lowenberg-DeBoer, J.: Evaluating the profitability of site-specific farming. JPA **11**(4), 439 (1998). https://doi.org/10.2134/jpa1998.0439

Sylvester, G.: Information and communication technologies for sustainable agriculture - Indicators from Asia and the Pacific, Bangkok (2013)

Sylvester, G.: Success Stories on Information and Communication Technologies for Agriculture and Rural Development, Bangkok (2015). www.fao.org/3/a-i4622e.pdf

Tekin, A.B.: Variable rate fertilizer application in Turkish wheat agriculture: economic assessment. Afr. J. Agric. Res. **5**(8), 647–652 (2010). https://doi.org/10.5897/AJAR09.562

Thöni, A., Tjoa, A.M.: Information technology for sustainable supply chain management: a literature survey. Enterp. Inf. Syst. **11**(6), 828–858 (2017). https://doi.org/10.1080/17517575.2015.1091950

Timmermann, C., Gerhards, R., Kühbauch, W.: The economic impact of site-specific weed control. Precision Agric. **4**(3), 249–260 (2003). https://doi.org/10.1023/A:1024988022674

Umstatter, C.: The evolution of virtual fences: a review. Comput. Electron. Agric. **75**(1), 10–22 (2011). https://doi.org/10.1016/j.compag.2010.10.005

UN-ESCAP: Information and communications technology for food security and sustainable agriculture in the knowledge economy (2008). http://www.unescap.org/sites/default/files/CICT_1-E.pdf

UNEP: Sustainable food systems programme (2018). http://web.unep.org/10yfp/programmes/sustainable-food-systems-programme. Accessed 11 June 2018

United Nations: The Millennium Development Goals Report 2015, New York (2015a). http://www.un.org/millenniumgoals/2015_MDG_Report/pdf/MDG2015rev(July1).pdf

United Nations: Transforming Our World: The 2030 Agenda for Sustainable Development. Resolution adopted by the General Assembly on 25 September 2015, New York (2015b). http://www.un.org/ga/search/view_doc.asp?symbol=A/RES/70/1&Lang=E

van Marrewijk, M., Hardjono, T.W.: European corporate sustainability framework for managing complexity and corporate transformation. J. Bus. Ethics **44**(2), 121–132 (2003). https://doi.org/10.1023/A:1023335414065

Verdouw, C.N., Wolfert, J., Beulens, A.J.M., Rialland, A.: Virtualization of food supply chains with the internet of things. J. Food Eng. **176**, 128–136 (2016). https://doi.org/10.1016/j.jfoodeng.2015.11.009

Walter, A., Finger, R., Huber, R., Buchmann, N.: Smart farming is key to developing sustainable agriculture. Proc. Natl. Acad. Sci. **114**(24), 6148–6150 (2017). https://doi.org/10.1073/pnas.1707462114

Westermann, O., Förch, W., Thornton, P., Körner, J., Cramer, L., Campbell, B.: Scaling up agricultural interventions: case studies of climate-smart agriculture. Agric. Syst. **165**, 283–293 (2018). https://doi.org/10.1016/j.agsy.2018.07.007

Wognum, N., Bremmers, H.: Environmental transparency of food supply chains - current status and challenges. In: Global Perspective for Competitive Enterprise, Economy and Ecology - Proceedings of the 16th ISPE International Conference on Concurrent Engineering, pp. 645–652 (2009)

Wognum, P.M., Bremmers, H., Trienekens, J.H., van der Vorst, J.G.A.J., Bloemhof, J.M.: Systems for sustainability and transparency of food supply chains – current status and challenges. Adv. Eng. Inform. **25**(1), 65–76 (2011). https://doi.org/10.1016/j.aei.2010.06.001

Wolfert, J., Sørensen, C.G., Goense, D.: A future internet collaboration platform for safe and healthy food from farm to fork. In: 2014 Annual Global Conference SRII, San Jose, CA, USA, pp. 266–273 (2014)

Wolfert, S.: Future Internet of ICT in Agriculture, Food and the Environment (2015). https://www.wur.nl/web/file?uuid=11a80128-82b4-4c90-8980-ce3fd5cb9f83&owner=128fe7fd-b753-4930-91d5-3e9a3c1138b8. Accessed 11 Mar 2018

Wolfert, S., Ge, L., Verdouw, C., Bogaardt, M.-J.: Big data in smart farming – a review. Agric. Syst. **153**, 69–80 (2017). https://doi.org/10.1016/j.agsy.2017.01.023

World Bank: ICT in agriculture: connecting smallholders to knowledge, networks, and institutions, Washington, DC (2011). http://documents.worldbank.org/curated/en/455701468340165132/ICT-in-agriculture-connecting-smallholders-to-knowledge-networks-and-institutions

World Bank: ICT in agriculture – Resources (2012). https://web.archive.org/web/20160320050224/. http://www.ictinagriculture.org/content/resources. Accessed 29 June 2019

World Bank: ICT in Agriculture: Connecting Smallholders to Knowledge, Networks, and Institutions, Updated edition, Washington DC (2017). http://documents.worldbank.org/curated/en/522141499680975973/pdf/117319-PUB-Date-6-27-2017-PUBLIC.pdf

Zarco-Tejada, P., Hubbard, N., Loudjani, P.: Precision Agriculture: An Opportunity for EU Farmers - Potential Support with the CAP 2014–2020 (2014). http://www.europarl.europa.eu/RegData/etudes/note/join/2014/529049/IPOL-AGRI_NT%282014%29529049_EN.pdf. Accessed 14 Mar 2018

Green Synthesis and Characterization of Silver Nanoparticles Using Fresh Leaf Extract of *Aloe vera barbadensis Miller, Aloe vera* and *Sempervivum tectorum* and Its Antimicrobial Activity Studies

Anera Kazlagić[1(✉)], Amela Lagumdžija[1], Berina Borovac[1], Saud Hamidović[1], Odej Ali Abud[2], and Enisa Omanović-Mikličanin[1]

[1] Faculty of Agriculture and Food Sciences, University of Sarajevo, Zmaja od Bosne 8, 71000 Sarajevo, Bosnia and Herzegovina
e.omanovic-miklicanin@ppf.unsa.ba

[2] Department of Radiology, Clinical Center University of Sarajevo, Bolnička 25, 71000 Sarajevo, Bosnia and Herzegovina

Abstract. New advances in technology have inspired the development and application of new products in food industry, which brings to new challenges featuring food safety. Demands from consumers for more natural foodstuff are increasing, which puts aside conventional methods for controlling the microbial growth, such as thermal processing or adding chemical preservatives. New methods are being highlighted, due to features such as controlling microbial growth without damaging the product. Nowadays, nanotechnology is considered as promising technology for the food industry in general, as well as for new methodologies in the entire agri-food chain.

The use of plant extract while synthesizing metal nanoparticles represents an eco-friendly and promising alternative to traditional chemical methods. In this study, a new green method of preparing silver nanoparticles (AgNPs) using fresh leaf extract of *Aloe vera barbadensis Miller*, *Aloe vera* and *Sempervivum tectorum* has been proposed. Antimicrobial activity study was assessed using Kirby-Bauer test, commonly known as disc diffusion method. AgNPs demonstrated antimicrobial activity against *Salmonella spp.* but showed no effect on *Escherichia coli*. This study indicates that AgNPs synthesized this way could be a new potential candidate for use in agricultural, biological, and pharmaceutical applications for fighting infections caused by microorganisms.

Keywords: Green synthesis · Silver nanoparticles · Spectrophotometry · Antimicrobial activity · *Aloe vera* · *Sempervivum tectorum*

1 Introduction

One of the most active areas of research in modern material sciences is the field of nanotechnology. Nowadays, nanotechnology is making an impact in all spheres of human life, especially in biomedical devices and biotechnology [1].

M. Brka et al. (Eds.): AgriConf 2019, IFMBE Proceedings 78, pp. 335–341, 2020.
https://doi.org/10.1007/978-3-030-40049-1_42

Nanotechnology has an essential role in agriculture and food production. Nanoparticles (NPs) exhibit different properties and various possibilities of application. In compassion to other materials, nanomaterials are making headway in almost all sectors from production to consumption [2].

Nanoparticles (size 1–100 nm) have numerous applications in various fields such as food industry, pharmaceuticals, agriculture etc. [3].

In comparison to other metal nanoparticles, silver has been used in numerous applications because it retains antimicrobial activity against pathogens. It is utilized in the form of pure free metal or as a compound. Synthesis of silver nanoparticles (AgNPs) is of high interest to scientists because of their broad application in the various fields. Due to the production of non-eco-friendly by-products while synthesizing AgNPs, the development of a more ecologically safe green methods is progressing as an important branch of nanotechnology [4, 5].

Green synthesis of NPs is considered as an eco-friendly approach, due to utilization of the environmental friendly chemicals. In our previous research, formation of AgNPs was secured using gallic acid which acted as a reducing and stabilizing agent simultaneously [6]. In this research, we used Sodium dodecyl sulfate (SDS) to prevent aggregation of newly synthesized AgNPs. Nanoparticles synthesized using green technology have diverse natures, with greater stability and appropriate dimensions since they are synthesized using a one-step procedure [7]. Utilization of plants as a green source for nanoparticles synthesis is most commonly practiced as plants are the hub of a wide range of phytochemicals which can act as reducing and stabilizing agent for NPs synthesis [8]. In fact, in this method, there is no need to use high pressure, energy, temperature, and toxic chemicals [9].

2 Materials and Methods

2.1 Materials

Silver nitrate $AgNO_3$ Merck pro analysis, Sodium dodecyl sulfate $NaC_{12}H_{25}SO_4$

Sigma Aldrich, redistilled water, *Aloe vera barbadensis Miller* leaves, *Aloe vera* leaves, *Sempervivum tectorum* leaves.

2.2 Microorganisms

Escherichia coli and *Salmonella spp.* (biochemically isolated and tested by using I test, MC test, VP test, C test, Urea test, H2S test) were obtained from the river Lasva, Bosnia and Herzegovina.

2.3 Preparation of *Aloe vera barbadensis Miller* and *Aloe vera* Extract

The plant leaves were washed with tap water first, to remove any visible impurities, and after that with redistilled water. Leaves were left to dry in air, and the following step was cutting them into small pieces. This size reduction step facilitates in increasing the extraction rate, because of increased surface area.

In a 250 mL beaker, a 100 mL of redistilled water and 10 g of *A. vera barbadensis Miller* and in another beaker, the same amount of *A. vera* leaves was mixed and left in water bath for 20 min, on 80 °C. After the extract was cooled down on room temperature, the filtration was performed, following the centrifugation which was set on 12 000 rpm, for 15 min. 5 mL of 1 mM SDS ($NaC_{12}H_{25}SO_4$) solution was added to the final extract of *Aloe vera*, due to its stabilization necessity. AgNPs synthesized using *Aloe vera barbadensis Miller* showed great stability, so in this case there was no need for SDS addition.

2.4 Preparation of *Sempervivum tectorum* Extract

Preparation of *Sempervivum tectorum* leaves includes washing with tap water and afterwards with redistilled water, leaving them to dry in air and cutting them into small pieces. 20 g of *S. tectorum* was mixed with 100 mL of redistilled water, and heated on 80 °C for 10 min. After the extract was cooled down on room temperature, the filtration was performed, also following the centrifugation for 15 min (12 000 rpm).

5 mL of 1 mM SDS solution was added to the final *S. tectorum* extract.

2.5 Synthesis of AgNPs Using *Aloe vera barbadensis Miller*, *Aloe vera* and *Sempervivum tectorum* Extract

10 mL of plant's extract was slowly, drop by drop added to 90 mL of 1 mmol/L $AgNO_3$ solution on a magnetic stirrer. The observation of pale brown colour after 24 h of reaction indicated the formation of AgNPs.

2.6 Characterization of Synthesized AgNPs Using UV/VIS Spectroscopy

UV-visible spectroscopy is one of the most widely used techniques for structural characterization of silver nanoparticles. To verify reduction of silver ions, the solution was scanned in the range of 200–600 nm in a spectrophotometer (Carry, 1E, UV-VIS) using a quartz cuvette with redistilled water as the reference. The following step was evaluation of maximum absorbance of synthesized nanoparticles.

3 Antibacterial Activity of Silver Nanoparticles

3.1 Disc Diffusion Method

In order to determine the antimicrobial activity of the synthesized silver nanoparticles, the Kirby-Bauer test, commonly known as disc diffusion method, was used. Pure bacterial cultures (*Escherichia coli* and *Salmonella spp.*) were swabbed uniformly across a culture plates. A filter paper disks, impregnated with the AgNPs were then placed on the surface of the Mueller-Hinton agar. The next step was incubation of the dishes for 24 h at 37 °C.

4 Results and Discussion

Silver nanoparticles were synthesized according to the previously described method, the colloidal solution turned pale brown (Fig. 1), indicating the formation of AgNPs.

Fig. 1. Silver nanoparticles synthesized using *Aloe vera barbadensis Miller extract (1), Aloe vera* extract *(2)*, and *Sempervivum tectorum* extract (3)

The absorption spectrum (Fig. 2) of the pale yellow-brown silver colloids prepared by plant extract reduction showed a surface plasmon absorption band with a maximum of 423 nm for AgNPs made of *Aloe vera barbadensis Miller* and *Aloe vera* extract, and 440 nm for AgNPs made using *Sempervivum tectorum* extract, indicating the presence of Ag nanoparticles.

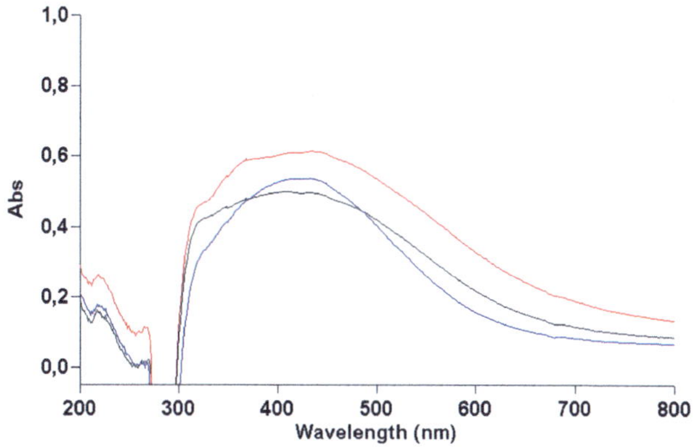

Fig. 2. UV/VIS spectrum of AgNPs using *A. vera barbadensis Miller* (green line, λ_{max} 423 nm), AgNPs using *A. vera* (blue line, λ_{max} 423 nm) and AgNPs using *S. tectorum* (red line, λ_{max} 440 nm).

Table 1. Diameters of inhibition zone (Φ) for *Escherichia coli* and *Salmonella spp*, for Disc diffusion method; * indicates presence of 5 mL of SDS

Sample	*Escherichia coli*	*Salmonella spp.*
No.	Φ (mm)	Φ (mm)
Disc diffusion		
1	0	18
2*	0	10
3*	0	16,5

Results obtained with this method were compared with the standardized values for antibiotic Doxycycline for bacteria *Escherichia coli* and *Salmonella spp.* (Table 2).

Table 2. Zone of growth inhibition for both *Escherichia coli* and *Salmonella spp.* (*Enterobacteriaceae*)

Zone of growth inhibition			
Doxycycline	R	I	S
Enterobacteriaceae	≤ 10	11–13	≥ 14

In Table 2, **R, I and S category** have been shown. **S category** states that the treatment with antibiotics is very successful, **I category** states that there is also possibility of successful inhibition of microorganism growth, and **R category** presents resistant microorganism groups, which does not show any sensitivity.

According to Disk diffusion method, and in comparison to Doxycycline, AgNPs synthesized using *A. vera B.M extract* had great, the highest potential of all examined samples, to inhibit the growth of *Salmonela spp.* (18 mm diameter). This AgNPs have no potential to inhibit the growth of *Escherichia coli* (Table 1).

AgNPs synthesized using *A. vera* extract with addition of 5 mL of SDS, showed sensitivity on *Salmonella spp.* (10 mm diameter), and not activity at all on *Escherichia coli,* which is also presented in Table 1.

AgNPs synthesized using *S. tectorum* extract with addition of 5 mL of SDS, showed great potential (16 mm) to inhibit the growth of *Salmonela spp., and* no potential at all to inhibit the growth of *Escherichia coli.* (Table 1).

5 Conclusion

A rapid green method of synthesizing stable AgNPs has been developed by using fresh leaf extract of *Aloe vera B.M, Aloe vera* and *Sempervivum tectorum.* Results obtained in the studies support the idea of using these plant extract as a reducing agent to produce stable silver nanoparticles in a green manner. While assessing the antimicrobial effect on *Salmonella spp.* and *Eschericia coli,* we came to the conclusion that AgNPs synthesized in this green, eco-friendly way, have influence on inhibiting *Salmonella spp.* grow and multiplication, but no effect was showed on *Eschericia coli.*

Conflict of Interest. The authors declare that there are no conflicts of interest regarding the publication of this paper.

References

1. Parida, K., Bindhani, B.K., Nayak, P.: Green synthesis and characterization of gold nanoparticles using onion (Allium cepa) extract. World J. Nano Sci. Eng. **1**, 93–98 (2011)
2. Mishra, M., Dashora, K., Srivastava, A., Fasake, V.D., Nag, R.H.: Prospects, challenges and need for regulation of nanotechnology with special reference to India. Ecotoxicol. Environ. Saf. **171**, 677–682 (2018)
3. Ashraf, J.M., Ansari, M.A., Khan, H.M., Alzohairy, M.A., Choi, I.: Green synthesis of silver nanoparticles and characterization of their inhibitory effects on AGEs formation using biophysical techniques. Sci. Rep. **6**, 20414 (2016)
4. Kalaiselvia, D., Mohankumara, A., Shanmugama, G., Nivithab, S., Sundararaj, P.: Green synthesis of silver nanoparticles using latex extract of Euphorbia tirucalli: a novel approach for the management of root knot nematode, Meloidogyne incognita. Crop Prot. **117**, 108–114 (2018)
5. Tippayawat, P., Phromviyo, N., Boureoy, P., Chompoosor, A.: Green synthesis of silver nanoparticles in aloe vera plant extract prepared by a hydrothermal method and their synergistic antibacterial activity. Peer J. **4**, e2589 (2016).
6. Kazlagić, A., Omanović-Mikličanin, E., Hamidović, S.: Synthesis, characterization and antimicrobial activity of silver nanoparticles. In: Karabegović, I. (ed.) New Technologies, Development and Application II. NT 2019. Lecture Notes in Networks and Systems, vol. 76. Springer, Cham (2020)
7. Zhang, Y., Yang, D., Kong, Y., Wang, X., Pandoli, O., Guo, G.: Synergetic antibacterial effects of silver nanoparticles aloe vera prepared via a green method. Nano Biomed. Eng. **2**(4), 252–257 (2010)

8. Happy, A., Soumya, M., Kumar, S.V., Rajeshkumar, S., Sheba, R.D., Lakshmi, T., Nallaswamy, V.D.: Phyto-assisted synthesis of zinc oxide nanoparticles using Cassia *alata* and its antibacterial activity against *Escherichia coli*. Biochem. Biophys. Rep. **17**, 208–211 (2019)
9. Joghee, S., Ganeshan, P., Vincent, A., et al.: Ecofriendly biosynthesis of zinc oxide and magnesium oxide particles from medicinal plant *Pisonia grandis* R. Br. leaf extract and their antimicrobial activity. BioNanoSci. **9**, 141–154 (2019)

Effects of Stabilizing Agent on the Green Synthesized Silver Nanoparticles and Its Antimicrobial Activity Studies

Anera Kazlagić[1(✉)], Ediba Mršo[1], Odej Ali Abud[2], Berina Borovac[1], Saud Hamidović[1], and Enisa Omanović-Mikličanin[1]

[1] Faculty of Agriculture and Food Sciences, University of Sarajevo, Zmaja od Bosne 8, 71000 Sarajevo, Bosnia and Herzegovina
e.omanovic-miklicanin@ppf.unsa.ba
[2] Department of Radiology, Clinical Center University of Sarajevo, Bolnička 25, 71000 Sarajevo, Bosnia and Herzegovina

Abstract. In this study, a new green method of preparing silver nanoparticles (AgNPs) using fresh leaf extract of different types of onion has been proposed. The effects of stabilizing agent on the green synthesized AgNPs has been investigated, as well as the antimicrobial activity using Disk diffusion method. AgNPs demonstrated antimicrobial activity against *Salmonella spp.*

Keywords: Green synthesis · Stable silver nanoparticles · Spectrophotometry · Antimicrobial activity

1 Introduction

Nanotechnology is considered as one of the key enabling technologies recognized by European Union. Novel nanomaterials and nanotechnology are making headway in almost all sectors from production to consumption. This scientific field opens new insights in various sectors of food packaging, animal husbandry, environment and industry application making it one of the most intriguing industrial phenomena of the modern era. Nanoparticles have unique beneficial electrical, optical, magnetic, anti-inflammatory, wound-healing and antimicrobial properties, which enables them to target the disease most effectively unlike their bulk form [1]. Nanomaterials can be synthesized by different methods such as chemical, physical, and biocompatible method. Most chemical or physical methods likely cause environmental pollutions and/or waste problems, since the chemical procedures involved in the synthesis of nanomaterials generate a large amount of hazardous by-products [2].

To address the challenge of making technology more sustainable, eco-friendly and of precise application, nanotechnology is emerging up as a technology of choice.

Looking to the eco-friendly properties of nanotechnology, it is rapidly stepping up in the role of being considered for development of safer pesticides formulations as a substitute to the conventional pesticides [3]. The alternatives for quick plant protection was being searched and the idea of green revolution started stepping in many parts of the world. During the green revolution, extensive use of chemicals was done for crop

M. Brka et al. (Eds.): AgriConf 2019, IFMBE Proceedings 78, pp. 342–348, 2020.
https://doi.org/10.1007/978-3-030-40049-1_43

protection and production. Realizing the adverse impact on environment and people, scientists started working on alternate technologies which were eco-friendly and sustainable. Use of biological control for pest management found its way in the mainstream farming. In the present time, there is a shift from macrotechnology to nanotechnology due to its precision and sustainability. The development of nanotechnology in conjunction with biotechnology has significantly expanded the application domain of nanomaterials in various fields. Nanotechnology has the potential to revolutionize the agriculture with new tools to enhancing the ability of plants to absorb nutrients. Nanoparticles have interactions at molecular level in living cells and nanotechnology in agriculture involves nanoparticles so that these particles impart some beneficial effects to the crop [4].

Green chemistry which aims to reduce or eliminate substances hazardous to human health and the environment in the design, development and implementation of chemical processes and products is becoming much more important [5].

The three main steps in the preparation of nanoparticles (NPs) to be evaluated from a green chemistry perspective are: (a) selection of the solvent medium used for the synthesis, (b) selection of an environmentally benign reducing agent and (c) selection of a non-toxic material for the stabilization of the nanoparticles [6]. Green synthesis includes a clean, safe, eco-friendly, and environmentally non-toxic method of NPs synthesis, without using high pressure, energy, temperature, and toxic chemicals [7]. Synthesis of metal and metal oxide nanoparticles through green routes and environmentally benign method has been focused by researchers to avoid the use of toxic chemicals making the process and the final product eco-friendly and non-toxic in nature. Production of nanoparticles under nontoxic green conditions is of vital importance to address growing concerns on the overall toxicity of nanoparticles for medical and technological applications [8, 9]. The use of nanoparticles in growth of plants and for the control of plant diseases is a recent practice where nanoparticles are used as additive (e.g. silver nanoparticles) [10].

Synthesis of silver nanoparticles (AgNPs) is of high interest to the scientific community because of their broad application in the cosmetic, therapeutic and pharmaceutical industry [11, 12]. The synthesis of metal and semiconductor nanoparticles is an expanding research area due to the potential applications for the development of novel technologies. Generally, nanoparticles are prepared by a variety of chemical methods which are not environmentally friendly, but there is a fast, convenient and extracellular method for the synthesis of silver nanoparticles by reducing silver nitrate with the help of onion (*Allium cepa*) extract [13].

2 Materials and Method

2.1 Materials

Silver nitrate - Merck pro analysis, Sodium dodecyl sulfate - Sigma Aldrich, redistilled water, *Allium cepa* (onion bulbs, scale leaves), *Allium cepa* (green leaves), *Allium sativum*.

2.2 Microorganisms

Escherichia coli and *Salmonella spp.* were obtained from the river Lasva, Bosnia and Herzegovina. Biochemical isolation and various tests were performed. (I test, MC test, VP test, C test, Urea test, H2S test).

2.3 Preparation of *Allium cepa* (Green Leaves) Extract

The plant leaves were washed with tap water, and after that with redistilled water. Green leaves were left to dry in air, and then cut into small pieces. This size reduction step facilitates in increasing the extraction rate, because of increased surface area.

In a 250 mL beaker, a 100 mL of redistilled water and 10 g of leaves were mixed and left in water bath for 20 min, on 50 °C. After the extract was cooled down on room temperature, the filtration was performed, following the centrifugation which was set on 6 000 rpm, for 15 min.

2.4 Preparation of *Allium cepa*, Onion Bulbs (Scale Leaves) Extract

After washing the scale leaves with water, and dying them in air, they were cut into small pieces.

In a 250 mL beaker, a 100 mL of redistilled water and 25 g of leaves were mixed and left in water bath for 10 min, on 50 °C. After the extract was cooled down on room temperature, the filtration was performed, following the centrifugation set on 6 000 rpm during 15 min.

2.5 Preparation of *Allium sativum* Extract

For the first step of preparation, *A. sativum* was prepared the same way as *A. cepa*. Total amount of 20 g of dried *A. sativum* was mixed with 200 mL redistilled water in a 500 mL beaker and left in water bath for 15 min, on 50 °C. After the extract was cooled down on room temperature, and after that filtrated, following the centrifugation which was set on 6 000 rpm, for 15 min.

2.6 Synthesis of Silver Nanoparticles

Various amount (5–10 mL) of each extract was slowly, drop by drop added to 90 mL of 1 mmol/L $AgNO_3$ solution on a magnetic stirrer. The observation of brownish color after 24 h of reaction indicated the formation of AgNPs.

2.7 Addition of SDS

Three replicate samples of AgNPs were prepared. First one was free of SDS, second and third one contained 1 and 5 mL of 0,1 M SDS, respectively. SDS was added slowly in the samples, drop by drop.

3 Characterization of Synthesized AgNPs Using UV/VIS Spectroscopy

UV-visible spectroscopy is one of the most widely used techniques for structural characterization of silver nanoparticles. To verify reduction of silver ions, the solution was scanned in the range of 200–600 nm in a spectrophotometer (Carry, 1E, UV-VIS) using a quartz cuvette with water as the reference. The following step was evaluation of maximum absorbance of synthesized nanoparticles.

4 Analysis of the Antibacterial Activity of Silver Nanoparticles

4.1 Disc Diffusion Method

Antimicrobial activity of the AgNPs was determined by using disc diffusion method, effects on *Salmonella spp.* and *Escherichia coli.*

Using the sterile swabs, 4–5 colonies were picked up from the bacterial broth.

The next step was preparation of the suspension which consist of investigated bacterial culture, *Escherichia coli* and *Salmonella spp.* After that, the equal distribution of bacterial suspension across the whole Mueller – Hinton nutrient agar media was made. That uniform layer of bacteria was achieved by rotating the plate while continuing swabbing. The next step was putting on the discs, approximately 15–22 cm apart from each other. The dishes were 24-hours incubated at 37 °C.

5 Results and Discussion

Green synthesis of silver nanoparticles was performed according to the method described in this paper. The solution turned brown, indicating the formation of silver nanoparticles. The absorption spectrum showed a surface plasmon absorption band with a maximum of 432 nm, showed on Fig. 1, (AgNPs synthesized using *A. cepa* - green leaves), and 425 nm (AgNPs synthesized using *A. cepa* - onion bulb scale leaves), showed on Fig. 2. There was no maximum of absorption for *A. sativum* showed.

UV/VIS spectrum was preformed 24 h after synthesis and after 10 days to examine the stability of synthesized AgNPs with and without SDS addition.

The addition of SDS for *A. cepa* green leaves did not take step, due to their great stability for 10 days, but for other two samples, the addition was necessary, due to aggregation tendency.

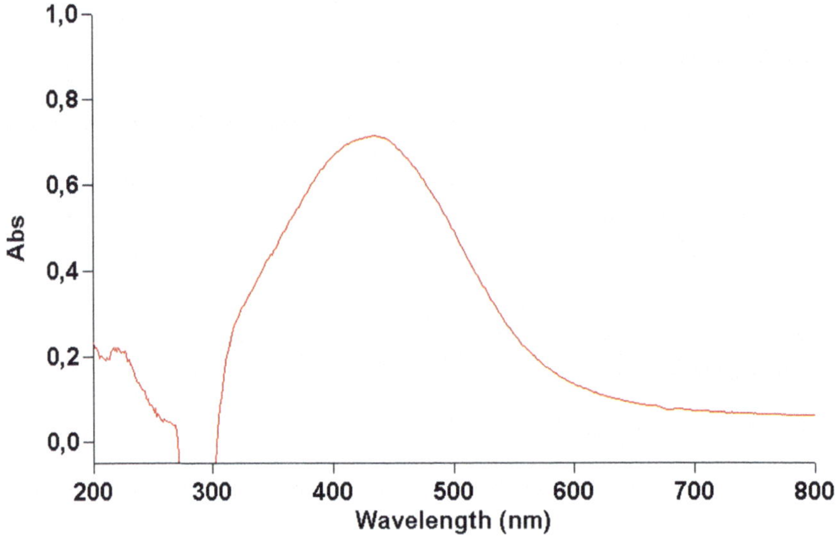

Fig. 1. *A. cepa* (green leaves) - stable without SDS after 10 days. (λ_{max} 432 nm)

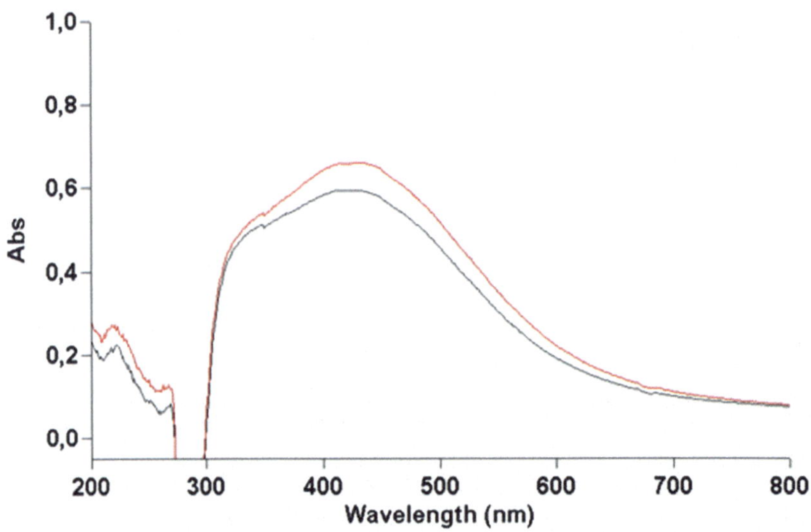

Fig. 2. *A. cepa* (onion bulbs, scale leaves) with the presence of 3 mL (blue line) and 5 mL (red line) of SDS (λ_{max} 425 nm)

A. cepa (onion bulbs) silver nanoparticles, prepared without SDS, are stable for about 10 h.

Results obtained with this method were compared with the standardized values for antibiotic Ampicillin-Sulbactam for bacteria *Escherichia coli* and *Salmonella spp* (Table 2).

Table 1. Diameters of inhibition zone (Φ) for *Escherichia coli* and *Salmonella spp.* for Disc diffusion method, where 1 represents *A. cepa* green leaves, and 2 represents *A. cepa* scale leaves

Sample	*Escherichia coli*	*Salmonella spp.*
No. **Sample**	Φ (mm)	Φ (mm)
1	/	11,5
2 + 5 mL SDS	/	12,5

Table 2. Zone of growth inhibition for both *Escherichia coli* and *Salmonella spp.* (*Enterobacteriaceae*)

Zone of growth inhibition			
Ampicillin-Sulbactam	R	I	S
Enterobacteriaceae	≤ 11	12–14	≥ 15

According to Disc Diffusion method, both AgNPs synthesized using *A. cepa (green leaves)* extract without SDS addition, and using *A. cepa (*scale leaves*)* extract with addition of 5 mL of SDS, showed intermediate sensitivity on *Salmonella spp.* (11,5 mm, 12,5 mm diameter respectively), and not activity at all on *Escherichia coli* in comparison to Ampicillin-Sulbactam, which is also presented in Tables 1 and 2.

6 Conclusion

A new rapid, green and eco-friendly method of preparing silver nanoparticles (AgNPs) using fresh leaf extract of different types of onion has been proposed. *A. cepa* (scale leaves) prepared AgNPs have tendention to aggregate overnight, and the addition of SDS as stabilizer helps them to remain stable.

The results of antimicrobial study clearly demonstrated that the silver nanoparticles inhibited the growth and multiplication of *Salmonella spp*. This intermediate antibacterial activity was observed at very low total concentrations of silver (0,001 mol/L).

Other methods and techniques, such as SEM (scanning electron microscope) and/or TEM (transmission electron microscopy) are needed in further research, in order to help determine size, shape and distribution of prepared nanoparticles.

These nanoparticles may be applicable in biomedical therapies, in the development of novel technologies, in agro-food sector with significant attention to nanoparticle's safety and toxicity.

References

1. Gunalan, S., Sivaraj, R., Rajendran, V.: Green synthesized ZnO nanoparticles against bacterial and fungal pathogens. Prog. Nat. Sci. Mater. Int. **22**, 693–700 (2013)
2. Zhang, M., Liu, M., Prest, H., Fischer, S.: Nanoparticles secreted from ivy rootlets for surface climbing. Nano Lett. **8**, 1277–1280 (2008)

3. Koirala, P., Dhakal, S., Tamrakar, A.S.: Pesticide application and food safety issue in Nepal. J. Agric. Environ. **10**, 128–132 (2009)
4. Kotegooda, N., Munaweera, I.: A green show release fertilizer composition based on urea-modified hydroxyapatite nanoparticles encapsulated wood. Curr. Sci. **101**, 43–78 (2011)
5. Poliakoff, M., Anastas, P.: A Principled Stance. Nature **413**, 257 (2001). https://doi.org/10.1038/35095133
6. Mishraa, M., Dashorab, K., Srivastavaa, A., Fasakea, V.D., Nag, R.H.: Prospects, challenges and need for regulation of nanotechnology with special reference to India. Ecotoxicol. Environ. Saf. **171**, 677–682 (2019)
7. Jeong, S., Yeo, S., Yi, S.: Antibacterial characterization of silver nanoparticles against E. coli ATCC-15224. J. Mater. Sci. **40**, 5407 (2005)
8. Hardman, R.: A Toxicologic review of quantum dots: toxicity depends on physicochemical and environmental factors. Environ. Health Perspect. **114**(2), 165–172 (2006). https://doi.org/10.1289/ehp.8284
9. Lewinski, N., Colvin, V., Drezek, R.: Cytotoxicity of nanoparticles. Small **4**(1), 26–49 (2008). https://doi.org/10.1002/smll.200700595
10. Zheng, L., Hong, F., Lu, S., Liu, C.: Effect of nano-TiO2 on strength of naturally aged seeds and growth of spinach. Biol. Trace Elem. Res. **104**(1), 83–91 (2005)
11. Bhattacharya, R., Mukherjee, P.: Biological properties of "naked" metal nanoparticles. Adv. Drug Deliv. Rev. **60**, 1289–1306 (2008). https://doi.org/10.1016/j.addr.2008.03.013
12. Bhumkar, D.R., Joshi, H.M., Sastry, M., Pokharkar, V.B.: Chitosan reduced gold nanoparticles as novel carriers for transmucosal delivery of insulin. Pharm. Res. **24**, 1415–1426 (2007). https://doi.org/10.1007/s11095-007-9257-9
13. Saxena, A., Tripathi, R.M., Singh, R.P.: Biological synthesis of silver nanoparticles by using onion (*Allium* cepa) extract and their antibacterial activity. Dig. J. Nanomater. Biostruct. **5**(2), 427–432 (2010). Amity Institute of Nanotechnology, Amity University, Noida – 201301, U.P, India

Application and Impact of Nanotechnology in Sport

Maid Ćibo(✉), Adis Šator, Anera Kazlagić, and Enisa Omanović-Mikličanin

Faculty of Agriculture and Food Sciences, University of Sarajevo, Zmaja od Bosne 8, 71 000 Sarajevo, Bosnia and Herzegovina
maid_cibo@hotmail.com, adissaca@hotmail.com, {a.kazlagic,e.omanovic-miklicanin}@ppf.unsa.ba

Abstract. Nanotechnology has been widely used in the last decade. Today, nanotechnology brings many benefits in our daily lives.

The term "nano" in theory appeared much earlier in the same application in various areas, due to not having adequate equipment nor technology that could explore such "small" particles.

In addition to all possible applications, nanotechnology, has one of the most important application and that is sports. Improvement of sports equipment by nanoparticles in many sports it brought to a clear development in the performance of athletes, achieving better results and quality of new equipment.

In this review, we are approaching nanotechnologies and sports, the possibility of applying nanomaterials such as carbon nanotubes, fullerenes, carbon nanoparticles, silicon dioxide nanoparticles and others to create sports equipment in various sports such as cycling, golf, tennis, skiing, surfing. The impact of nanotechnology, as a new technology in sports has been briefly explained.

Keywords: Nanotechnology · Application · Nanoparticles · Nanomaterial · Sport

1 Introduction

Nanotechnology is a branch of research that has been gaining momentum in the broad application of its principles and products in recent years. The application of nanotechnology research varies from different areas such as the prevention and treatment of diseases to advanced and improved electronic devices. Nanotechnology has even found its application in a wide range of sports. In the field of sports, nanotechnology has proven to be very useful and has the potential to improve the broad range of aspects of the sport world.

Innovations donate not only multiple advantages in sports, but also at the level of sports quality, reducing injury opportunities. Nanotechnology in sports offers many advantages and huge potential for improving sports equipment. Benefits such as endurance of the equipment, strength, and a sense of comfort and safety for sportsmen, reduced the likelihood of injuries and increased agility compared to the equipment used earlier (Fig. 1).

M. Brka et al. (Eds.): AgriConf 2019, IFMBE Proceedings 78, pp. 349–362, 2020.
https://doi.org/10.1007/978-3-030-40049-1_44

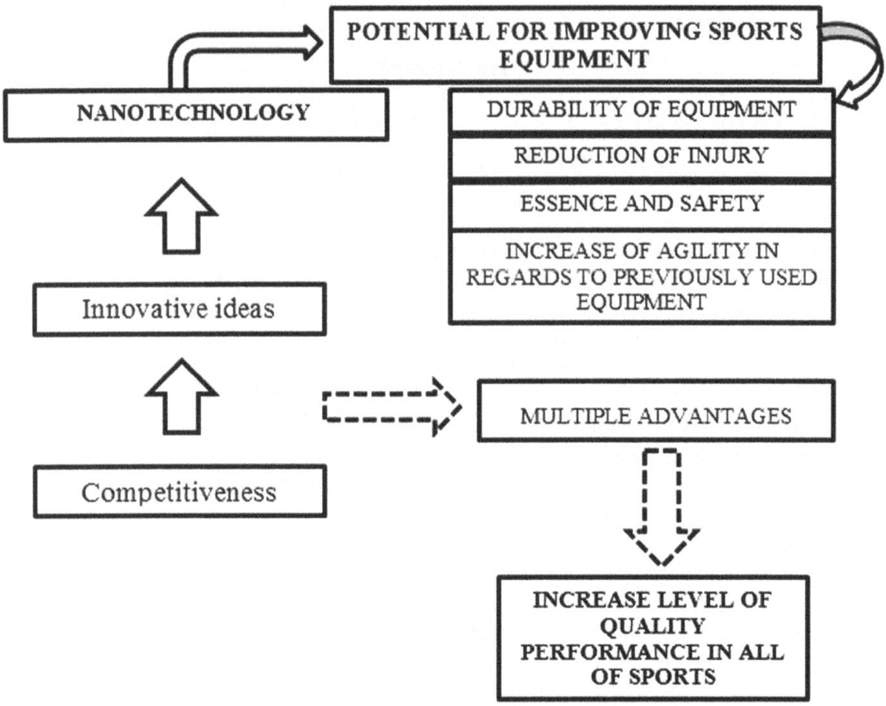

Fig. 1. Application, benefits of nanotechnology in sport utilization [1]

The most commonly used nanomaterials are carbon nanotubes, silicon dioxide nanoparticles, fullerenes etc. There are various advantages that these materials provide in various sports such as golf, tennis, skiing, cycling etc. [1].

2 Nanotechnology in General

In July, 1990, the first international nanotechnology meetings was held in Baltimore, which indicated the birth of nanotechnology [2]. Nanotechnology is defined as science, engineering, and technology conducted at the nanoscale, which is about 1 to 100 nm.

At the Californian Technological Institute, during a meeting of the American Physical Society, Richard Feynman's physicist statement "There's a lot of places at the bottom", many ideas and concepts, based on nanoscience and nanotechnology, have been launched. The physicist Richard Feynman gave a description of the process in which scientist could manipulate and control individual atoms and molecules. The era of modern nanotechnology begin in 1981, over a decade after explorations of ultra precision machining, when a scanning tunnel microscope was developed, which can see individual atoms and molecules. Ultimately, nanotechnology has a potential to improve sciences, such as chemistry, biology, physics, materials science and engineering [3].

Figure 2 presents a schematic, sequencing both science and commercial technologies. The sequence begins with scientific and technological discoveries in instruments (such as scanning tunneling microscopes or STM and atomic force microscopes or AFM) and nanomaterials (such as buckminster fullerenes and carbon nanotubes), forming the basis for a value chain. Complements, intermediate goods include nanocoatings and composites manufactures and nano-core processing. The end of the value chain shows a broad final products. The boundaries between these positions in the value chain overlap [4, 5].

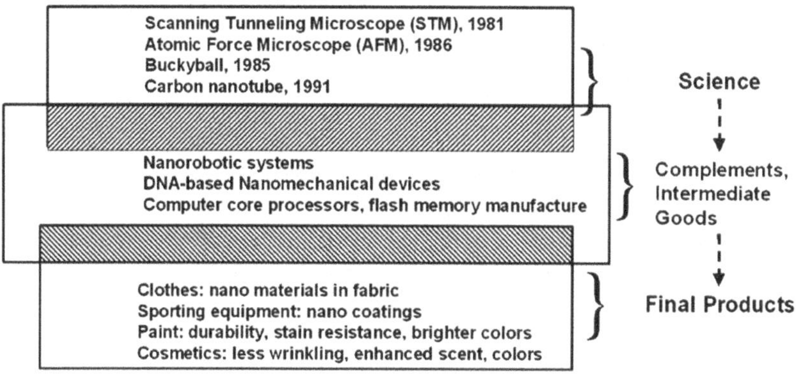

Fig. 2. Potential sequence of science, intermediate goods, and final products in nanotechnology [5]

2.1 Advantages of Using Nanotechnology in Production Chain - Improving Sports Equipment

The degree of competitiveness in sports has been remarkably impacted by nanotechnology like any other innovative idea in materials science. Within the niche of sport equipments, nanotechnology offers a number of advantages (Fig. 3) and huge potential to improve sporting equipments making athletes safer, comfortable and more agile than ever [6].

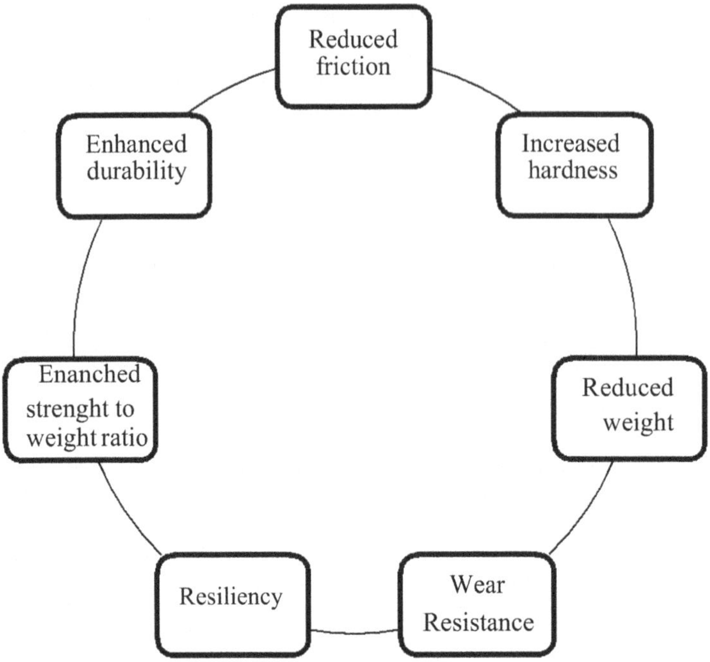

Fig. 3. Nanotechnology advantages in sports equipment [6, 7]

2.2 Advantages of Nanomaterial Uses in Sports Equipment

Nanomaterials in different sport sections have brought many benefits to the sport equipment [8]. Today's sport venues have been dramatically changed through application of nanomaterials in the construction of sport stadiums from their roof and floors to walls and glasses [9–11]. Table 1 shows nanomaterials that have been found to be applied in a particular sport and thus bring various benefits.

Table 1. Nanomaterials and advantages of its incorporation in sporting equipments [11]

Nanomaterial	Sport	Benefits
Carbon nanotubes	Tennis/Badminton	Increase stiffness, strength, consistency, endurance, elasticity, impact, rejection power and vibration control of racquets
	Golf	Reduce weight, lower spin of clubs
	Kayaking	Enhance abrasion/crack resistance and easy paddling in kayaks
	Archery	Better vibration control in arrows
Silica nanoparticles	Tennis/Badminton	Increase stability, power and durability of racquets
	Skiing	Decrease torsion index and facilitate transition in skins
	Fly-fishing	Enhance hoop and flex strength of rods
Fullerenes	Tennis/Badminton	Reduce weight and twisting o racquets frames
	Golf	Facilitate flexible club whipping
	Bowling	Reduce chipping and cracking of balls
Nano clay	Tennis/Golf	Increase resiliency and bounce of balls
	Water craft	Reduce weight and enhance speed of water-boats
Carbon nanofibers	Cycling	Reduce weight and increase stiffness of bicycles
Nano-titanium	Tennis/Badminton	Resist deformation, increase strength and durability of racquets, transmit more power to shuttlecock/ball, more accurate shots
Nano-nickel	Golf	Increase moment of inertia and stability of clubs
Carbon nanoparticles	Road Racing	Decrease rolling resistance

3 Types of Nanomaterials Used in Sports

Application of nanotechnology in sport includes mainly used nanomaterials such as carbon nanotubes, silicon dioxide nanoparticles, fullerenes, carbon nanoparticles, bringing multi-properties in sport equipment.

For Example, Scholler, a Swiss company, has developed a nano-based technology to produce clothing with optimal balance of comfort, air permeability, wind and water resistance and self-cleaning property for extreme cold weather sports such as mountaineering and ski [12].

3.1 Carbon Nanotubes

The carbon nanotube (CTN), discovered by Iijima in 1991, is one of the allotropes of carbon, apart from diamond, graphite, and fullerenes [13]. Carbon nanotubes are cylindrical nanostructures and currently the most durable and hardest material, making it the most used material for making sports equipment. The carbon nanotube is a structure consisting of carbon hollow cylindrical atoms, which are closed at the ends by semi-fullerene structures [14].

The simplest or single-walled carbon nanotube (SWCNT) is formed by rolling up a sheet of a honeycomb network of carbon atoms, or graphene, seamlessly into a tubular form (Fig. 4) [15, 16].

Individual carbon nanotubes can align themselves by pi-pi stacking and Van der Waals interactions. In more chemical terms, carbon nanotubes bond due to sp^2 hybrid carbon atoms. These bonds are stronger than the alkanes and provide CNTs which are extremly high in strength [17].

The wrapping of the graphene sheets is represented by a pair of indices (n, m) where n and m denote the number of unit vector in the crystal structure of graphene as shown in Fig. 4.

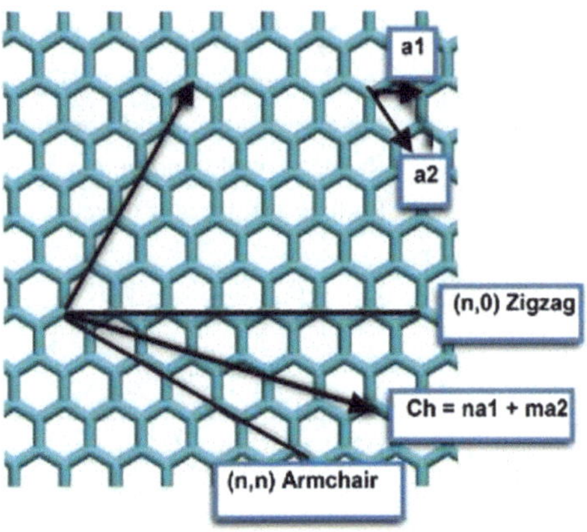

Fig. 4. The (n, m) naming scheme of nanotubes can be thought of as a vector C in an infinite graphene sheet; this will describe how to roll up the sheet to form the nanotube. Also, a1 and a2 denotes the unit vector for the graphene [18].

CNT, the most commonly used material in nano-enhanced sports equipment, have a higher specific strength and specific stiffness than other conventional materials.

They are 100 times stronger, but 6 times lighter than steel and rigid as a diamond, making them an ideal component for making sports equipment where low weight and high strength have primary importance. CNTs are used to create super-strong steering wheels for mountain bikes, durable tennis rackets and ultra-lightweight bicycle frames [19, 20].

3.2 Silicon-Dioxide Nanoparticles

Silicon dioxide is a material used to produce glass characterized as a bad heat conductor. Silicon dioxide coatings make the skeleton of aerogel. Aerogels of silicon dioxide are made of nanoparticles of aerated airborne nanoparticles. As a result, this substance is mostly made up of air [6, 21].

Since both air and silicon dioxide have very low thermal conductivity, they are excellent insulators. These features make nano-aerogels one of the currently the best heat insulators [22].

3.3 Fullerenes

Fullerene is an allotropic carbon modification that looks like a hollow sphere. Spherical fullerenes look like a football ball. Macroscopic quantities of fullerenes with C60 molecules, as well as other fullerenes, began to be mass produced after 1990, when Kratschmer and Huffman discovered the method of preparation by electric arc. Great enthusiasm in fullerenes research has revealed that doped fullerenes exhibit superconductive properties (1991, doped with potassium) [23, 24].

3.4 Carbon Nanoparticles

Carbon nanoparticles have widely been applied in sport such as road racing. The electrical conductivity of polymer materials, like carbon nanoparticles, can be considerably improved by incorporation of conductive fillers.

The latest discovery from Rice University is the result of research that lasted for more than ten years, and despite this, it has caused quite a shock. Rice scientists, the US Air Force and the Israeli Technion Institute discovered that they produced a new type of carbon microtubes (CNTs), or microfibers that look similar to textile fibers, or carry heat and electricity such as metal wire. In a journal, scientists have described the process of industrial production that has ended and whose characteristics in many ways go beyond existing commercially available high performance materials [11, 25–27].

4 Application of Nanotechnology in Sports

As already been said, nanotechnology has found great application in most sports, and the largest application in terms of enhancing equipment in golf, tennis, swimming, racing and cycling. Nanotechnology has a big potential to improve a broad range of aspects of the sports world [28, 29].

4.1 Nanotechnology in Golf

By enhancing equipment, nanotechnology changes the way golf plays. A newly designed golf stick with a long, narrow part of the rod covered with nanomaterials enables us to reduce weight and increase the strength by ensuring the speed of the golf ball head. The Wilson company predominates in the use of nanotechnology in sports not only by using nanomaterials in golf clubs, but also by mixing nanomaterials with graphite fibers [6].

Electrodeposition is performed using an electrochemical cell. A negatively charged metal is dissolved and metal atoms redeposit on the positively charged club head (Fig. 5).

Dense layer of a nano grained metal on the surface of the club head leads to a lighter and stronger club face (Fig. 6).

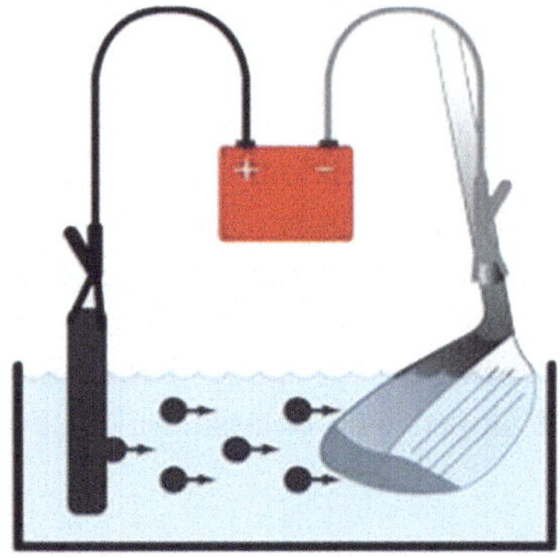

Fig. 5. Perform electrodeposition using an electrochemical cell [30]

Fig. 6. Dense layer of a nano grained metal on the surface of club head [30]

The Fig. 7 shows how nanometers fill small gaps in the rod, making the gap, stronger and more consistent.

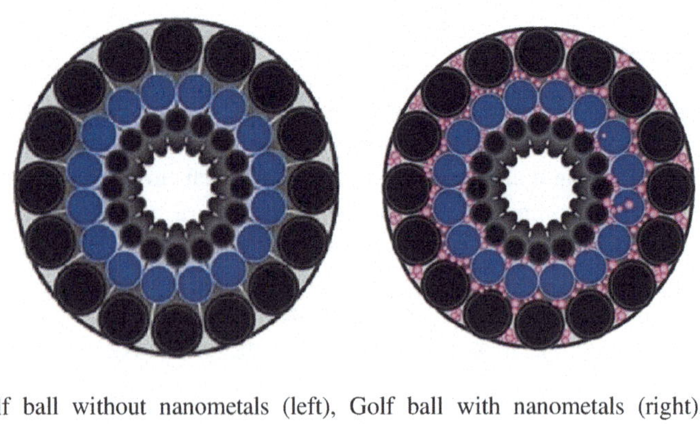

Fig. 7. Golf ball without nanometals (left), Golf ball with nanometals (right). Gray areas presents gaps. Pink spots presents nanomaterial [31].

4.2 Nanotechnology in Tennis

Tennis is a main example of how nanotechnology is having an interesting impact on sports equipment. "Wilson" has been able to create tennis rackets that are twice more stable compared to conventional rackets. According to present research, this rackets are up to 22% more powerful.

As a result of increasing the quality and performance capability can also lead to big speed increases in what is already an extremely fast paced game. This is accomplished by bonding microscopic balls of butyl rubber with clay particles. This mixture is then applied to the inner layer of the ball, but still flexible boundary that keeps the gas inside the tennis ball much longer. A visual explanation of the results of this new airtight boundary is found below in Fig. 8.

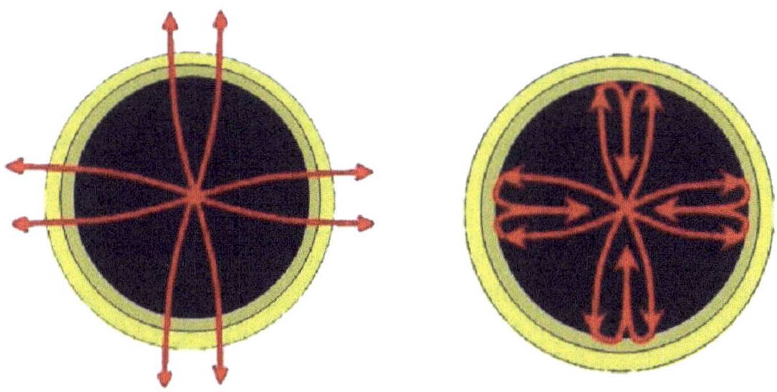

Fig. 8. Ordinary tennis ball (left). New "Wilson" tennis ball coated with gas barrier (right) [32]

Nanotechnology is the cause of many changes in tennis starting from tennis rackets which gain additional strength with the help of nannies to new tennis balls that prevent the release of gases from the core. In 2002 "Babolat" created a racetrack with graphite interlaced with nanoparticle and was one of the first tennis equipment manufacturer who experimented with nanotechnology.

This new material has resulted in the creation of a very light rectangle that still had extraordinary power [32].

For example, Roger Federer has won many tournaments including Wimbledon using rackets enhanced with nanotechnology and nanoparticles of silicon oxide. These rackets are more stable, stronger, more power when hitting balls than other nano-enhanced rackets.

4.3 Nanotechnology in Swimming

One of the first examples of how nanotechnology has changed the world of sports took place at the 2008. Olympics when sports equipment manufacturers introduced a new, watertight, waterproofing suit and ensures greater navigability. These swimsuits have led to skirmishes of nearly 200 world records, as they have a number of advantages over ordinary, swimsuit outfits. In February 2008, Speedo launched the latest generation of the LZR Racer [33].

Nano innovation is the 'LZR Racer Swimsuit' which is now used by most professional swimmers because of its revolutionary aerodynamics [34].

For example, American swimmer Michael Phelps was able to gain eight gold medals while breaking seven World records at the same time. He was accompanied by 25 others who broke World records at this Olympic Game. Though their amazing athleticism can't be doubted, neither can the effect of their high-tech swimwear on these World record races [35, 36].

4.4 Nanotechnology in Sports Racetrack

Like swimming, where friction reduction was an important milestone, nanotechnology also found its applications in car racing. The Nanotec - USA company has implemented Nano Bionics treatment on Nish Sports' engines and has thus enabled cars to go much faster. It has been possible to reduce fuel consumption by 50%, and the cars could develop a speed that was 60 km/h higher than the previous maximum speed. Nano Bionics treatment makes the surface smooth to apply smoothly. During this treatment, nanoparticles are combined with paint and the construction of the carboys, adhering to the nanopore of the material. Nano Biotics clever particles are synchronized, forming new structures, and also makes it easier to maintain the surfaces to be applied. The Millers' Oill automotive fuel company has significantly improved engine protection. This is achieved by reducing friction, which results in increased engine power and more economical fuel use [37].

4.5 Nanotechnology in Bicycling

One of the first uses of nanomaterials in cycling was in 2006. When Floyd Landis was riding a bicycle with built-in carbon nanotubes, he won the famous World Tour Tour de France. This was the first example of the use of nanomaterials in standard bicycle design [38].

BMC, the Swiss company producing the Landis bicycle, in co-operation with Easton Sports USA, used carbon nanotubes to make bicycles because they are 200 times stronger and 5 times elastic than steel and less than half the size of aluminum. The use of composite fibers to produce a product with desired properties implies a lot of effort.

In this case, the BMC bicycle has reduced weight and increased power by replacing carbon fibers with solid, lighter carbon nanoparticles. The bicycle wheel that drove Landis weighed less than 1 kg but was strong enough to withstand a run of almost 90 h [38, 39].

4.6 Nanotechnology in Athletics

Applications in athletics including materials such as nano silver are non-toxic, smelless, non-metamorphic, without decomposition and has no stimulation to human skin. Countries, like China, has added nano silver in the development of sports shoes to make them antibacterial [5]. Nano fiber plays a vital role in the friction brake, especially to ball players, starting and braking. In addition, people are constantly upgrading nano fiber's elasticity and reducing its weight. In that way we can improve the elastic potential energy and variable rate of shoes greatly [39].

Also, Toho rayon has exploited a radioactive care fabric which can emit α, β, γ rays. They grind rare elements such as uranium to nanoscale and then add these elements into viscose to turn the compound to fibers just so they can make cloth with this kind of new material. The rays it emitted will make the atoms and molecules that it passed through become electric ions [9].

That's why the T-shirts, underwear, socks, protective belts, bellybands, sheets and cushions made of this radioactive care fabric has health care functions, such as treatment of shoulder pain, low back pain, limb cold and anti-inflammatory analgesic, diuretic, promoting sprain and contusion, trauma and treatment of neuralgia, etc. [40, 41]. For example, "Adidas" worked with Olympic 400 m runner Jeremy Wariner for more than two years to create the revolutionary 'Adidas Lone Star spike', which is named after Wariner's home state of Texas [42].

4.7 Nanotechnology in Surfing

In 2006 Honolulu's Oceanit Laboratories Inc. was the first producer of a surfboard with embedded nanomaterials. It is the first "nano" surfboards in the world where tiny particles are found in epoxy resin coatings, and make the board stronger than the normal epoxy resin [43].

Andy Jordan, a surfer from Sea City has come up with a water-repellent solution, using nano products to slim the surf board. It is defined as a plate enhancer that meets all the irregularities at the bottom of the surfboard [5].

These products consist of millions of particles of silver, glass, ceramics and diamond particles that are molecularly bound. The surface applied to the surfboard is about 30 nm [30].

5 Market Potential

In 2011, the global sports equipment market was around US$ 64.9 billion and is expected to grow at 12.2% to achieve a value of US$ 72.8 billion by the year 2016. The segment of equipment for sports balls with 23.7% of total market value represents the largest market share. The major markets for sports equipment are Japan, France, UK, Germany and USA. USA commands 40.6% share of the market [44]. Among various nanomaterials, CNTs are the most prominent nanomaterial being used in nano-enhanced sporting equipments such as golf shafts, rackets, skis, snowboards, etc. In 2011, the global market of CNTs (various grades) was around US$ 192 million and is expected to grow at a CAGR of 22.4% to reach US$ 527 million by 2016 [45]. Sporting equipments is one of the prominent sectors of CNTs market and constitute around 14% share of total annual consumption of CNTs, which is likely to be converted into a more significant figure in the near future. Nanomaterials may make some of the sports equipment perhaps many times better. Sports equipments becomes better selling items in the market and even 10–20% more enhanced in terms of performance/durability. However, despite the various advantages proposed, there is not a myriad of nanotechnology-enhanced sports equipment in the market as the technology is highly expensive and complicated to deal with [46].

The Project on Emerging Nanotechnologies, set up in part by the Woodrow Wilson International Center for Scholars, is maintaining a nanotechnology consumer products inventory. Unlike the March 2006. database release in October 2007, there were 580 products in the database, representing an increase of 175% (Fig. 9) [47].

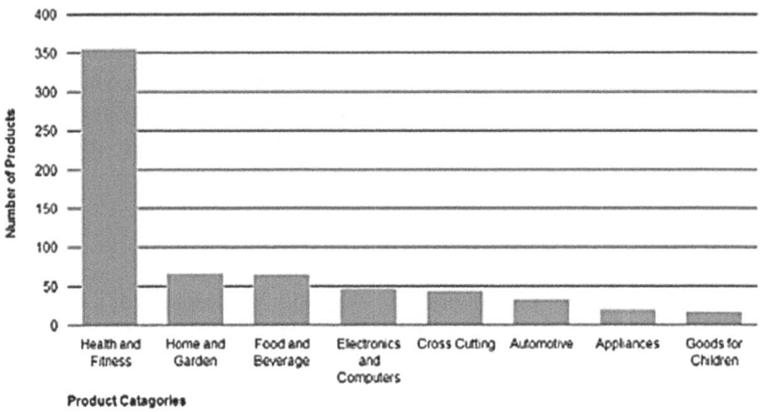

Fig. 9. Products by category [47]

The biggest category, with 61% of products, is health and fitness which includes clothing, cosmetics and sunscreens. This category also includes sports equipment that includes improved tennis balls and stronger rackets. These are examples of embedded nanotechnology in the same way that carbon nanotubes could be used to reinforce car bodies for extra safety and strength [47, 48].

6 Conclusion

Nanotechnology is a new technology that has influenced every sphere of our everyday life, and currently most nanotechnology products are to be found in the sports.

Nano-enhanced sports equipment is much more advanced and superior in terms of strength, rigidity and durability compared to conventional sports equipment. Various companies come up with more innovative ideas for further application of nanotechnology to improve sports equipment. Many prominent athletes choosing nano-built sports equipment to enhance their performance. However, since this is a set of technologies, every athlete cannot afford nano-enhanced sports equipment, and if any competitor has not the same sports equipment, it would be unjust and difficult to judge the natural sports abilities and the human spirit.

Therefore, the ultimate goal to be achieved is shifting the level of performance instead of increasing the cost of high-tech sports equipment. In addition, in order to address possible toxicity problems associated with nano-enhanced sports equipment, proper monitoring of their production processes, period of use and disposal is required to prohibit the release of nanoparticles due to degradation or any other reason.

Investments in sports equipment created of various nanomaterials are increasing, confirming a large difference in the share of previously used sports equipment.

References

1. Omanović-Mikličanin, E., Badnjević, A.: Uvod u nanotehnologiju. University of Sarajevo, Faculty of Agriculture and Food Sciences (2017)
2. Du, W. (ed.): Informatics and Management Science IV. Springer, London (2013)
3. Forloni, G.: Responsible nanotechnology development. J. Nanopart. Res. **14**(8), 1–17 (2012). [1007]
4. Roco, M.: Nanoscale science and engineering: unifying and transforming tools. AIChE J. **50** (5), 890–897 (2004)
5. Youtie, J., Iacopetta, M., Graham, S.: Assessing the nature of nanotechnology: can we uncover an emerging general purpose technology? J. Technol. Transf. **33**(3), 315–329 (2007)
6. Gong, Z.G.: Nanotechnology application in sports. Adv. Mater. Res. **662**, 186–189 (2013)
7. Patra, J.K., Gouda, S.: Application of nanotechnology in textile engineering: an overview. J. Eng. Technol. Res. **5**, 104–111 (2013)
8. Song, Z.Q., Cai, Y.T.: Application of nano-materials in sports engineering. Adv. Mater. Res. **602–604**, 281–284 (2012)
9. Li, B.F.: Design of sports field based on nanometer materials. Appl. Mech. Mater. **340**, 366–369 (2013)
10. Su, Q.F.: Analysis of new materials in competitive sports. Appl. Mech. Mater. **539**, 925–927 (2014)
11. Zhao, H.E., Shen, F.: The applied research of nanophase materials in sports engineering. Adv. Mater. Res. **496**, 126–129 (2012)
12. Miah, A.: Rethinking enhancement in sport. Acad. Sci. **1093**, 301–320 (2006)
13. Iijima, S.: Helical microtubules of graphitic carbon. Nature **354**, 56–58 (1991)
14. De Volder, M.F., et al.: Carbon nanotubes: present and future commercial applications. Science **339**, 535–539 (2013)
15. Iijima, S., Ichihashi, T.: Single-shell carbon nanotubes of 1-nm diameter. Nature **363**, 603 (1993)
16. Jacobs, M.M., et al.: Precarious promise: a case study of engineered carbon nanotubes. Desalination **336**, 97–109 (2014)
17. Rahman, M., Asiri, M.A.: Carbon Nanotubes. King Abdul Aziz University, Jeddah (2018)
18. De Graef, M., McHenry, M.: Structure of Materials. Cambridge University Press, Cambridge (2007)
19. Grebler, S., Fries, R., Simko, M.: Carbon nanotubes- part I: introduction, production, areas of application. NanoTrust Dossiers **22**, 1–5 (2012)
20. Deborah, D.L.C.: Composite Materials: Science and Applications. Springer, London (2010)
21. Kolbe, P.: UH at ForeFront of nanotechnology. UH Today Fall Rev. **06**, 1 (2007)
22. Lian, Y., Wang, X.: Novel Nanomaterials for Biomedical, Environmental and Energy Applications. Elsevier, Amsterdam (2019)

23. Wang, P., Wang, J.Y.: Development and application of nanotechnology in sports. Adv. Mater. Res. **918**, 54–58 (2014)
24. Huanh, J.Y., et al.: Superplastic carbon nanotubes. Nature **439**, 281 (2006)
25. Wood, J.: The top ten advances in materials science. Mater. Today **11**, 40–46 (2008)
26. Shea, C.: Future management research directions in nanotechnology: a case study. J. Eng. Tech. Manag. **22**, 185–200 (2005)
27. Jiang, Y., Wang, Y.: Study of sport devices based on nanomaterial. Appl. Mech. Mater. **329**, 75–78 (2013)
28. Tannenbaum, R.: Advances in materials and self-assembly. In: Proceedings of Advanced Technology and the Future of Manufacturing. Georgia Institute of Technology (2005)
29. Meng, Y.L., Zhu, J.Q.: Application of nanomaterial in sports and its safety research. Appl. Mech. Mater. **387**, 36–39 (2013)
30. Harifi, T., Montazer, M.: Application of nanotechnology in sports clothing and flooring for enhanced sport activities, performance, efficiency and comfort: a review. J. Ind. Text. **46**(5), 1147–1169 (2016)
31. Kimel, R.A.: NACK Network. NanoScience and Nanotechnology: How the Smallest Building Blocks are Impacting Life Today and Tomorrow. Pennsylvania State University (2018). https://nanohub.org/resources/27769
32. Lim, Y., Jang, J., Choi, Y.: Development of golf shaft with carbon nanotubes and marketing strategy. Int. J. Softw. Eng. Appl. **8**(12), 165–170 (2014)
33. Berry-Helmlinger, L.: Boulder firm uses nanotechnology to improve tennis game. Denver Bus. J. (2002). http://www.bizjournals.com/denver/stories/2002/12/02/story6.html?page=all
34. Weidong, Y., Yan, H., Wang, K., et al.: Nanotechnology in competitive sports. Sports Sci. Res. **3**, 27 (2006)
35. Xiaoxiang, L.: Olympics and science: innovation of sports apparel and equipment. Cent. South Univ. Press **4**, 119–122 (2005)
36. Mukhopadhyay, A., Midha, V.K.: A review on designing the waterproof breathable fabrics part I: fundamental principles and designing aspects of breathable fabrics. J. Ind. Text **37**, 225–262 (2008)
37. Dong, C.C., MA, C.: Research on co-evolution of technology and sports based on application of high technology. J. Shenyang Sport Univ. **3**, 16 (2008)
38. Saidur, R., Lai, Y.K.: Nanotechnology in vehicle's weight reduction and associated energy savings. Energy Educ. Sci. Technol. Part A-Energy Sci. Res. **26**(2), 87–101 (2013)
39. Maney, K.: Nanotech could put a new spin on sport. USA Today. (2004). http://www.usatoday.com/sports/2004-11-17-nanotechnology-sports_x.htm
40. Ji, Y.: The applied research of nano-material in athletic facilities and sports equipment. Adv. Mater. Res. **459**, 398–401 (2012)
41. Uttam, D.: Active sportswear fabrics. Int. J. IT Eng. Appl. Sci. Res. **2**, 34–40 (2013)
42. Pliakas, A.: Application of nanotechnology for multifunctional sportswear fabrics. Ph.D. Thesis, University of Leeds, UK (2011)
43. Fugao, C.: LSMZ series single guide type nano functional shoes. Research **32**(5) (2010)
44. Zhu, Z.X.: Research on nano materials tennis rackets. Adv. Mater. Res. **507**, 75–78 (2012)
45. Sports Equipment: Global Industry Guide. MarketLine (2012)
46. Global Markets and Technologies for Carbon Nanotubes. BCC Research (2012)
47. Opportunities for Nanomaterials in Sporting Applications: Trend, Forecast and Competitive Analysis, Research and Market (2008–2013)
48. Project on Nanotechnologies: A Nanotechnology Consumer Products Inventory. Woodrow Wilson Center for International Scholars (2005). http://www.nanotechproject.org/index.php?id=44

Author Index

M. Brka et al. (Eds.): AgriConf 2019, IFMBE Proceedings 78, pp. 363–365, 2020.
https://doi.org/10.1007/978-3-030-40049-1

The manufacturer's authorised representative in the EU is Springer Nature Customer Service Centre GmbH, Europaplatz 3, 69115 Heidelberg, Germany. If you have any concerns regarding our products, please contact ProductSafety@springernature.com

Printed and bound by CPI Group (UK) Ltd, Croydon, CR0 4YY
25/09/2026
02222853-0002